# Quaternary Period in Saudi Arabia

## 1: Sedimentological, Hydrogeological, Hydrochemical, Geomorphological, and Climatological Investigations in Central and Eastern Saudi Arabia

A Cooperative Research Project
of the University of Petroleum and Minerals, Dhahran
and the Austrian Academy of Sciences, Vienna

Edited by

**Saad S. Al-Sayari**
**and Josef G. Zötl**

Springer-Verlag Wien New York

Dr. Saad S. Al-Sayari
Department of Earth Sciences, University of Petroleum and Minerals, Dhahran, Saudi Arabia

Prof. Dr. Josef Georg Zötl
Department of Hydrogeology, Technical University, Graz, Austria, and Quaternary Commission, Austrian Academy of Sciences, Vienna, Austria



Printed in Austria

With 101 Figures and 1 Pullout

Library of Congress Cataloging in Publication Data. Main entry under title: Quaternary period in Saudi Arabia. Bibliography: p. Includes index. Contents: 1. Sedimentological, hydrogeological, hydrochemical, geomorphological, and climatological investigations in central and eastern Saudi Arabia. 1. Geology, Stratigraphic-Quaternary. 2. Geology-Saudi Arabia. I. Al-Sayari, Saad S. II. Zötl, Josef. III. Jāmi 'at al-Bitrūl wa-al-Ma 'ādin. IV. Akademie der Wissenschaften, Vienna. QE696.Q34. 551.7'9'09538. 78-1708.

ISBN 3-211-81448-5 Springer-Verlag Wien-New York
ISBN 0-387-81448-5 Springer-Verlag New York-Wien

## Foreword

The publication of this first volume on the Quaternary Period in Saudi Arabia marks a significant milestone in the study of this Kingdom's basic geology. It also represents a major achievement in international cooperation in scholarship.

The study of the Quaternary Period of Saudi Arabia was first proposed by Austrian scientists and by the officers of the Austrian Academy of Sciences early in 1972. A meeting was arranged between representatives of that Academy and of the University of Petroleum and Minerals, in Vienna, during May and June of that year, and tentative agreement was reached on the details of this cooperative venture. An exchange of letters, 15th June 1972 from the Austrian Academy of Sciences and 21th June 1972 from the University of Petroleum and Minerals, formally recorded the basis of the agreement and arranged for the first visit to Saudi Arabia of the initial research team and accompanying officials from Austria. This team was headed by Professor Dr. JOSEF ZÖTL, of the Department for Hydrogeology, Technical University of Graz, the initial project director.

Many scientists, administrators, and officials have contributed to the success of this fundamental research project. Their number is too extensive to mention in this Foreword, but the principal contributors have been listed in the body of this text in conjunction with the specific areas of investigation in which they participated. The enormous contribution of Professor Dr. ZÖTL, however, merits exceptional commendation. His was the initial conception of the study. His participation in the field studies and later with the reduction and interpretation of data was an inspiration to all his collaborators. His unflagging enthusiasm ensured high morale and surmounted all difficulties. If a single name may be associated with this important research, it must be that of Professor Dr. ZÖTL.

The study of the Quaternary Period of Saudi Arabia is remarkable not only for its contribution to basic scientific knowledge, but also as a demonstration of successful international cooperation in the area of open research.

The agreement between the two organizational partners in the Project, the Austrian Academy of Sciences and the University of Petroleum and Minerals, was not a contract. It was consciously designed to be an understanding between peers, in their mutual self interest and enforceable only because of their mutual respect.

Thus, each party to the agreement supplied those elements in specialized manpower, equipment, laboratory facilities, and support logistics which it was best fitted to contribute. Transportation outside the Kingdom and the

Middle East was supplied by the Academy; that within the Kingdom and Middle East was provided by the University. Hospitality for visiting scientists in Europe was supplied by the Academy; in Saudi Arabia by the University. Data was reduced in Europe or in Saudi Arabia depending upon the laboratory best suited for the study. Where equipment was contributed, it was supplied in kind and its value determined by the cost to the contributing partner, thus avoiding the troublesome problem of widely different costs for the same item in Europe and Arabia. Each party maintained its separate accounts, each according to its own accounting rules for its own contribution to the total project.

The agreement provided that all findings of the investigation would be considered "open research", and as such would be published and made a contribution to science. Publication would be made periodically and not upon final completion of the entire project, so that new discoveries might be reported on a timely basis and made immediately useful to contemporary students of the geology of this area. Copyrights would be held by the University and the Austrian Academy of Sciences, which have transferred the copyrights of this first volume to Springer-Verlag Wien thus providing for an adequate presentation of the scientific results. The quality of scholarship was ensured by the established reputation of the Austrian Academy of Sciences and the rapidly growing reputation of the University of Petroleum and Minerals.

It is, therefore, with quiet pride that we present this first result of the work of many scientists from many countries who have spent the past five years in its preparation. In a world troubled by dissension and materialism, it is an achievement of some merit to record this fundamental research on the silent eternal geology of this Kingdom, motivated only by the selflessness of scholarship.

Dhahran, March 1978 BAKR ABDULLAH BAKR

# Contents

# List of Authors

AL-SAYARI, SAAD S., Chairman, Department of Earth Sciences, University of Petroleum and Minerals, Dhahran, Saudi Arabia.

BECKMAN, RICHARD J., Los Alamos Scientific Laboratory, University of California, Los Alamos, NM 87545, U.S.A.

CHAPMAN, RANDOLPH W., Professor, Department of Earth Sciences, University of Petroleum and Minerals, Dhahran, Saudi Arabia.

FELBER, HANS, Oberrat, Institut für Radiumforschung und Kernphysik der Österreichischen Akademie der Wissenschaften, Boltzmanngasse 3, A-1090 Wien, Austria.

HÖTZL, HEINZ, Professor, Institut für Geologie, Universität Karlsruhe, Kaiserstraße 12, D-7500 Karlsruhe, Federal Republic of Germany.

JOB, CARLO, Professor, Institut für Balneologie, Universität Innsbruck, Fritz-Pregl-Straße 3, A-6020 Innsbruck, Austria.

JOHNSON, DONALD H., Professor, Department of Earth Sciences, University of Petroleum and Minerals, Dhahran, Saudi Arabia.

KAMAL, MARWAN R., Professor, Department of Chemistry, University of Petroleum and Minerals, Dhahran, Saudi Arabia.

KRÄMER, FRIEDRICH, Akad. O.-Rat, Institut für Geologie, Universität Karlsruhe, Kaiserstraße 12, D-7500 Karlsruhe, Federal Republic of Germany.

LIPPOLT, HANS J., Professor, Laboratorium für Geochronologie, Universität Heidelberg, Im Neuenheimer Feld 234, D-6900 Heidelberg, Federal Republic of Germany.

MAURIN, VIKTOR, Professor, Institut für Geologie, Universität Karlsruhe, Kaiserstraße 12, D-7500 Karlsruhe, Federal Republic of Germany.

MCCLURE, HAROLD A., Arabian American Oil Company, Box 2376, Dhahran, Saudi Arabia.

MOSER, HERIBERT, Professor, Institut für Radiohydrometrie der Gesellschaft für Strahlen- und Umweltforschung mbH München, Ingolstädter Landstraße 1, D-8042 Neuherberg, Federal Republic of Germany.

PAK, EDWIN, Institut für Radiumforschung und Kernphysik der Österreichischen Akademie der Wissenschaften, Boltzmanngasse 3, A-1090 Wien, Austria.

PIERSON, GARY O., Professor, Department of Chemistry, University of Petroleum and Minerals, Dhahran, Saudi Arabia.

RAMSAY, JOHN B., Professor, Department of Chemistry, University of Petroleum and Minerals, Dhahran, Saudi Arabia.

RAUERT, WERNER, Institut für Radiohydrometrie der Gesellschaft für Strahlen- und Umweltforschung mbH München, Ingolstädter Landstraße 1, D-8042 Neuherberg, Federal Republic of Germany.

SCHYFSMA, ERNEST, Professor, Department of Earth Sciences, University of Petroleum and Minerals, Dhahran, Saudi Arabia.

STICHLER, WILLIBALD, Institut für Radiohydrometrie der Gesellschaft für Strahlen- und Umweltforschung mbH München, Ingolstädter Landstraße 1, D-8042 Neuherberg, Federal Republic of Germany.

ZÖTL, JOSEF G., Professor, Abteilung für Hydrogeologie, Technische Universität Graz, Rechbauerstraße 12, A-8010 Graz, Austria.

# Introduction

The Kingdom of Saudi Arabia, covering an area of more than two million square kilometers, represents the largest part of the Arabian Peninsula. Despite its enormous size, its geology is relatively well-known. There exist not only a geological map at a scale of 1 : 2,000,000 (U.S. Geological Survey and Arabian American Oil Company, 1963), but also quadrangle sheets of 1 : 500,000 for the entire area (see references). However, little attention has been paid to the Quaternary Period, as it has had little bearing on prospecting for petroleum and ores, and only local importance for water supply projects.

The Quaternary Period, be it the Pleistocene or Holocene, shows an alternation of warm and cold climates in Europe and North America, the course of which has been very intensively studied. The change of climate was caused by a worldwide displacement of climatic zones resulting in various periods, each single period being characterized by its own oscillations.

During the cold periods of the Pleistocene, large regions in the middle and northern latitudes of Europe and North America were covered with ice sheets up to 2,000 meters thick, the so-called continental glaciers or "inland-ice", whereas the valley systems of the Alps or the Rocky Mountains were filled with large valley glaciers. At the borders of the inland-ice and the lower ends of the valley glaciers, large terminal moraines still bear witness to Pleistocene glaciation and its phases.

Of considerable interest are studies of Quaternary geology of the periglacial regions, such as the development of the Baltic Sea and of the large accumulation terraces downstream from the Pleistocene glaciers. Thus, in addition to general geological problems, the Quaternary geology includes studies of erosion and sedimentation caused by climatic conditions. As a consequence of the large ice masses accumulated during the cold periods of the Pleistocene in the northern regions the average sea level was lower than during the interglacials. Thus, in the Mediterranean during the Würm period, the sea level was about 100 meters lower than it is today, in contrast to the warm periods which partly showed a higher sea level than at present. Results of these fluctuations include marine abrasion terraces along the coastal areas at various elevations, changes of underground drainage caused by the different base levels of erosion, increased terrestrial sedimentation into shallow marine basins, and alternating phases of erosion and accumulation in the river systems. Precise information has been gained through studies of these phenomena in the coastal areas of Florida, and especially around the Mediterranean in Africa, Europe, and the Middle East.

There can be no doubt about the simultaneous nature of these climatic phenomena during the Holocene and Late Pleistocene in Europe and the Mediterranean area. Yet, no full agreement has been reached so far as to the number of cold periods during the Pleistocene. Likewise, estimates of the total duration of the Quaternary have changed from about one million to two million years and more (cf., chapters 3.1.5., 3.2.).

The terminology that has been introduced for the Quaternary is summarized in Table 51. It gives a comparison of the classifications and nomenclature used for marine niveaus (coastal terraces) of the Mediterranean and for the climatic periods in Europe and North America.

Starting from the Mediterranean area, the studies of the Quaternary gradually were extended to the adjoining southern and eastern regions where scientists are still working in the Sahara, Niger, Tchad and Tibesti, in Iran, and in Pakistan.

The research results point to climatic variations of different kinds. In these southern and eastern areas the glacials of Europe were partly rainy periods, so-called "pluvials". In addition these phases displayed a very different character in the arid African region on the one hand, and in Asia on the other. The great interest of the Arabian subcontinent—where up to 1972 only few investigation of Quaternary geology were completed, except for some parts in the north and east (Mesopotamia, Basin of Damascus, Arabian Gulf)—lies in the fact that it is situated between these two regions. Thus, it can be assumed that research work in Saudi Arabia will provide a key for understanding the differences between Quaternary pluvials in North Africa and climatic changes in the monsoon areas of South and Southwest Asia.

Therefore, the study of Saudi Arabia's Quaternary Period is one of the fascinating tasks in the field of the natural sciences and of great importance because of the size of the country. For the country itself, these studies serve as the basis for subsequent specialized and local investigations which also might be necessary for a successful development of its economy.

Research is being carried out according to the following plan:

a) Field campaigns in the Arabian Shelf, i.e., the basin of sedimentary rocks from the eastern border of the Shield to the west coast of the Arabian Gulf, crossing the large cuesta landscape of central Saudi Arabia. The general geological map of 1 : 2,000,000 (U.S. Geological Survey and Arabian American Oil Company, 1963) reflects the large river systems which once crossed the central and eastern part of the peninsula during the Late Pliocene and the Early Pleistocene.

b) Field campaigns in the western part of the country, the so-called Arabian Shield which consists of north-south-trending mountains of crystalline and volcanic rocks now rising to altitudes of 3,000 meters, with young deformations and large faults along the graben of the Red Sea with its narrow coastal plain.

A description of these different areas is followed by the discussion of the results obtained. In the first volume, after an introduction to the general features of the country (geology, geomorphology, and climate) the main topic comprises the description of the field campaigns carried out in 1973, 1974, and 1975 and their results concerning the features of the Quaternary Period of the Arabian Gulf coastal area, its hinterland, and the cuesta region of central Saudi Arabia.

The second volume, already in preparation, will bring the results of the studies in the Arabian Shield and the coastal plains of the Red Sea. A special chapter of the second volume will be devoted to a comparison of the results of investigations carried out in Saudi Arabia with those which are known from research in the Sahara, Ethiopia, Iran, Pakistan, and the Mediterranean and its surroundings.

## Acknowledgements

The scientists of the Saudi-Austrian team who have carried out these investigations wish to express their gratitude to the sponsors of this project. Special thanks go to H. E. Dr. BAKR ABDULLAH BAKR, Rector of the University of Petroleum and Minerals in Dhahran, and to Professor Dr. HERBERT HUNGER, President of the Austrian Academy of Sciences. The team also wishes to thank Professor Dr. E. SCHMID, Vice-President of the Austrian Academy of Sciences, and the members of the Academy, Professor Dr. E. W. PETRASCHECK, formerly Chairman of the Academy's Commission for the International Geological Correlation Programme, Professor Dr. H. ZAPFE, present Chairman of the Academy's Commission for the International Geological Correlation Programme, and Professor Dr. J. FINK, Chairman of the Academy's Commission on Quaternary Problems. Lastly, acknowledgements are due to the Austrian Fund for the Promotion of Scientific Research, which generously supported this project through grants required to cover the costs of air transport and various analyses carried out in Austria. The special enthusiasm and support of DDr. ELMAR WALTER, Ministry of Science and Research in Vienna, was decisive for the project's start.

Particular thanks must be expressed to the University of Petroleum and Minerals for bearing the local costs of the project as well as the organizational responsibility for the work in Saudi Arabia. Many of its staff members, e.g., Dr. ABDUL AZIZ GWAIZ, Vice Rector, Dr. ROBERT KING-HALL, Senior Adviser, and Dr. MARWAN E. KAMAL, former Dean of Sciences, secured the success of the project through their initiative and ready assistance.

Vehicles, field equipment, and auxiliary personnel were furnished by the Institute for Applied Geology at the KING ABDUL-AZIZ University in Jeddah. Thanks are due to Dr. AHMED AL-SHANTI, Director of the IAG, and SHEIKH GHAZI SULTAN, Deputy Minister of Mineral Resources of the Ministry of Petroleum and Minerals, who supported the field work in many ways.

Valuable contributions to the project were made by the Department of Geology of the Ministry of Agriculture and Water, Ar Riyadh, permitting the research team to consult its extensive files of basic data. Thanks are due to MUSTAFA NOORY, Director General of the Department of Geology, and its former Chief Geologist, Dr. G. ÖTKUN. Thanks are extended to T. DINÇER, formerly Hydrologist at the Ministry of Agriculture and Water for his personal assistance in Ar Riyadh, for his advice concerning field excursions, and for useful discussions of the results of isotopic measurements.

The U.S. Geological Survey (Jeddah) and the Arabian American Oil Company (ARAMCO) were co-operating in an amicable way in many cases.

# 1. General Information on the Arabian Peninsula

## 1.1. Geology

(R. W. Chapman)

### 1.1.1. Introduction

The Arabian Peninsula, lying between Africa and western Asia, has been of academic interest to geologists since the end of the last century in relation to rifting in the Red Sea, separation of Arabia and Africa, continental drift, sea floor spreading, and plate tectonics. It has been of interest also for the part it appears to have played in the folding and elevation of the Zagros Mountains in western Iran and the Oman Mountains during the Alpine orogeny. Since the 1930's, close attention has been focused on the stratigraphy, structure, and economic geology of the eastern part of the peninsula, near the Arabian Gulf, the site of the world's greatest petroleum deposits. More recently, from about 1950 to the present, the Saudi Arabian Directorate General of Mineral Resources and the United States Geological Survey have pursued an intensive program of mapping and mineral exploration in the Precambrian shield in western Arabia in an attempt to locate mineral deposits. The results of much of the above work have already been published and have led to a good, although incomplete, understanding of the geology and tectonic history of the Arabian Peninsula.

The present account is a summary of Arabian Peninsula geology written mainly for those who are unfamiliar with this interesting and important region. The account begins with a brief preliminary description of the structure and history of the peninsula and the surrounding region. Then, the two structural provinces of the penisula, the Arabian Shield and the Arabian Shelf, are described in some detail. Following this, a brief description is given of the Zagros Mountains, the Oman Mountains, and the Red Sea, all three of which are closely related geologically to the peninsula. Finally, the economic geology of the penisula is considered [1].

### 1.1.2. Structure and Geologic History of the Peninsula

The Arabian Peninsula is a huge crustal plate composed of ancient sedimentary and volcanic rocks, deformed and metamorphosed and injected by

[1] The Arabian American Oil Company (ARAMCO) provided the geologic map (Fig. 1) that accompanies this account and granted permission to publish it. The writer wishes to express his appreciation to ARAMCO for their generosity. No geographic names are shown on the map, but in the text the locations of cities and towns and certain specific points are given by latitude and longitude. The reader should refer to a geographic map of the Arabian Peninsula for the locations of other geographic features.

plutonic intrusions. In Precambrian time, long before the formation of the Red Sea, the peninsula was attached to Africa as a part of the African (Nubian) Shield. In the late Precambrian its surface was deeply eroded and peneplaned.

At the beginning of the Cambrian, a great sedimentary basin or geosyncline (the Tethys) had developed north and east of Arabia in the area now occupied by Turkey, northern Iraq, and southwestern Iran. Throughout Paleozoic, Mesozoic, and early Cenozoic times many thousands of meters of sediment accumulated in this deep, slowly-sinking trough. Between the Tethys and the Arabian Peninsula lay broad epicontinental seas. These spread widely over the eastern parts of the peninsula, depositing on it a relatively thin succession of almost flat-lying Paleozoic, Mesozoic, and early Cenozoic strata. Inland of the shoreline, continental sediments were laid down. From time to time, in response to slight vertical movements, the shoreline moved back and forth across parts of the peninsula, eroding some sediments and producing unconformities.

In the late Cretaceous, orogenic movements all the way from Gibraltar to the Island Arcs of the Pacific Coast heralded the destruction of the ancient Tethys. The thick sequence of marine strata in the geosyncline was buckled into great folds and sliced by great overthrusts. This deformation constituted the first stage of the Alpine orogeny which was to give rise to the mighty Alpine-Himalayan mountain chain. The second stage, or climax, began in the late Tertiary when, as a result of powerful deep-seated forces, the deformed rocks of the geosyncline started to rise. The Arabian Peninsula itself was little affected by this uplift except for being tilted slightly toward the eastern Arabian Gulf region where sinking and sedimentation continued. This second stage of the orogeny culminated in the formation of the Taurus, Zagros, and Oman Mountains.

In the middle Tertiary, while these events were in progress, the Arabian plate split away from the African Shield along the Red Sea trough. It then began moving slowly northeastward, impinging on the edge of the great Asian plate in Iran and perhaps even sliding beneath the latter. This separation of Arabia from Africa was accompanied by extensive volcanism along the western edge of the peninsula. Throughout the Paleozoic and Mesozoic and during the Tertiary orogeny, the Arabian plate remained relatively stable, and its cover of shelf sediment was barely disturbed.

Today, the peninsula can be divided into two structural provinces. A western province, known as the Arabian Shield, is a part of the Precambrian crustal plate, generally exposed but locally covered by Tertiary volcanic rocks. An eastern province, called the Arabian Shelf, consists of a thick sedimentary sequence covering the plate. The geology of these two provinces will now be described in some detail.

## 1.1.3. Arabian Shield

### *1.1.3.1. General nature*

The Arabian Shield is an ancient land mass that occupies a trapezoidal area of about 770,000 square kilometers in the western part of the peninsula along the Red Sea (Fig. 1). Its relatively smooth, slightly-arched surface is a peneplain now sloping very gently toward the north, northeast, and east under a thin sedimentary cover in eastern Arabia. The framework of the shield consists of Precambrian gneiss and metamorphosed sedimentary and volcanic rocks that were folded and faulted in two major orogenies and then intruded by granites. Following peneplanation in late Precambrian time, the surface was covered in part at least by lower Paleozoic basal sands. In the Tertiary great floods of basic lava poured out onto the surface. The principal rocks and geologic events are discussed more fully below.

### *1.1.3.2. Basement rocks*

The basement rocks consist of: 1. Rocks that have been termed "basement gneiss" and 2. bedded rocks consisting of metamorphosed sediments and volcanic rocks. The basement gneiss is believed to represent the oldest rocks exposed in the shield. Some massive bodies of gneiss occupy diapiric structures apparently squeezed upward through younger schistose rocks along fault planes. Gneissic boundaries are sometimes sheared. Other bodies show intrusive characteristics suggesting palingenesis and regeneration. Most of the rock types are orthogneisses derived from diorite, quartz diorite, granodiorite, or granite and most belong to the amphibolite facies. The basement gneiss is definitely of Precambrian age (R. J. FLECK et al., 1976).

The bedded rocks, mainly of sedimentary and volcanic origin, have been metamorphosed to the greenschist facies. They occur either in separate basins or closely associated with plutonic bodies, either resting on their flanks or occurring in roof pendants. These rocks have been folded and strongly compressed and show linear structures trending generally north, northeast, and northwest. There are several formations and many are very thick. For example, the Hali schist, consisting of various types of schist together with marble, quartzite, slate, conglomerate, and greenstone, is 9,000 meters thick. Another formation, the Baish greenstone, locally pillow structured and spilitic, includes metadiorite, metagabbro, and amphibolite and a subordinate amount of slate. The total measured thickness is 12,000 meters. The great variety of rock types in the bedded Precambrian formations is well shown by the Halaban andesite which attains a thickness of 10,000 meters and contains: 1. various types of volcanic rocks, 2. sills and dikes of diorite, diabase, gabbro, and serpentine, and 3. quartzite, arkose, greywacke, conglomerate, marble, and phyllite.

### *1.1.3.3. Plutonic rocks*

Scores of stocks and batholiths are scattered throughout the Arabian Shield. These plutonic bodies, principally granitoid and granitic, have been grouped as follows: 1. pretectonic granitoid rocks, 2. syntectonic granitoid rocks, 3. late tectonic granitic rocks, and 4. posttectonic intrusive rocks.

The pretectonic granitoid rocks are granodiorite and related quartz

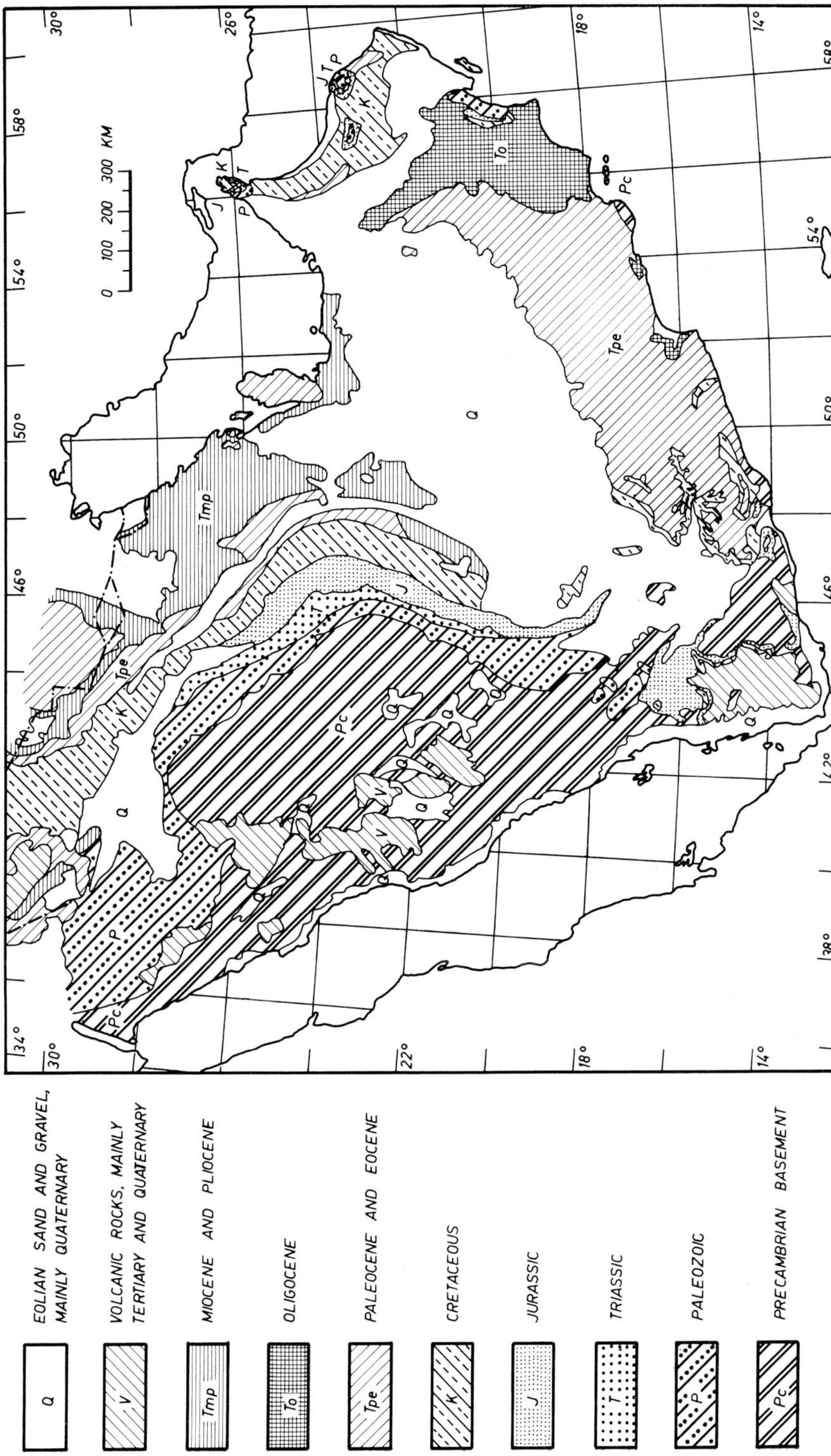


Fig. 1. Generalized geologic map of the Arabian Peninsula

diorite, well exposed in large and small plutons near Mecca and to the northwest, east, and southeast. Rb-Sr age determinations on these rocks average about 1,000 million years and indicate that the latter predate Al Hijaz orogeny (described later).

The syntectonic rocks are mostly diorite or granodiorite emplaced during Al Hijaz orogeny. They are gneissic, migmatitic, commonly sheared and somewhat metamorphosed, with zones of schlieren and xenoliths. Radiometric dating indicates that they are between 650 and 800 million years of age.

The late tectonic granitic rocks, emplaced in the late stages of Al Hijaz orogeny, are calc-alkalic and generally massive, but in some places show metasomatic layering. Ages range from 600 to 650 million years.

Finally, the posttectonic intrusions, which followed the Najd orogeny (described later), are alkaline or peralkaline and commonly characterized by sodic amphiboles and pyroxenes. They range from gabbro to granite and from syenite to silicic albite granite. They occur as stocks, pipes, and batholiths. Their ages range from 490 to 600 million years.

The grouping of the granitic rocks into syntectonic, late tectonic, and posttectonic stages is based on their radiometric ages, their structural forms and textures, and on their general magmatic trend.

### *1.1.3.4. Orogenies*

The fold-fault pattern of the shield, together with certain stratigraphic relationships, suggests that there were two major orogenic cycles, each with more than one period of folding and igneous activity, subsequent to the intrusion of the pretectonic granitoid rocks some 1,000 million years ago. The older of the two, which was more intense and more widespread both areally and timewise, is called Al Hijaz orogeny. This cycle involved folding and the intrusion of syntectonic granitoid rocks about 660 million years ago. East-west compression was dominant so that strongly folded and faulted beds now crop out in belts trending north-northeast. Associated faults in the southern part of the shield trend generally northerly whereas those in the west-central part trend generally northeasterly. In the metamorphism that accompanied the folding and intrusion, the rocks were changed to the amphibolite facies.

The second or Najd orogeny represents a younger period of mountain building. The effects of this younger and shallower deformation may be seen in the long northwesterly-trending left-lateral faults that have displaced some rock units as much as 240 kilometers. Most of the posttectonic alkalic rocks described earlier were probably emplaced late in this orogeny or after the fault movement. The Najd orogeny seems to have involved at least two periods of igneous activity, interrupted by long quiet intervals and intermittent fault movements.

It appears that Al Hijaz and Najd orogenies resulted from different forces. The older Al Hijaz folding and intrusion may have occurred at greater depth, whereas the younger Najd orogeny may reflect shears from shallower movement.

### *1.1.3.5. Peneplanation and sedimentation*

At the close of Precambrian time, the shield was deeply eroded to a peneplain with small scattered monadnocks. It was then depressed and sedimentation began so that now lower Paleozoic sedimentary rocks overlap the basement rocks along its northern and eastern edges. The lowest formation, the Saq sandstone of Cambrian and Ordovician age and probably of terrestrial origin, rests unconformably on the eastern edge of the shield and forms outliers in its central part.

### *1.1.3.6. Tertiary volcanism*

Volcanism was widespread in western Arabia during the Tertiary Period. The oldest lavas, the so-called "Trap Series", rest on Cretaceous clastics in Yemen where they are closely associated with rifting of the Red Sea. They include plateau basalt, andesite, trachyte, and rhyolite flows interbedded with tuff and nonmarine sedimentary rocks. Most are probably Oligocene to Miocene in age, but the lowermost flows may be late Cretaceous or Eocene.

Northward across the shield and out onto the Eocene rocks east of Jordan, thick effusions of basalt and andesite cover vast areas. These effusions have been subdivided on the basis of K-Ar age determinations, and they range in age from Oligocene (22.5–36 million years) to Holocene (0–1.8 million years). Thus, volcanic activity continued right up to the present time. The Miocene and Holocene flows in western Saudi Arabia consist of alkali olivine basalt like those of the Trap Series.

## 1.1.4. Arabian Shelf

### *1.1.4.1. General nature*

The Arabian Shelf, the second of the two major structural provinces, lies to the east of the shield where it forms about two thirds of the peninsula. Its foundation is a part of the same Precambrian plate that makes up the shield. On this rigid peneplaned basement lies a sequence of continental and shallow-water marine sedimentary rocks, ranging in age from Cambrian to Pliocene. These dip gently away from the shield and into a number of deep sedimentary basins.

### *1.1.4.2. Distribution of sedimentary rocks*

In central Arabia, Paleozoic, Mesozoic, and lower Cenozoic sedimentary rocks are exceptionally well exposed in a great curved belt along the eastern margin of the shield. The beds dip gently and uniformly northeast, east, and southeast away from this ancient land mass, reflecting the gently-arched surface of the buried basement. Dips range from about one degree in the older units to less than half a degree in the Upper Cretaceous and Eocene beds. These low, constant dips characterize this part of the shelf which is known as the Interior Homocline. Along the exposed part of the homocline is a series of parallel strike escarpments facing westward, each capped with resistant limestone. Intervening valleys and plains are floored with sandstone and shale. The scarps offer an excellent opportunity to study the stratigraphic sequence and to trace the rock units.

In northwestern Arabia mainly lower Paleozoic sedimentary rocks are exposed. These show constantly changing strikes as they bend around the northern end of the shield south and west of the Great Nefud sand area in the general vicinity of lat 28° 00′ N, long 38° 00′ E. A basin on the north side of the homocline contains a thickened section of Upper Cretaceous to Tertiary sedimentary rocks. Tertiary and Recent volcanic rocks cover much of the area.

On the southern edge of the peninsula a broad band of Paleocene and Eocene rocks extends along the homocline from the eastern part of the Republic of Southern Yemen across Dhufar and into Oman and the eastern Rub' Al Khali. East of this band is a large area of Oligocene sedimentary rocks.

East of the homocline is a very broad area of low relief that includes Ar Rub' Al Khali and much of northeastern Arabia. This is the Interior Platform, a surprisingly flat structural feature separated from the homocline by an abrupt break-in-slope. Sedimentary rocks in this region are essentially horizontal, but their thickness increases gradually eastward and beneath the Arabian Gulf may exceed 10,000 meters. Paleozoic and Mesozoic strata are buried by Tertiary and other late deposits, and do not reappear at the surface until they reach the Oman Mountains and the far side of the Arabian Gulf.

### *1.1.4.3. Description of sedimentary rocks*

The sedimentary section in Saudi Arabia above the Precambrian has been divided into eight major divisions, each characterized by a particular lithology and separated by significant breaks in the geologic record (R. W. POWERS et al., 1966, p. D 6). These divisions are described briefly below, beginning with the oldest. It will be noted that the Oligocene rocks are not included in the descriptions.

#### *Lower Paleozoic clastic rocks*

These range from Cambrian through Lower Devonian and are mainly coarse clastic types with some thin carbonate beds in the uppermost part. In northwestern Saudi Arabia, at least 1,900 meters of these rocks are exposed. Here they extend southeastward from the Jordan border, curve around the northern margin of the shield and disappear eastward under the sands of the Great Nefud. Equivalent sedimentary rocks appear again south of the Great Nefud where they are gradually overlapped by Permian beds and finally wedge out completely west of Ar Riyadh.

The lower third of the lower Paleozoic section is almost exclusively white, buff, and red, medium- to coarse-grained cross-bedded sandstone of terrestrial origin. This is followed by a thick section of graptolite-bearing shale and cross-bedded sandstone, and then by a capping of 300 meters of shale, sandstone, and limestone.

Some beds have been positively dated by graptolites and brachiopods as Lower Ordovician, Silurian, and Devonian, but no Cambrian fossils have been found. It is believed, however, that the lower part of the section is equivalent to certain Jordanian units of known Cambrian age.

In southwestern Arabia a wedge of coarse-grained sandstone, more than 900 meters thick, occurs between the Precambrian basement and the Upper Permian beds. The precise age of this sandstone is not known.

The Paleozoic units rest, with marked unconformity, on the peneplaned surface of the Precambrian shield, but the time represented by this break is not known. Along the unconformity at the top of the section, Middle Devonian through middle Permian rocks are missing.

### *Permian and Triassic clastic rocks*

The Permian and Triassic clastic rocks consist of nearly 1,000 meters of alternating nonmarine and marine clastic units with some thick calcareous sections. They are exposed in a wide belt that swings around the eastern margin of the shield. At the base is limestone and dolomite alternating with gypsiferous shale. Above this is unfossiliferous nonmarine shale, and then a thick section of limestone interbedded with sandstone and gypsiferous shale. At the top are brightly colored clastic beds containing petrified wood, suggesting that the unit is mainly continental. Ammonites and other mollusks, and abundant spores and pollens from equivalent layers indicate that these beds are Upper Permian to Upper Triassic. The base of the section rests on Precambrian, Cambrian through Early Devonian, and middle Permian beds and marks an important unconformity. At the top of the section is another distinct unconformity.

### *Lower and Middle Jurassic clastic and carbonate rocks*

In central Arabia these rocks are represented by some 500 meters of shallow-water marine shale and carbonate. In the vicinity of Ar Riyadh and Dhurma (lat 24° 38′ N), thin basal sandstone is followed by alternating highly fossiliferous limestone and shale. Southward, along the outcrop, shale and carbonate gradually give way to sandstone which is apparently of continental origin.

The Lower Jurassic rocks rest on Upper Triassic sandstone with distinct disconformity. However, the change from Middle to Upper Jurassic is believed to take place without interruption or with only slight disconformity.

### *Upper Jurassic and early Lower Cretaceous carbonate rocks*

These rocks, all shallow-water limestones, are well exposed in central Arabia where they form a series of scarps roughly parallel to the eastern edge of the shield. A section of more than 850 meters has been measured near

Ar Riyadh (lat 24° 38′ N). Here, calcarenitic limestone predominates although thick beds of calcarenite are common.

The basal 200 meters of Upper Jurassic section is a lithographic limestone with abundant corals and stromatoporoids. This massive unit stands up in sharp relief to form the great scarp of Jebal [1] Tuwayq which trends roughly north-south for more than 1,000 kilometers across the Arabian Peninsula. At the top of the Upper Jurassic is a unit of massive anhydrite, the Hith Formation. Much of this evaporite has been removed by solution, so that most of the area where it might have cropped out is now a solution-collapse zone.

The boundary between Upper Jurassic and older beds is conformable north of lat 19° 00′ N and unconformable south of it. The boundary between the Upper Jurassic-early Lower Cretaceous carbonate rocks and overlying strata involves two unconformities.

*Late Lower Cretaceous clastic rocks*

This division consists mainly of continental sandstone with a basal limestone unit. The belt of sandstone is very narrow near lat 26° 00′ N but widens rapidly to the south, maintaining a uniform width from lat 24° 00′ N to lat 20° 30′ N where it is covered by younger rocks. The maximum thickness of the sandstone is about 700 meters. The basal limestone unit has a measured maximum thickness of 20 meters and crops out as a discontinuous scarp from lat 25° 30′ N to lat 20° 35′ N. The upper and lower boundaries of this division are obvious unconformities.

*Middle Cretaceous clastic rocks*

These rocks are represented by a thin persistent sandstone unit with layers of interbedded shale, dolomite, and limestone. Most of the sandstone is brown and cross bedded; the shale is green, red, and purple. The sandstone crops out as a narrow curved band from about lat 20° 00′ N to the eastern edge of the Great Nefud (lat 28° 00′ N), a distance of about 1,100 kilometers. Patches of these rocks also crop out in the southwestern Rub' Al Khali and north of the Great Nefud near Sakakah (lat 30° 00′ N). South of lat 26° 00′ N the thickness of the Middle Cretaceous section is very uniform and averages 40 meters. The unit thickens northwestward, however, and at Sakakah it measures 250 to 300 meters. The lower and upper contacts of the Middle Cretaceous clastic section are well-marked unconformities.

*Upper Cretaceous to Eocene carbonate rocks*

These are widely exposed in a belt in the central and northern parts of the Arabian Peninsula. South of lat 24° 00′ N the width of outcrop is relatively constant and the belt follows the trend of the older units. Northward the belt cuts across the strike of the older units and gradually widens until near Sakakah (lat 30° 00′ N) the width of outcrop of the Upper Cretaceous rocks alone is 160 kilometers, and younger units extend into Iraq.

[1] Jebal (mountains) is the plural of Jabal (mountain) in Arabic.

The Upper Cretaceous and Eocene rocks are mostly limestone and dolomite with thin layers of marl and shale. At the northern and southern ends of the belt, the Upper Cretaceous section is gradually replaced from the bottom up by sandstone. The lower contact of the Upper Cretaceous to Eocene carbonate section is marked by a strong angular unconformity in some places, but appears to be merely a time line in others. The upper contact, however, is a widespread, persistent unconformity.

Cretaceous rocks are not exposed east of the main outcrop belt, but Paleocene and/or Eocene rocks appear again near Abqaiq, Dhahran, Qatar (near lat 26° 00′ N, long 50° 00′ E), and in the southeastern Rub' Al Khali. Also, a broad band of Eocene rocks is exposed along the southern edge of the peninsula in the eastern part of the Republic of Southern Yemen and in Dhufar.

*Miocene and Pliocene clastic rocks*

These strata are mostly nonmarine and have a thickness of only 200 to 300 meters. In the Eastern Province they consist of sandy limestone, calcareous sandstone, sandy clay, and sandstone. Fresh-water fossils are sparse and of no use in dating, but some middle Miocene marine molluscs and echinoids have been found in the middle part of the sequence. A similar but thicker section of Miocene and Pliocene sandstone, sandy marl, and sandy limestone covers Ar Rub' Al Khali. These rocks are all nonmarine clastics and attain a thickness of as much as 600 meters. Miocene and Pliocene rocks everywhere rest unconformably on Eocene or older beds. Quaternary and Recent sand and gravel are distributed erratically on the Miocene and Pliocene beds.

### *1.1.4.4. Structure of the shelf*

The Arabian Shelf can be divided into several subprovinces: 1. an Interior Homocline, everywhere marginal to the shield, 2. an Interior Platform bordering the homocline, and 3. several basins, adjacent to the platform, that have collected thicker sedimentary deposits.

*Interior Homocline*

The Interior Homocline extends from Tabuk (lat 28° 22′ N, long 36° 35′ E), and Al Jawf (lat 29° 49′ N, long 39° 52′ E), in the northern part of Saudi Arabia, southward along the eastern edge of the shield to the Republic of Southern Yemen and then swings eastward into Oman. In the north its width averages 400 kilometers, but west of Ar Rub' Al Khali its width is 600 kilometers and in Dhufar and Oman it is 300 kilometers. It is a region in which the Paleozoic, Mesozoic and Cenozoic strata dip very gently, i.e. about 1° 00′ in the Permian and Triassic beds and 0° 20′ in the Upper Cretaceous and Eocene beds. For the most part the beds dip off the shield so that their strikes tend to parallel the shield boundary. Along the southern edge of the peninsula, however, the beds dip northward into Ar Rub' Al Khali basin.

The sharp change in strike of the Paleozoic, Mesozoic, and Cenozoic strata in the homocline near lat 24° 00′ N is caused by the so-called central Arabian arch. This structure is a gentle ridge or swell in the basement rocks that affects the distribution and attitude of all the sedimentary rocks from the basement up. The arch extends all the way to Qatar. On the homocline it separates the northern from the southern Tuwayq escarpments. The arch is apparently a residual high, that originated in Middle or Late Cretaceous time, between the periodically sinking northern Arabian Gulf and Ar Rub' Al Khali basins (described later), and not an independent positive feature. This view is supported by the presence of an arc of tensional structures (grabens and troughs) near the crest of the arch north of lat 24° 00′ N.

The Hail arch is a similar structural feature of the homocline that has caused an abrupt shift in strike of the sedimentary rocks around the northern end of the shield in the vicinity of Hail (lat 27° 33′ N, long 41° 42′ E).

### *Interior Platform*

The Interior Platform borders the Interior Homocline on the east. It is an area, varying greatly in width, in which a systematic eastward dip of the strata no longer prevails. This flat structural feature is separated from the homocline by an abrupt break-in-slope. The platform is widest (400 kilometers) in the vicinity of Qatar where it extends as far east as the coast of Iran. Around the western and southern sides of Ar Rub' Al Khali the width ranges from 100 to 200 kilometers. Structure drill and seismic data indicate that the sedimentary rocks on the platform are thick and unusually flat. However, several major north-south anticlinal axes rise above the level of the platform and some of the larger ones include the important oil fields of Al Ghawar, Abqaiq, and Al Qatif, all in the area just west of Qatar. The origin of the anticlines is uncertain, but there is reason to believe that they are related to some sort of horstlike uplift at great depth.

### *Basins*

Four major sedimentary basins, containing thick sedimentary deposits, lie on the Arabian Shelf adjacent to parts of the Interior Platform: 1. Ar Rub' Al Khali basin, 2. northern Arabian Gulf Basin, 3. Ad Dibdibah basin, and 4. As Sirhan-Turayf basin.

Ar Rub' Al Khali basin trends northeast from the center of Ar Rub' Al Khali desert almost to the coast of Iran. Its average width is 300 kilometers. It appears to be primarily a Tertiary feature in which Paleocene, lower and middle Eocene, and upper Tertiary sedimentary rocks thicken toward the center. The northern Arabian Gulf basin at the head of the Arabian Gulf, and the Dibdibah basin lying just west of it, are smaller, and their nature and history are less well known. In northwestern Saudi Arabia just south of Turayf (lat 31° 40′ N, long 38° 38′ E), is the Sirhan-Turayf Basin, a synclinal depression filled with a thick sequence of Upper Cretaceous and Eocene sediments. Volcanic rocks cover much of these sediments. The structure of the basin is confused by complicated faulting, but it is postulated that the basin developed as a sag in a structurally flat area during Late Cretaceous and Eocene times.

## 1.1.5. Mobile Belt

### *1.1.5.1. General nature*

The Zagros Mountains of Iran and the Oman Mountains lie in the mobile belt of the Alpine orogeny. They consist of deposits that were laid down in the geosyncline during its lifetime and subsequently deformed and uplifted. The geology of these mountains differs greatly from that of the Arabian Shield and Arabian Shelf.

### *1.1.5.2. Zagros Mountains*

The nearly flat-lying Paleozoic, Mesozoic, and Cenozoic sedimentary rocks of the Arabian Shelf extend beneath the Arabian Gulf and rise up again on the opposite side in the Zagros Mountains. In a strip about 50 kilometers wide along the northeast side of the Arabian Gulf these rocks have been gently deformed into elongate, symmetrical folds parallel to the main mountain range. This strip of gentle folds, which has not been uplifted appreciably, constitutes the Zagros Mountains foreland. It is of interest to note that the rich oil fields of southwestern Iran and eastern Iraq are in the anticlines of this foreland. Behind the foreland is a belt about 225 kilometers wide in which the strata have been deformed into larger and tighter folds. The axes of these trend northwest throughout most of the belt, but become east-west near the eastern end of the Arabian Gulf. Eastward this belt of folding finally passes into a region of overthrusts about 50 kilometers wide. Here the rocks have been greatly deformed and large nappes have been pushed forward toward the southwest. The strongly folded and faulted sedimentary rocks described here have been thrown up into high mountains where their eroded edges now reach elevations of over 4,000 meters. Farther northeast, Precambrian basement rocks like those exposed in the Arabian Shield appear at the surface again.

### *1.1.5.3. Oman Mountains*

The Oman Mountains are of the Alpine type and completely different from the other mountains of Arabia. The highest tectonic unit visible in the Oman Mountains is the complex Semail nappe, composed of ophiolites. Beneath this nappe are other imbricated nappes of Paleozoic and Mesozoic geosynclinal sediments (carbonates and cherts) without any regional metamorphism. Gently deformed Mesozoic rocks, generally dipping westward, lie on the flanks of the mountains and are almost entirely covered by flat-lying Tertiary sediments.

Ophiolites, such as those that appear in the Semail nappe, are rather unusual rocks, and their presence here is of considerable interest. The Oman ophiolites are an assemblage of igneous and metamorphic rocks, believed by some to be polygenetic. They can be grouped into: 1. older ultrabasic types (peridotites), probably pre-geosynclinal (pre-Middle Permian), and 2. gabbroic-basaltic types covering the whole geosynclinal cycle (Middle Permian lower Upper Cretaceous). The ophiolites are believed to have come from the earth's mantle. They are now associated with rocks that accumulated on the geosyncline floor possibly as early as Middle or Late Permian time but certainly by early Late Cretaceous time. Field relations suggest to some geologists that large amounts of the plutonic rocks were actually expelled from depth in a solid state. During the Alpine orogeny, the emplaced ophiolite mass and its associated rocks were transported westward into their present position along a low angle thrust plane.

It is of interest to note how the Oman Mountains trend northward along the Oman line, directly into the Zagros Mountains. The structure and tectonic history of the whole Oman region are very intriguing, but these aspects of the geology are complex and rather controversial and beyond the scope of this paper.

## 1.1.6. Red Sea

### *1.1.6.1. Topography*

The Red Sea is an elongate trough extending NW-SE from the Sinai Peninsula to Bab Al Mandab and separating the Arabian Peninsula from the continent of Africa. Throughout most of its length of over 1,800 kilometers its opposing shorelines are remarkably parallel in plan. In the north the width of the sea is only 175 kilometers but southward it increases to a maximum of 350 kilometers near Jizan (lat 16° 55′ N, long 42° 35′ E), and from there it decreases to a minimum of 30–40 kilometers at Bab Al Mandab (lat 12° 35′ N, long 43° 30′ E). Inland of the shorelines is a narrow coastal plain, from one to 40 kilometers wide, which is backed by a prominent escarpment 1,500–3,000 meters high. The escarpment marks the uplift of the margins of the Arabian and African Shields, and is the structural edge of the Red Sea rift area.

The Red Sea area can be divided into two main physiographic units: 1. the shelf area which consists of the coastal plain and the Red Sea marginal shelves, and 2. the trough area which consists of the Red Sea trough and the axial trough. The shelf area is separated from the deeper trough area by a steep break-in-slope that drops from 200–300 meters to 600–1,000 meters. The shelf is very narrow in the north but widens to the south, whereas the trough area is wide in the north and narrows southward until it nearly disappears at Bab Al Mandab. The axial trough, which is very narrow and deep, lies along the axial zone of the Red Sea. Near lat 21° 00′ N, isolated deeps within the axial zone are more than 2,000 meters below sea level.

### *1.1.6.2. Sediments*

The oldest sediments in the Red Sea are of Upper Cretaceous age. These, as well as younger Eocene and Oliogocene sandstones and marls, are scant and have been found only in scattered areas. A thick sequence of middle and upper Miocene evaporites, exceeding 3,500 meters in thickness in the southern Red Sea, overlies the Eocene-Oligocene strata. Above the evaporites are limestones and reefs of Pliocene to Recent age. Thus, it appears that the Red Sea dates mainly from the Miocene.

### *1.1.6.3. Structure*

Structurally, the Red Sea is a graben along the crest of an anticline that formed in the Arabian-African Shield. The inner margins of the shield apparently underwent considerable uplift (2,000–3,000 meters) and now form prominent scarps at the edge of the Red Sea rift. A zone, 1–2 kilometers wide, of high-angle and tensional faults concealed by coastal sediments, lies at the foot of the escarpments. On the seaward side of this zone the basement has been step faulted downward in blocks, and now lies beneath the shelf area at depths between 2,000 and 3,000 meters below sea level. The basement is covered by a thick sequence of sediments.

Three sets of faults seem to have controlled the development of the Red Sea: 1. a NW-SE set which determined the major lines of weakness and is associated mainly with step faulting, 2. a NNE-SSW, or Gulf of Al Aqabah, set which is mainly one of wrench faulting and which appears to have determined the resultant direction of separation of Arabia and Africa, and 3. a WNW-ESE set which is a major fault trend in the Precambrian basement and which caused many irregularities in the coastlines.

### *1.1.6.4. Tectonics*

The Red Sea is a part of a much larger tectonic system associated with Tertiary orogenic movements. It is connected to the Levantine fault via the Gulf of Al Aqabah, to the East African rift system via the Danakil Depression, and to the mid-oceanic ridges of the world via the Gulf of Aden and the Carlsberg Ridge. It is generally believed that Arabia separated from Africa along the Red Sea by crustal thinning and sea floor spreading. Evidence for this comes from several sources: 1. numerous geophysical measurements and surveys, 2. interpretations of satellite photographs, and 3. paleomagnetic data.

The process began in late Cretaceous-Eocene time with the doming of the Arabian-African Shield and the formation of a crack along the crest of this elongate dome. In late Eocene-Oligocene time NE-SW tension caused normal step faulting and rifting of the crack and associated monoclinal downwarping of the edges. During this time plateau basalts (Trap Series) were extruded in the south, and perhaps the northeast separation of Arabia and Africa began. During the Miocene the Red Sea was relatively quiet, but in late Miocene-Pliocene time renewed activity caused the breaking of the basement crust and the start of sea floor spreading. Today, movement is still in progress and involves: 1. transcurrent movement in a N-NE direction,

2. anticlockwise movement of Arabia relative to Africa, 3. E-W sea floor spreading with infusion of basic magma in the newly-formed axial trough, and 4. lava effusion in the marginal zones.

## 1.1.7. Economic Geology

### *1.1.7.1. Oil and gas*

The enormous oil and gas reserves of the Arabian Gulf region are well known. At the end of 1974, the estimated crude oil reserves, in millions of barrels, of the various countries of the Arabian Gulf region were as follows: Saudi Arabia 181,219, Kuwait 81,450, Iran 66,000, Iraq 35,000, Abu Dhabi 30,000, Qatar 6,000, Oman 6,000, Dubai 2,420, and Bahrain 336. Saudi Arabia's oil reserves are the greatest of any country in the world, more than twice those of the U.S.S.R., its closest competitor. Moreover, Saudi Arabia is the world's third greatest oil producing country.

Most Saudi oil and gas come from gentle north-south anticlines developed in the shelf sedimentary rocks along the western edge of the Arabian Gulf. The anticlines lie both onshore and offshore. One of these structures, 250 kilometers long, gives rise to the huge Al Ghawar field, the largest oil field in the world. About 85 per cent of Saudi Arabia's petroleum comes from the D Member of the Arab Formation (Upper Jurassic) which is characterized by a high percentage of porous calcarenite beds.

### *1.1.7.2. Mineral deposits*

The Arabian Peninsula is less well endowed with mineral deposists of commercial value than with petroleum. However, intensive investigations are under way in the shield area and elsewhere, and these have resulted in some rather significant mineral findings. Although in the past none of these deposits has been considered to be of commercial grade or size, current mineral prices may make some of them attractive to mining companies.

Large tonnages of phosphate rock have been discovered near Thaniyat Turaif (lat 29° 45′ N, long 38° 10′ E), and Turayf (lat 31° 41′ N, long 38° 40′ E) in northern Saudi Arabia. The content of $P_2O_5$ is low, but parts of these deposits are believed suitable for open pit mining methods, and the phosphate can be upgraded by milling.

Hematite has long been known in Wadi Fatimah (lat 21° 30′ N, long 39° 30′ E) and in the Wadi As Sawawin area (lat 28° 00′ N, long 36° 00′ E), but has not been mined as yet. Large deposits of massive pyrite and pyrrhotite occur in the Wadi Wassat area (lat 18° 20′ N, long 44° 10′ E), and in Wadi Qatan (lat 18° 07′ N, long 44° 10′ E), but these minerals are of no value in Arabia at the present time. The discovery of nickel in the Wadi Qatan sulfides could have a bearing eventually on the value of this deposit.

The Arabian Peninsula is dotted with localities in which gold and silver have been mined off and on since ancient times. However, the Mahd Ath

Thahab deposit (lat 23° 50′ N, long 40° 50′ E) is the only one that has been productive in recent times. From 1939 to 1954 it produced 765,768 ounces of gold and 1,002,029 ounces of silver. Known reserves in stopes and in the quarry are believed to be significant. The Mahd Adh Dhahab deposit is considered as a major producer of gold and silver, and its potential is currently being investigated. The findings thus far are encouraging. The gold-lead-zinc deposit at Samrah (lat 24° 15′ N, long 44° 20′ E) is being reevaluated, and despite its small size, the rich ore should make it of commercial value. The well-known gold-silver-zinc-copper deposit at Al Amar (lat 23° 45′ N, long 45° 00′ E) is being drilled and studied to determine its potential.

Other deposits of possible commercial value include: 1. Jabal Said (lat 23° 45′ N, long 40° 58′ E) which contains copper, zinc, gold, and silver, 2. As Safra (lat 24° 10′ N, long 41° 50′ E) which contains copper and silver, and 3. Jabal Dhaylan area (lat 37° 10′ N, long 25° 45′ E) which contains lead, zinc, and copper.

Mineral deposition is active in the deeps of the Red Sea, and the sediments and overlying hot stratiform brines are rich in lead, zinc, and copper. Upwelling solutions from below the sea floor are considered to be the source of both the heat and metals. The mineralization is located in the area where sea floor spreading is most active and where there may be a fault zone. Much exploration has been done in the deeps, and the findings show clearly that vast mineral resources are present here. No doubt these will be mined at some future date when deep sea mining methods have been improved and there is need for the metals.

## 1.2. Geomorphology

(R. W. Chapman)

### 1.2.1. Introduction

Geomorphic studies in the Arabian Peninsula have lagged far behind geologic studies. Published accounts of work that has been done are few and scattered widely throughout the literature so that it is practically impossible to gain a comprehensive picture of the geomorphology of the peninsula. This brief account, mainly descriptive, is designed to acquaint the reader with what is known about Arabian landforms and the processes that have caused them. It is hoped that this will enable him to understand and evaluate better the Quaternary events described in this volume.

The geomorphology is considered by provinces, beginning in the northeast on the Arabian Gulf and working westward, then southward, and then eastward to the Oman Mountains, in a counter-clockwise direction. Since the landforms of many areas are not well known, the descriptions of some of the provinces are brief and lacking in detail.

It must be remembered that although the present climate of the peninsula is generally hot and arid and most regions are desert, this has not always been so. During the several Pleistocene pluvials, for example, rainfall was heavier than now and temperatures somewhat lower. It can be expected that such different climatic conditions would have left their imprints on the landforms [1].

### 1.2.2. Arabian Gulf Coastal Region

The eastern edge of the Arabian Peninsula, from the southern end of the Arabian Gulf northwestward through Iraq and into Syria, lies in a rectangular depression formerly known as Mesopotamia. In the northern part of this depression are the Tigris and Euphrates Rivers which have their headwaters in the Taurus and Anti-Taurus Ranges in Turkey. On the east there is a pronounced scarp separating the depression from the highlands of Kurdistan, but on the west the ascent to the Syrian Plateau is less abrupt. North of Baghdad the region is an open undulating plain with low hills; south of Baghdad it constitutes the flood plain and deltas of the two great rivers. The southern part of the depression includes the western half of the Arabian Gulf and a narrow coastal strip of the Arabian Peninsula. This coastal strip is called the Arabian Gulf coastal region.

The Arabian Gulf is a broad, shallow epicontinental sea with a maximum depth of 110 meters. The shore is irregular, low, and sandy and the water has many shoals, so that changes in tide and wind direction cause the waterfront to shift back and forth for several kilometers. Consequently, the low coastal flats are covered by the sea periodically. The elevation of the Arabian Gulf coastal region rises gradually inland at the rate of about one meter per kilometer.

Salt flats or sabkhahs are common along the shoreline all the way from Kuwait to the southern end of the Arabian Gulf. Sabkhat Matti, southeast of Qatar, is the largest of these with an area of about 6,000 square kilometers. These low, exceedingly flat features are composed of layers and mixtures of sand, silt, mud, and salts to a depth of several meters. Intense evaporation commonly brings up salts that form a thin surface crust. The surface of the sabkhahs is maintained at the level to which moisture rises above the static water table. During dry weather, the sabkhah surface becomes firm and smooth and is often covered with mud polygons. Some sabkhahs are now found many kilometers inland at elevations as high as 150 meters. It is not known whether they formed at these higher elevations or whether they were elevated after their formation.

Along the northwestern shore of the Arabian Gulf, from Al Jubayl northward to Kuwait Bay, are low rolling plains covered with a thin mantle of sand. The roots of bushes and grass hold the sand in hummocks and form an irregular, bumpy terrain called dikakah. South of Al Jubayl is a wide belt of drifting sand and large dunes which widens southward and, beyond Al Hofuf, merges with the sand area known as Al Jafurah. Those parts of

[1] The geomorphic map (Fig. 2) that accompanies this account was drawn originally by Clarence J. Beauchamp to whom the writer is greatly indebted. It was prepared from a base map kindly furnished by the Arabian American Oil Company.

the Arabian Gulf coastal region not covered with sand or sabkhahs are barren rock terrain developed on Eocene, Miocene, and Pliocene limestones.

In the fertile oases of Al Qatif and Al Hasa (around Al Hofuf) are many wells and artesian springs with abundant fresh water which make these two

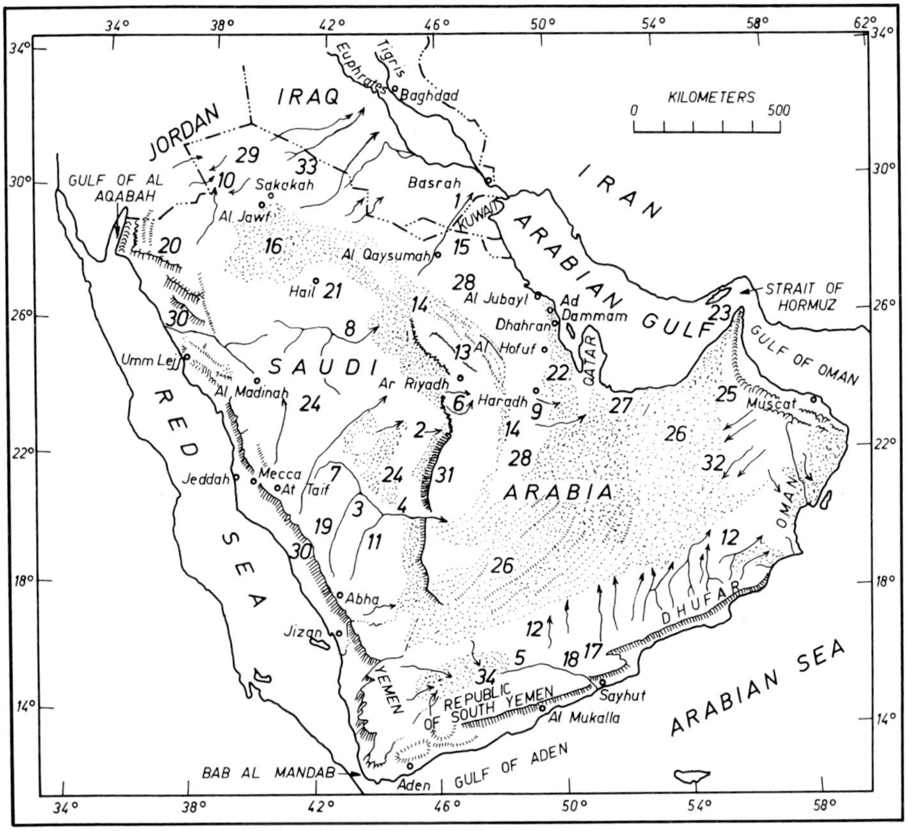


Fig. 2. Map of the Arabian Peninsula showing the principal geomorphic features. Numbers indicate the locations of the wadis and landforms listed in the following explanation.

Wadis: *1* Wadi Al Batin, *2* Wadi Birk, *3* Wadi Bishah, *4* Wadi Ad Dawasir, *5* Wadi Hadhramawt, *6* Wadi Nisah, *7* Wadi Ranyah, *8* Wadi Ar Rimah, *9* Wadi As Sah'ba, *10* Wadi As Sirhan, *11* Wadi Tathlith

Land Forms: *12* Aden Hinterland, *13* Al 'Aramah Escarpment, *14* Ad Dahna, *15* Ad Dibdibah, *16* Great Nefud, *17* Hadhramawt Plateau, *18* Hadhramawt Valley, *19* Al Hijaz Plateau, *20* Hisma Plateau, *21* Jabal Shammar, *22* Al Jafurah, *23* Musandam Peninsula, *24* Najd Pediplain, *25* Oman Mountains, *26* Ar Rub' Al Khali, *27* Sabkhat Matti, *28* As Summan Plateau, *29* Syrian Plateau, *30* Tihamah, *31* Tuwayq Escarpment, *32* Umm As Samim, *33* Al Widyan, *34* Yemen Highlands

areas important producers of dates and other agricultural products. Water infiltrates the upturned edges of aquifers in the higher country to the west and flows eastward through the aquifers by gravity.

A low-level bench, containing abundant shells and serpulite reefs and standing 1.5 to 2 meters above sea level, follows the coastline from the vicinity of Kuwait to a point south of Qatar and then appears again along the southern shore of the Arabian Gulf. Another, higher marine deposit, known as the Bahr Formation and consisting of calcarenite and quartz sand interbedded with dune sand, forms low cliffs along much of the shoreline. It lies at various heights ranging from slightly below sea level to 150 meters above sea level. These two deposits indicate higher stands of the Arabian Gulf in former times.

An interesting and significant feature of the Arabian Gulf coastal region is a great gravel plain in Ad Dibdibah area southwest of Kuwait. This triangular-shaped plain has its apex near Al Qaysumah and it spreads northeastward almost to the Tigris and Euphrates Valleys. It is composed of cobbles and pebbles of quartz, igneous and metamorphic rocks, and limestones with finer admixed sediments. The cobbles and pebbles decrease in size toward the northeast. Ad Dibdibah gravel plain is interpreted as a delta of the Wadi Ar Rimah-Wadi Al Batin drainage system which swept rock debris from the crystalline uplands and sedimentary steppes of the northern Najd during one of the pluvials.

Another similar deltaic gravel plain has its apex in Wadi As Sah'ba near Haradh. From this point divergent gravel trains fan out toward the Arabian Gulf. The present channel of Wadi As Sah'ba, blocked with sands of Al Jafurah, crosses the plain and enters the Arabian Gulf south of Qatar. Foreset beds of sands and silts have been found along the margins beneath the gravels.

The Arabian Gulf coastal region west of Al Hofuf is bounded by a prominent east-facing escarpment extending several kilometers north and south. This cliff is cut in the eastern margin of As Summan Plateau (described later), and the elevation of its base ranges from about 150 meters to about 200 meters. The escarpment has not been studied in detail but it is believed now that it is the result, in part at least, of wave erosion during a higher stand of sea level.

Thus, the elevated shell bench and Bahr formation, the high level sabkhahs, the two delta-like gravel plains, and the prominent, possibly wave-cut, escarpment in the Arabian Gulf coastal region, all suggest one or more higher stands of sea level.

### 1.2.3. As Summan Plateau

Bordering the Arabian Gulf coastal region on the west in Saudi Arabia is a long, flat, hard-rock plateau, from 80 to 250 kilometers wide, trending north-south. This is As Summan Plateau. Along its western margin the elevation of As Summan averages about 400 meters, and along its eastern edge about 250 meters. Thus, it slopes eastward approximately 0.7 meter/kilometer. The surface of the plateau is flat and barren and probably mainly of depositional origin. The eastern edge is a prominent escarpment indented by ancient stream valleys and possibly developed in part by marine erosion.

Beyond the escarpment, isolated buttes and mesas project prominently into the coastal lowlands. The bedrock is permeable, flat-lying sandy limestone (Hofuf Formation) of late Miocene or Pliocene age, riddled with sink holes, solution cavities, and caves (karst topography). Beneath the plateau, west of Al Hofuf, lies the Arabian American Oil Company's gigantic Al Ghawar oil field, the largest in the world.

As Summan Plateau is actually a prong of the larger Syrian Plateau to the north. The whole physiographic unit extends from a point about 300 kilometers south of Al Hofuf northwestward across northeastern Saudi Arabia and western Iraq and into Syria where it abuts against the Trans-Jordan Highlands. There it attains its maximum width of 650 kilometers. In Iraq, Syria, and northern Saudi Arabia the plateau is a vast desert of gravel and rock plains underlain by Cretaceous and Tertiary sedimentary rocks and Pliocene basalt. It is only slightly warped, and there is no sharp break between it and the higher Trans-Jordan Highlands. From Sakakah northward to the junction of the borders of Saudi Arabia, Jordan, and Iraq, the plateau attains an elevation of nearly 900 meters and forms an important drainage divide. On the east side of this divide a well-coordinated system of dendritic wadis, known as Al Widyan, carries drainage eastward onto the plains of Iraq and into the Euphrates River. On the west side, drainage flows into Wadi As Sirhan, a great depression, over 300 kilometers long, from 30–50 kilometers wide, and 300 meters below the plateau surface. Prior to the construction of the Trans-Arabian pipeline, Wadi As Sirhan served as a major trade route between the Mediterranean region and central Arabia.

### 1.2.4. Eolian Sand Areas

Approximately one third of the Arabian Peninsula is covered by mobile sand in one form or another. In the northern part of the peninsula, the term "nefud" is applied to areas of deep sand and wind-built dunes. These areas include: 1. The Great Nefud, 2. Al Jafurah, 3. Ad Dahna, and 4. a group of sand bodies along the western side of the Tuwayq Escarpment. In the southern part of the peninsula is another area of eolian sand, the Great Arabian Desert or Ar Rub' Al Khali (Empty Quarter). Since all these areas have characteristics in common, they will be considered together.

The Great Nefud, in the northwestern part of the peninsula between Hail and Al Jawf, is shaped like a giant hand with long fingers pointing toward the west. It occupies a broad basin of low relief, Al Jawf-Sakakah Basin, and covers an area of approximately 57,000 square kilometers. To the east and north is the Syrian Plateau, and to the west and south are the Hisma Plateau and Jabal Shammar, respectively. The Great Nefud is a vast expanse of reddish sand and rolling dunes with sparse vegetation, a region without streams and without oases.

Ad Dahna is one of the most distinctive physiographic features of the Arabian Peninsula. It is a long narrow belt of shifting sand and dunes extending nearly 1,300 kilometers in a broad arc from the Great Nefud in the north to Ar Rub' Al Khali in the south. It lies between As Summan Plateau on the east and the cuesta region on the west. The sands of Ad Dahna are bright red-orange due to a coating of iron oxide on the quartz grains. In the early morning and late afternoon sunlight they take on a soft pinkish-purple hue. Although water is scarce, Ad Dahna is a favourite grazing ground for camels in winter and spring.

East of As Summan Plateau and south of Al Hofuf is an area of drifting sand called Al Jafurah, the north end of which lies within the Arabian Gulf coastal region. Al Jafurah is narrow here but widens rapidly to the south where it eventually merges with the sand area of Ar Rub' Al Khali. In contrast with Ad Dahna sands, Al Jafurah sands are buff to tan in color.

A fourth area of eolian sand, roughly parallel to Ad Dahna, follows along the western edge of the Tuwayq Escarpment from the Great Nefud southward to the vicinity of Wadi Ad Dawasir. The sand is not continuous but occurs in small elongate patches strung out in a chain. It tends to lie in low areas and its distribution and movement appear to be influenced in part by the high escarpment.

In southern Arabia is the immense, forbidding area of Ar Rub' Al Khali or Empty Quarter. The sand here lies in a huge basin bounded by the Oman Mountains on the east, the Hadhramawt Highlands and the Aden Hinterland on the south, and Al Hijaz Plateau on the west. This sand region has an approximate length of 1,200 kilometers, a maximum width of about 650 kilometers, and a total area of 600,000 square kilometers, making it the largest continuous body of sand in the world. The sand is of red-orange color and of medium to fine grain like that of the Great Nefud and Ad Dahna. In the eastern part of Ar Rub' Al Khali is an enormous group of sabkhahs interspersed among the sand dunes. More than half of these are elongate in a WNW-ESE direction. By far the largest is Umm As Samim with a length of over 100 kilometers. Most of Ar Rub' Al Khali is uninhabited except for some Bedouins who move into the fringe areas during rains to take advantage of the pasturage, and others who frequent such places as the northeastern section, where water wells are common.

Geologists recognize four classes of eolian sand terrain in the Arabian Peninsula:

I. *Transverse.* Predominantly simple and compound barchan dunes in areas of more mobile sand and/or simple rounded ridges, oriented transverse to the prevailing wind direction.

II. *Longitudinal.* Primarily dikakah (bush- or grass-covered sand) and various types of undulating sand sheets, in general characterized by elongation of the individual forms parallel to the prevailing wind direction, often partly stabilized by sparse vegetation.

III. *'Uruq.* Various forms of long, nearly parallel, sharp-crested narrow sand ridges and dune chains separated by broad sand valleys usually including elements of class I sand terrain. These forms are the resultant of a system of two dominant wind directions and are identical to the sayf dunes of north Africa.

IV. *Sand mountains.* Dominated by large sand massifs commonly cresting 50 to 300 meters above the substratum, often with superimposed dune patterns consisting of various types of complex barchans. Common forms are giant barchans spanning several kilometers from horn to horn, giant sigmoidal and pyramidal sand peaks as well as other less common peak forms, and giant oval to elongate sand mounds.

In the Great Nefud, classes II and III are by far the most common, but in the eastern part there is some terrain of class IV. In Ad Dahna, Class III is dominant although the other three classes also occur. Al Jafurah sand terrain is all of class I, and along the western side of the Tuwayq Escarpment the main classes are III and IV. Along the northern edge of Ar Rub' Al Khali classes I and II are most common, but throughout the central and western parts class III predominates. The eastern part of Ar Rub' Al Khali has mostly class IV terrain.

The dune shapes found in Ar Rub' Al Khali are typical, and they illustrate well the great variety of shapes present in all the sand areas. Dunes are crescent-shaped (barchans), dome-shaped, hook-shaped, and star-shaped. Some are sigmoidal and others are simple transverse types. Many dunes are long, single or parallel ridges ('uruq) with heights of 100 meters or more and lengths of 20 to 200 kilometers. Many others are huge complex pyramids or "sand mountains", attaining heights of up to 300 meters. In the eastern Rub' Al Khali are hundreds of magnificent red sand mountains, each one completely surrounded by sabkhah flats.

Most of the sand in nefuds and in Ar Rub' Al Khali comes from secondary and tertiary sources such as the Paleozoic and Mesozoic sandstones and the alluvium derived from all the older formations. A certain amount is, no doubt, blown in from Syria and Jordan to the north. The primary source of most of the sand is the large granite batholiths underlying the Arabian Shield.

Unfortunately, little is known about the movements of eolian sand in Arabia. This information can come only from a careful study of 1. regional wind patterns, 2. sources of sand supply, and 3. characteristics of local topography.

### 1.2.5. Cuesta Region

This region in central Arabia takes its name from several prominent north-south escarpments that sweep gracefully around the eastern edge of the crystalline shield in the general vicinity of Ar Riyadh. The escarpments are cuestas formed on the upturned edges of sedimentary beds dipping very gently toward the east. The zone of escarpments is bounded on the east by Ad Dahna and on the west by the central plateau region, and its width is over 250 kilometers.

The Tuwayq Escarpment west of Ar Riyadh is the largest. It is 800 kilometers long; its average elevation above sea level is 840 meters, and its maximum elevation about 1,500 meters. The top is capped by Upper Jurassic limestone and it rises about 240 meters above the plains to the west. Al 'Aramah

Escarpment, northeast of Ar Riyadh, is approximately 250 kilometers long, and its summit has an elevation above sea level of some 540 meters. It is less conspicuous than the Tuwayq Escarpment because it stands only 120 meters above the plain to the west. Al 'Aramah Escarpment is topped by Upper Cretaceous limestone. Other lower and shorter escarpments occur west of the Tuwayq Escarpment and roughly parallel to it. It is within the low, elongate depressions between the escarpments that the small nefuds, described earlier, are found.

In general, resistant limestones form the summits of the cuestas, and less resistant rocks, mainly sandstones, underly the intervening plains. Where soluble strata like anhydrite and gypsum were formerly present, their removal by solution is marked by large sinkholes and solution cavities at the base of the scarps. The tops of the cuestas are not true dip slopes; the limestone beds are truncated at very low angles by erosion surfaces sloping eastward toward the sandstone plains.

The Tuwayq Escarpment is transected by a number of major wadi channels that give rise to an interesting geomorphic problem. Examples of such channels are Wadi Al Hinuw, Wadi Ad Dawasir, Wadi Birk, and Wadi Nisah. The drainage along these is from west to east and it has been assumed that they were cut by eastward-flowing streams. However, although they are actually deep canyons throughout most of the cuesta, they are much wider at their western ends where they form V-shaped notches in the escarpment face.

The cuesta region is often called the Oasis Belt. Desert streams, many completely dry at the surface, flow in wadis from the higher plateau areas to the west, and enter the basins. Here they either appear as springs or cross the basins and become lost in the desert sands. Other, smaller streams flow down the dip slopes of the escarpments and contribute additional water to the basins. With its abundant springs and wells and its rich volcanic soil washed in from the west, the Oasis Belt is a region of considerable fertility.

### 1.2.6. Central Plateau Region

The immense area of elevated terrain in west-central Arabia is known as the central plateau region. This area is bounded on the east by the cuesta region, on the north by the Great Nefud and the Trans-Jordan Highlands, on the west by the mountains along the Red Sea coast, and on the south by the Hadhramawt Plateau and the Aden Hinterland. It has a trapezoidal shape and a width of 500–600 kilometers, and it is one of the larger geomorphic provinces of the peninsula.

The topography of the central plateau region was carved on an ancient land mass of crystalline rocks, the Arabian Shield. This land mass was peneplaned probably in Precambrian time, and at a much later date was elevated as a great plateau and tilted gently northeastward. The high western edge now forms the crest of the coastal mountains and marks the continental divide, the boundary between eastward and westward drainage.

The central plateau region can be divided into three physiographic subprovinces: 1. Al Hijaz Plateau, 2. Hisma Plateau, and 3. Najd Pediplain (Fig. 2). Al Hijaz Plateau is a triangular area in the southern part of the region that extends from At Taif southward to Abha where it is about 180 kilometers wide. This surface is highest along its western edge near Abha where the elevation is 3,133 meters. Northwestward, the elevation decreases to 2,386 meters at At Taif. The western edge of the plateau shows evidence of ramping and tilting along tension breaks and shears. Numerous benches between canyons indicate that the ramping was intermittent. Eastward-flowing wadis are entrenched 60–120 meters in their headwaters.

In northwestern Arabia is the Hisma Plateau, a ramped block comparable to Al Hijaz Plateau and elevated most at its southwestern corner. From the coastal mountains it slopes northeastward to a point about 100 kilometers distant.

Al Hijaz Plateau descends northeastward to the Najd Pediplain, a vast area nearly 600 kilometers wide, developed on Precambrian plutonic and metamorphic rocks. The pediplain is a mosaic of coalescing pediments and desert plains interrupted here and there by inselbergs, and its western part is dotted with lava fields of various sizes. Some are only a few hundreds of square kilometers in area, whereas the largest, Harrat Rahat, extends all the way from Mecca to Al Madinah and covers 24,000 square kilometers. In some places the lava flowed westward over Al Hijaz divide and down into the valleys toward the Red Sea. The older lavas were erupted from fissures, whereas the more recent ones were ejected from vents to form cinder cones. They range in age from Oligocene to Holocene, the middle Miocene and younger ones being more abundant. Clearly, the Najd Pediplain antedates the Oligocene volcanic eruptions. In fact, there is evidence that at least portions of it are much older, and represent parts of a Precambrian erosion surface that was buried and later exhumed with some modifications.

In general, the central plateau region has a gentle, gradual northeastward slope on which many drainage systems have developed. The largest of these is the Wadi Ar Rimah system which, through its extension, Al Batin, carries drainage all the way from its headwaters east of Al Madinah to the Arabian Gulf basin near Basrah. Another major drainage system consists of the three wadis, Ranyah, Bishah, and Tathlith which join to form Wadi Ad Dawasir.

Population is sparse throughout the central plateau region except in the northeast between the valley of Wadi Ar Rimah and the southern edge of the Great Nefud. In the vicinity of Jabal Shammar are scores of towns and villages clustered around the great oasis of Hail.

### 1.2.7. Mountains of Western Arabia

The mountains of western Arabia are often referred to as scarp mountains because they are carved in the steep western edge of the great eastward-tilted Arabian Shield. The northwesterly-trending escarpment is the result of Tertiary faulting that lowered the Red Sea coastal plain some 3,000 meters below the shield (Al Hijaz Plateau). The mountain belt runs the full length of the peninsula from the Gulf of Al Aqabah on the north to Bab Al Mandab on the south, and varies in width from 40–140 kilometers.

In the southern part of the peninsula the mountains rise boldly from the coastal plain and reach their highest elevation, 3,760 meters, in Yemen. From Yemen northward to the vicinity of Mecca, elevations decrease and knife-edge ridges and deep canyons are characteristic. The mountains are developed on folded and faulted Precambrian rocks and many of the ridges and valleys are related to structure. Between Jizan and At Taif they parallel the escarpment, occupying grabens and the crests of anticlines. In the vicinity of Jeddah and Mecca, one of these grabens has been eroded to a broad plain that serves as an important route of communication between the Red Sea and the interior of the peninsula.

From At Taif and Mecca northward, the height and ruggedness of the mountains decrease appreciably, but the width of the belt increases to a maximum northwest of Al Madinah. Here the scarp mountains are interrupted by block-fault mountains striking obliquely across them and into the Red Sea. North of the fault zone the mountains represent the dissected edge of the tilted block of the Hisma Plateau.

The height of the mountains in 'Asir and Yemen and their proximity to the monsoons of the Indian Ocean bring these areas the highest rainfall of any section of the Arabian Peninsula. As a result, the growing of coffee and other crops on terraced plantations up to 2,400 meters in elevation is the most important activity.

### 1.2.8. Red Sea Coastal Plain

The eastern coastal plain of the Red Sea is known as the Tihamah. It is bounded inland by the scarp mountains and seaward by the shelf area. Although very narrow in the north, it widens irregularly toward the south and attains a maximum width of 40 kilometers near Jizan.

The western portion of the Tihamah is a depositional surface, principally a coralline plain, about 3 meters above sea level. This grades upward into the higher eastern Tihamah which is largely a pediment cut on crystalline and Tertiary rocks and covered with alluvium, sands, and gravels. This pediment was cut after the last fault movement along the front of the scarp mountains. In the general area 175 kilometers northwest of Jizan, basaltic lava flows and cinder cones rest over the fault zone and on the adjacent coastal plain. Numerous drowned tributaries, reaching depths of 50 meters below sea level, cut through the seaward edge of the coastal plain and extend out onto the shelf. These must have formed at a time when sea level was lower and rainfall was greater.

On and near the coast north of Jeddah are remnants of a 20-meter raised surface. From Umm Lejj northward to the Gulf of Al Aqabah are other raised surfaces at elevations of 6, 10, 20, and 30 meters above sea level. These four elevated surfaces are on coralline rocks and have resulted from vertical movements along northwest-southeast faults.

On the whole, the Tihamah is a hot, dry plain, infertile and uncultivated except between Jizan and latitude 20° N. Because of the low, flat coral-bound coast, first-class natural harbors do not exist.

### 1.2.9. Mountains of Southern Arabia

The southern coast of the Arabian Peninsula, particularly along the Gulf of Aden, is backed by high mountains. From Bab Al Mandab eastward for 450 kilometers these mountains are cut in the southern edge of the Yemen Highlands, a plateau underlain by Precambrian basement rocks covered in part by Upper Cretaceous and Tertiary volcanic rocks. East of the Yemen Highlands, the coastal mountains are sculptured in the steep southern edge of the Hadhramawt Plateau, a great elevated block of Paleocene and Eocene strata dipping very gently northwestward into Ar Rub' Al Khali basin. The mountains here rise irregularly and precipitously to elevations averaging about 2,100 meters above sea level. One peak north of the town of Al Mukalla attains an elevation of over 3,300 meters. Elevations decrease eastward into Dhufar where they average about 1,000 meters. The Hadhramawt Plateau slopes more or less uniformly toward the east as far as Oman. Its abrupt termination on the south is believed to be the result of major faulting.

A long east-west valley, called the Tarim Trench or Hadhramawt Valley, cuts through the Hadhramawt Plateau for a distance of 800 kilometers. The Wadi Hadhramawt flows eastward along this valley from the Yemen Highlands to its mouth on the coast near Sayhut. Much of the valley is desert but there are some oasis towns surrounded by fields and date groves. It is a unique region, almost shut off from the rest of the world.

Behind the Hadhramawt Plateau is a flat featureless plain, a sort of transition belt between the high dissected plateau area and Ar Rub' Al Khali to the north. This is the Aden Hinterland, a region sloping gently eastward from high elevations at its western end almost to sea level where it abuts the folded belt of the Oman coast. Innumerable small wadis drain northeastward across the hinterland to become lost eventually in the sands of Ar Rub' Al Khali.

### 1.2.10. Oman Mountains

The Oman Mountains, known as Al Hajar, sweep in a narrow arc along the eastern edge of the Arabian Peninsula roughly parallel to the Gulf of Oman. These rugged mountains reach their maximum elevation of 3,000 meters in the central part of the range. Toward the north, the range becomes narrower and the peaks become lower until, at the Musandam Peninsula, they plunge beneath the Strait of Hormuz. Eastward of the central part, the range widens and elevations decrease and the mountains terminate abruptly against the Gulf of Oman. On their northeast flank the mountains slope steeply into the Gulf of Oman; on their southwest flank they descend with equal steepness toward the inland plateau. A network of wadis carries drainage from the mountains southwestward across this plateau into Ar Rub'

Al Khali. The Oman Mountains are sculptured in deformed sedimentary and igneous rocks, and they differ entirely from the mountains of western and southern Arabia.

The continental margin along the northeast side of the Oman Mountains is very steep and it drops quickly into the Gulf of Oman basin. Bathymetrical, geological, geophysical, and offshore drilling data all indicate that the Gulf of Oman is a young basin of subsidence. It is believed that the Oman Mountains started to rise in late Oligocene time, perhaps 30 million years ago, and that the Gulf of Oman started to sink at about the same time. A great amount of vertical movement was obviously involved since the highest mountain is 3,000 meters above sea level and the center of the Gulf is 4,000 meters deep.

Probably the major valleys in the Oman Mountains were initiated in the Miocene and suffered their greatest erosion in the Pliocene. The small canyon-like valleys may be Pleistocene in age like some of those elsewhere in Arabia.

The Musandam Peninsula, at the tip of Northern Oman, has been a part of the Arabian continental edge for about 200 million years. It has been the subject of recent intense investigation because of its remarkable drowned valleys and the information they may reveal about Quaternary changes of land level, sea level, and climate. The geomorphology of this area is closely controlled by lithology and structure. The highest peaks are usually synclinal masses of limestone and dolomite; the more gentle topography, including the interior parts of the drowned valleys, is underlain by soft shales. The coastline in many places is controlled by fault lines. Thrusts and normal faults are present, and vertical joints play an important part in coastal erosion. The east side of the peninsula is a down-faulted coastline, the downfaulting probably dating from Miocene-Pliocene to Recent.

Isthmuses, and narrow straits which must have been isthmuses before their rupture by the sea, are characteristic of this region. Probably these are the result of vertical movement and standstills in a desert environment. The inlets have a central flat area, 35–50 meters below sea level, which slopes up gradually at first, and then rapidly near shore. It is believed that the inlets may have been braided valleys before their submergence, but they have since been modified by coastal rock falls and the deposition of organic sediments.

No raised beaches have been identified with certainty although many stretches of coast display benches that truncate the bedding and could be of marine origin. About 75 kilometers south of the tip of the Musandam Peninsula, on the Arabian Gulf side, is a wave-cut terrace in solid rock, backed by an old coastline. This marine surface is between 100 and 200 meters above sea level. A similar terrace occurs on the Gulf of Oman side in the Oman Mountains south of Masqat. It can be recognized over a wide area but is not found in the Musandam Peninsula because it plunges northward into the sea.

It is generally agreed that the drowned coastline of Northern Oman is the result of subsidence combined with the Flandrian transgression. The evidence indicates tilting toward the northeast during the late Quaternary. It is estimated that the vertical displacement of the north and east coasts has exceeded 60 meters during the last 10,000 years.

# 1.3. Climate

(E. Schyfsma)

## 1.3.1. Introduction

The overall climate of the Arabian Peninsula falls within C. W. Thornthwaite's (1943, 1948) "arid province" and within W. Köppen's (1900) "dry climates" (arid or desert), except for parts of North and South Yemen, the

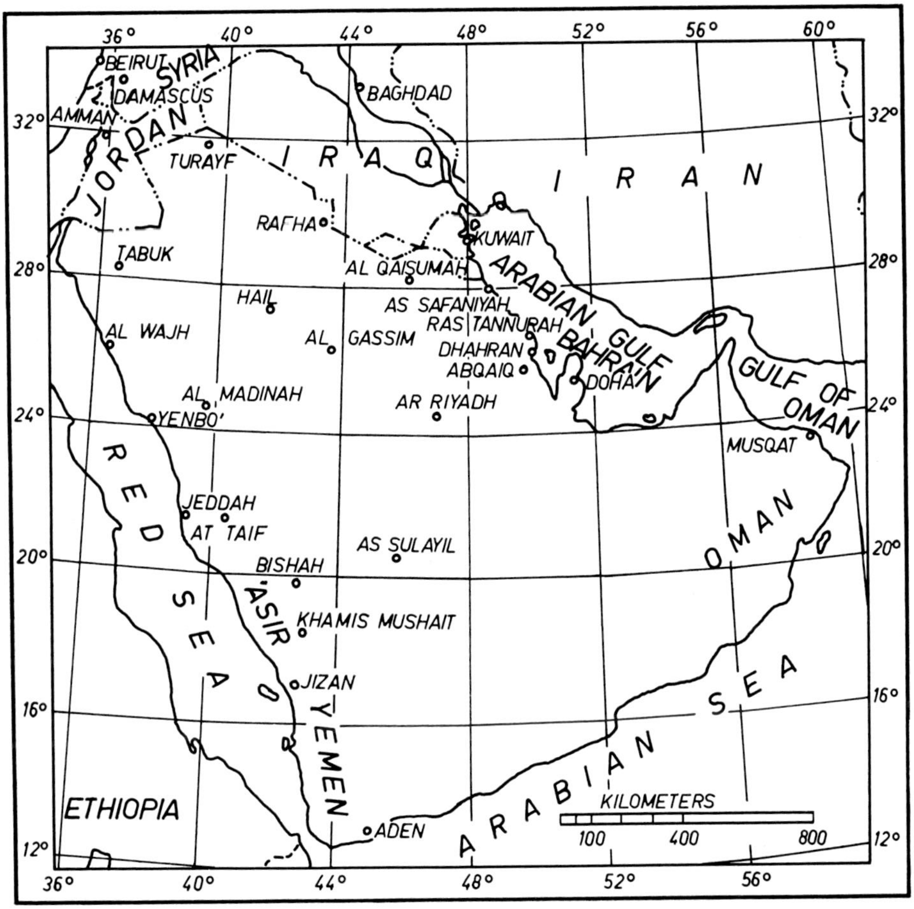


Fig. 3. Meteorological stations in the Arabian Peninsula

'Asir province of Saudi Arabia, and northeastern Oman (Fig. 3), where the temperatures are lower, the rainfall is greater than in the remaining part of the peninsula, and where the vegetation consists of mountain forest and woodland.

Deserts or areas with arid climates can be found from the equator to about 50° latitude both north and south. The Arabian Peninsula is situated approximately within lat 12° N and 32° N and long 36° E and 60° E. The total area amounts to about 3,100,000 square kilometers and falls well within the zone of deserts of the earth. As a matter of fact, it forms part of a virtually unbroken belt of desert environment that extends for some 9,300 kilometers from the Atlantic coast of North Africa to the Thar desert of northwestern India.

The number and distribution of weather observation stations in the area and the compiled data over the last few years are far from adequate for the accurate analysis of regional climatic data. Furthermore, bearing in mind that climatic regions are continuously changing in quality and distribution, the following climatological information on the Arabian Peninsula applies mainly to the situation during the last eight years. These climatic generalizations must be viewed with the nonuniform distribution of observation posts in mind. Figure 3 shows the distribution of meteorological stations in the peninsula from which data have been obtained. Table 1 gives the coordinates of these stations.

Table 1. *Locations of the meteorological stations providing the data used*

| Stations | Latitude north | Longitude east | Elevation in meters |
|---|---|---|---|
| Al Wajh | 26° 14′ | 36° 26′ | 19 |
| Yenbo' | 24° 07′ | 38° 03′ | 6 |
| Jeddah | 21° 30′ | 39° 12′ | 11 |
| Rafha | 29° 38′ | 43° 29′ | 440 |
| Jizan | 16° 54′ | 42° 33′ | 3 |
| Tabuk | 28° 24′ | 36° 35′ | 769 |
| Turayf | 31° 41′ | 38° 40′ | 824 |
| Hail | 27° 31′ | 41° 44′ | 988 |
| Al Madinah | 24° 31′ | 39° 42′ | 632 |
| At Taif | 21° 29′ | 40° 32′ | 1,457 |
| Khamis Mushait | 18° 18′ | 42° 48′ | 2,057 |
| As Sulayil | 20° 28′ | 45° 40′ | 612 |
| Ar Riyadh | 24° 42′ | 46° 44′ | 608 |
| Al Gassim | 26° 17′ | 43° 51′ | 645 |
| Dhahran | 26° 16′ | 50° 10′ | 21 |
| Bahrain | 26° 16′ | 50° 37′ | 2 |
| Kuwait | 29° 13′ | 47° 59′ | 55 |
| Bishah | 19° 58′ | 42° 40′ | 1,161 |
| Al Qaysumah | 28° 20′ | 46° 07′ | 356 |
| Dhahran (ARAMCO) | 26° 49′ | 50° 08′ | 104 |
| Ras Tannurah (ARAMCO) | 26° 42′ | 50° 05′ | 4 |
| Abqaiq (ARAMCO) | 25° 56′ | 49° 41′ | 105 |
| As Safaniyah (ARAMCO) | 28° 00′ | 48° 48′ | 5 |

The information has been compiled from various sources:

1. Monthly issues of "Monthly Climatic Data for the World", published by the U.S. National Oceanic and Atmospheric Administration, and sponsored by the World Meteorological Organization.

2. Valuable meteorological data have been collected by ARAMCO Engineering, Oceanographic-Meteorological Services, Survey Unit, as from

1935; these have been used especially for the ARAMCO stations in the Gulf coastal region (Dhahran, Ras Tannurah, Abqaiq, As Safaniyah).

3. Meteorological information has been published by the Meteorological Department of the Ministry of Defence and Aviation in Jeddah. The following issues were present in the Meteorological Department at Dhahran International Airport and have been used:

January–April 1972, October 1972–January 1975.

4. Hydrological Publications by the Ministry of Agriculture and Water, Department of Water Resources Development, Hydrology Division, Ar Riyadh, in which meteorological data are given on many stations in the Kingdom of Saudi Arabia over the years 1966–1974.

## 1.3.2. Temperatures

*Summer (June, July, August)*

During the summer months temperatures in the shade frequently exceed 48 °C over much of the central part of the Arabian Peninsula although the mean monthly temperatures, calculated over the years 1966 to 1974, remain well below such peak values.

The mean monthly temperatures for the inland stations range from 27.5° to 36.8 °C during the summer months. High stations show lower temperatures due to their elevation above sea level (Fig. 4).

Coastal stations show temperatures rarely exceeding 48 °C in the shade during the summer. Mean monthly temperatures along the Red Sea range from 27.9° to 33.5 °C during the summer months, whereas those along the Arabian Gulf are somewhat higher, ranging from 30.6° to 36.8 °C during the same period.

*Winter (December, January, February)*

During the winter months temperatures often drop below freezing in central and northernmost Arabia, but snow and ice are uncommon except in the highest places.

Mean monthly temperatures during the winter for the inland stations range from 4.5 °C (Turayf, January) to 20 °C (Al Madinah, February), these values representing the two extremes. Most of the winter mean monthly temperatures inland lie between 8° and 17 °C (Fig. 5).

In the coastal areas the temperature never drops below freezing. Mean monthly temperatures along the Red Sea coast range from 18.6° to 26.9 °C during the winter months; they are somewhat lower along the Arabian Gulf, ranging from 11° to 22.2 °C.

The winter temperatures along the coasts of the peninsula are thus higher due to the nearness of the sea than those occurring inland. This is especially the case with those along the Red Sea coast.

*Spring (March, April, May)*

During the months of March, April and May, the temperatures over the entire peninsula are rather pleasant with warm, sunny days and relatively cool nights.

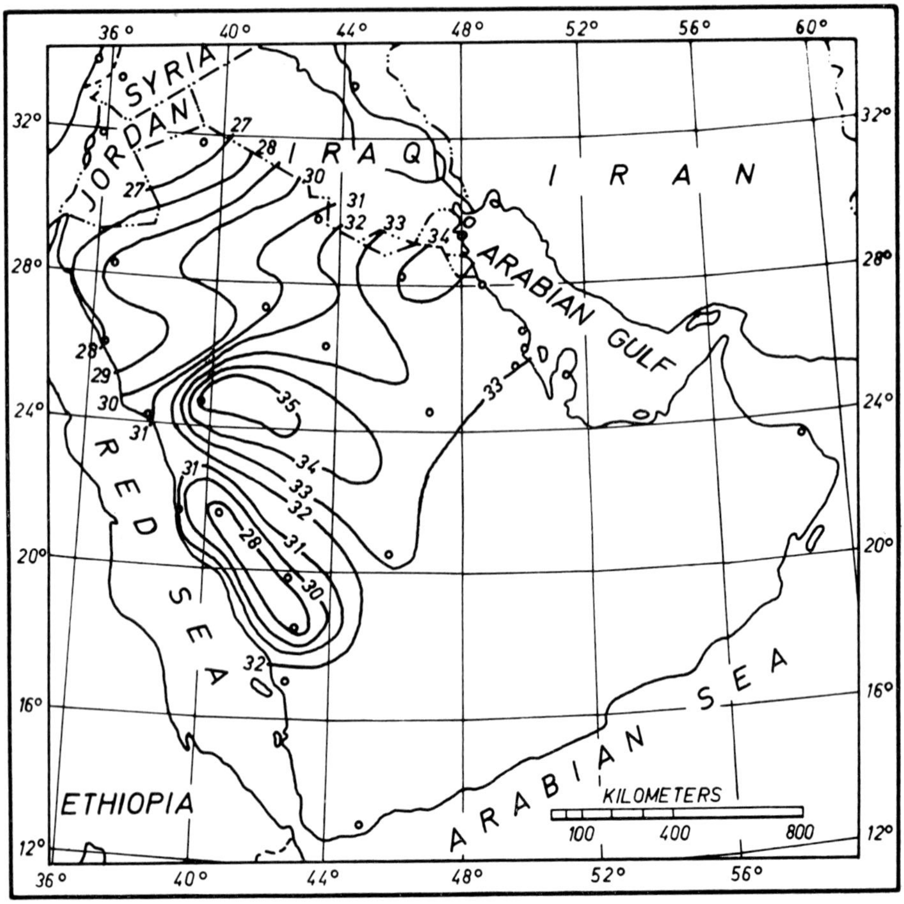


Fig. 4. Summer isotherms (mean values June, July, August 1966–1974)

The mean monthly temperatures for the inland stations range from 13.2 °C (Turayf, March) to 32.2 °C (Al Madinah, May) during the spring. The increase of the mean monthly temperatures from winter to summer for the inland stations is considerable: from 8°–17 °C for the winter, to 27.5° to 36.8 °C for the summer values, i.e., from 17 °C (winter maximum) to 27.5 °C (summer minimum).

Mean monthly temperatures for the coastal stations show a less drastic increase between winter and summer averages, especially for the stations along the Red Sea. Here, the mean monthly temperatures during the spring range from 21.5 °C (Al Wajh, March) to 32.9 °C (Jizan, May). The increase of the mean monthly temperatures during March, April, and May is slight: from 18.6°–27 °C for the winter, to 28°–33.5 °C for the summer, i.e., from 27 °C (winter maximum) to 28 °C (summer minimum).

Along the Arabian Gulf the winter and summer mean temperatures are

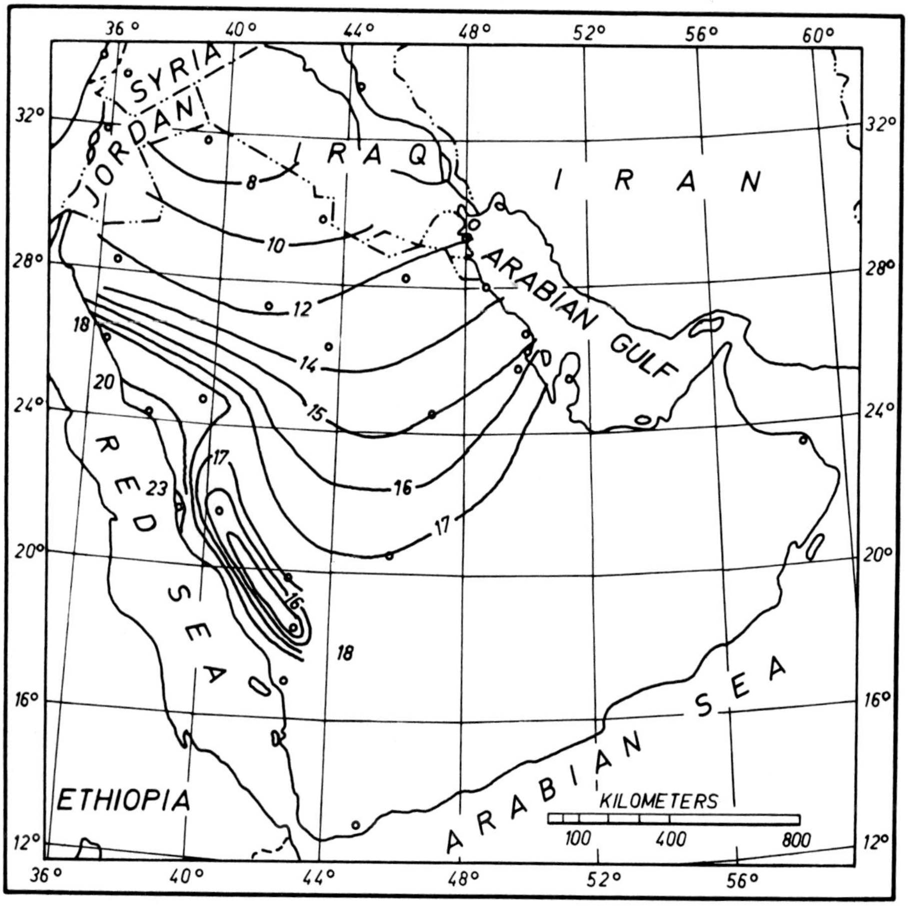


Fig. 5. Winter isotherms (mean values December, January, February 1966–1974)

more widely separated than along the Red Sea. Mean monthly temperatures of the Gulf coastal stations range from 18.7 °C (Kuwait, March) to 32.8 °C (Masqat, May) during the spring. The increase of the mean monthly temperatures during this season is fairly substantial: from 11°–22.2 °C for the winter, to 30.6°–36.8 °C for the summer, i.e., from 22.2 °C (winter maximum) to 30.6 °C (summer minimum).

*Fall (September, October, November)*

During the Fall, the temperatures over the entire peninsula are, like those of the spring, pleasant with cool nights and sunny, warm days.

The mean monthly temperatures for the inland stations range from 13.1 °C (Turayf, November) to 34.7 °C (Al Madinah and Abqaiq, September) during the fall. The higher stations such as Turayf (824 meters), Tabuk (769 meters), Hail (988 meters), Khamis Mushait (2,057 meters), show the lowest temperatures, respectively 13.1°, 17.5°, 16.7°, and 16.2 °C during the

Table 2. *Average mean monthly temperatures in degrees centigrades calculated over the years 1966–1974*

| Station | Jan | Feb | Mar | Apr | May | Jun | Jul | Aug | Sep | Oct | Nov | Dec |
|---|---|---|---|---|---|---|---|---|---|---|---|---|
| Al Wajh | 18.6 | 19.4 | 21.5 | 24.0 | 26.3 | 27.9 | 28.8 | 29.1 | 27.9 | 26.6 | 23.5 | 19.5 |
| Yenbo' | 19.2 | 21.0 | ? | ? | ? | 30.4 | 31.4 | 30.4 | 29.8 | 27.8 | 24.3 | 19.4 |
| Jeddah | 23.5 | 23.9 | 25.3 | 27.4 | 29.7 | 30.8 | 31.9 | 31.8 | 30.6 | 29.1 | 27.1 | 24.6 |
| Rafha | 8.4 | 10.4 | 16.6 | 22.1 | 29.5 | 32.0 | ? | 32.6 | 30.5 | ? | ? | ? |
| Jizan | 26.1 | 26.2 | 27.9 | 29.9 | 32.9 | 33.5 | 33.1 | 33.3 | 32.6 | 32.0 | 28.9 | 26.9 |
| Tabuk | 10.6 | 13.8 | 17.2 | 22.0 | 26.5 | 30.2 | 31.2 | 30.7 | 28.4 | 24.3 | 17.5 | 11.9 |
| Turayf | 4.5 | 7.8 | 13.2 | 17.5 | 23.4 | 27.5 | 28.1 | 27.6 | 24.8 | 23.4 | 13.1 | 6.7 |
| Hail | 10.1 | 12.5 | 17.0 | 21.2 | 28.1 | 30.8 | 31.7 | 31.8 | 29.5 | 24.6 | 16.7 | 11.2 |
| Al Madinah | 18.1 | 20.0 | 24.0 | 27.8 | 32.2 | 35.9 | 35.9 | 35.8 | 34.7 | 29.8 | 23.5 | 18.9 |
| At Taif | 14.8 | 16.2 | 19.4 | 21.9 | 24.9 | 28.3 | 27.9 | 28.1 | 27.1 | 22.5 | 18.5 | 15.5 |
| Khamis Mushait | 13.7 | 15.2 | 17.0 | 18.9 | 21.3 | 24.2 | 23.2 | 23.2 | 22.5 | 18.9 | 16.2 | 13.6 |
| As Sulayil | 15.5 | 19.2 | 22.4 | 28.5 | 31.5 | 33.7 | 33.7 | 34.8 | 32.5 | 27.2 | 20.3 | 16.4 |
| Ar Riyadh | 13.8 | 16.2 | 20.3 | 25.2 | 31.0 | 34.5 | 34.7 | 34.8 | 32.1 | 26.6 | 20.5 | 14.9 |
| Al Gassim | 12.2 | 14.8 | 19.5 | 23.6 | 29.2 | 32.9 | 33.9 | 33.7 | 31.6 | 25.7 | 19.5 | 13.1 |
| Dhahran | 14.0 | 15.8 | 20.1 | 24.6 | 30.4 | 33.3 | 34.4 | 34.0 | 30.9 | 27.2 | 21.3 | 14.8 |
| Bahrain | 16.4 | 16.7 | 20.9 | 25.0 | 29.5 | 32.7 | 33.5 | 34.4 | 32.7 | 29.3 | 23.9 | 17.0 |
| Kuwait | 11.0 | 12.8 | 18.7 | 24.6 | 30.2 | 35.5 | 36.8 | 36.7 | 33.8 | 27.9 | 18.8 | 12.3 |
| Bishah | 17.3 | ? | ? | 26.0 | 27.4 | 31.3 | 31.0 | 32.1 | 29.6 | 24.2 | 19.9 | 18.4 |
| Al Qaysumah | 10.7 | 16.8 | 18.3 | ? | 30.9 | 34.6 | 35.6 | 35.3 | 33.2 | ? | ? | 12.1 |
| Aden [1] | 24.4 | 25.0 | 26.1 | 28.3 | 30.6 | 32.2 | 31.1 | 30.0 | 31.1 | 28.9 | 26.7 | 25.0 |
| Masqat [1] | 21.7 | 21.7 | 25.0 | 28.9 | 32.8 | 33.9 | 32.8 | 30.6 | 30.6 | 29.4 | 26.1 | 22.2 |
| Abqaiq [2] | 15.8 | 17.2 | 21.8 | 26.3 | 32.0 | 35.3 | 36.8 | 36.8 | 34.7 | 28.7 | 22.6 | 17.2 |
| Dhahran [3] | 15.8 | 17.0 | 20.8 | 25.7 | 31.3 | 34.5 | 36.0 | 35.7 | 33.2 | 29.0 | 23.4 | 17.8 |
| Ras Tannurah [4] | 16.0 | 16.5 | 20.3 | 23.6 | 28.6 | 32.0 | 33.4 | 33.3 | 30.9 | 27.7 | 22.9 | 17.3 |

? Insufficient data.
[1] Information from G. R. Rumney, 1968 (period of reference unknown).
[2] Information from ARAMCO data from 1950 to 1970.
[3] Information from ARAMCO data from 1935 to 1974.
[4] Information from ARAMCO data from 1963 to 1972.

month of November. Al Madinah and Abqaiq registered the highest mean monthly temperature (34.7 °C) for the month of September. Apart from these extremes, all the mean monthly temperatures for the inland stations fall within the range of 18.6°–33 °C during the fall.

The decrease of the mean monthly temperatures from September through November is considerable: from 27.5°–36.8 °C for the summer temperatures of the inland stations to 8°–17 °C for the winter, i.e., from 27.5 °C (summer minimum) to 17 °C (winter maximum).

The decrease of mean monthly temperatures from summer to winter values is less for the coastal stations, especially for the Red Sea coast. Here, the mean monthly temperatures during the fall range from 23.5 °C (Al Wajh, November) to 32.6 °C (Jizan, September). The decrease of the monthly temperatures during the fall is slight: from 28°–33.5 °C for the summer, to 18.6° to 27 °C for the winter, i.e., from 28 °C (summer minimum) to 27 °C (winter maximum).

Along the Arabian Gulf the decrease of mean monthly temperatures from summer to winter values is more substantial than for the Red Sea stations. Mean monthly temperatures of the Gulf stations range from 18.8 °C (Kuwait, November) to 33.8 °C (Kuwait, September) during the fall. The decrease of the mean monthly temperatures during this season is from 30.6°–36.8 °C for the summer, to 11°–22.2 °C for the winter, i.e., from 30.6 °C (summer minimum) to 22.2 °C (winter maximum).

Isothermal maps for the summer and winter months have been made from the calculated values shown in Table 2. The maps of average summer (Fig. 4) and winter temperatures (Fig. 5) are prepared from data derived from Table 2.

### 1.3.3. Precipitation

The rainfall variability in the Arabian Peninsula is extremely high as is the case in all desert regions. Sporadic, sometimes very heavy rains occur during short, unpredictable intervals in very localized areas and are followed by long dry periods. It may happen, for instance, that heavy rain falls on Dhahran whilst Ad Dammam, 17 kilometers away, remains dry. The annual precipitation at one station shows great differences from year to year. Dhahran, for example, has an average annual rainfall of 72.4 millimeters, measured from 1935 to 1974 by the ARAMCO meteorological station, but registered 186.9 millimeters in 1974, and 5.3 millimeters only in 1946. Similar variabilities occurred at the other ARAMCO meteorological stations. The average annual precipitation of Ras Tannurah, measured from 1951 to 1974, is 88.1 millimeters, but in 1955 amounted to 215.1 millimeters, and in 1960 to 8.9 millimeters. Abqaiq has an average annual rainfall of 82.3 millimeters, measured from 1951 to 1974, but in 1955 had 175 millimeters recorded, and in 1965 only 7.1 millimeters. As Safaniyah has an average annual rainfall of 93 millimeters, measured from 1971 to 1974, but in 1972 recorded 159.5 millimeters and in 1973 only 30 millimeters. No rainfall was recorded during the months June through October.

To demonstrate the rainfall variability, G. R. Rumney (1968, p. 441) states: "at Khawr Sharja on the Trucial Coast of the Arabian Gulf, where 4.2 inches (106.7 millimeters) is normal, 4.3 inches (109.2 millimeters) have been recorded in twenty-four hours; at Bahrain, where 2.9 inches (73.7 millimeters) is the average, 2.8 inches (71.1 millimeters) have fallen in twenty-four hours."

Due to the rainfall variability, the value of average annual precipitations is debatable unless data over a considerable number of years and from many stations are used.

Table 3. *Average mean monthly and annual precipitations 1966–1974 in millimeters*

| Station | Jan | Feb | Mar | Apr | May | Jun | Jul | Aug | Sep | Oct | Nov | Dec | Average annual |
|---|---|---|---|---|---|---|---|---|---|---|---|---|---|
| Al Wajh | 1.1 | 2.4 | 4.2 | 1.7 | 0.8 | 0 | 0 | 0 | 0 | 0 | 6.3 | 6.4 | 20.9 |
| Yenbo' | 8.1 | 0.1 | 4.4 | 0 | 0 | 0 | 0 | 0 | T | 0 | 0.9 | 1.0 | 20.8 |
| Jeddah | 24.3 | 11.0 | 0.4 | 10.4 | 0.3 | 0 | 0 | 0 | 0 | T | 21.6 | 9.9 | 79.8 |
| Rafha | 24.3 | 17.1 | 45.6 | 3.2 | 4.8 | 0 | 0 | 0 | 0 | 0 | 6.2 | 26.2 | 114.5 |
| Jizan | 4.3 | 4.2 | 4.7 | 0.9 | 2.7 | 0 | 0.6 | 3.6 | T | 5.7 | 0.3 | 2.7 | 32.2 |
| Tabuk | 15.6 | 4.7 | 13.6 | 2.8 | 4.9 | 0 | 0 | 0 | 0 | 0 | 20.4 | 1.5 | 65.0 |
| Turayf | 21.6 | 18.2 | 15.7 | 0.7 | 0 | 0 | 0 | 0 | 0 | 13.9 | 26.3 | 8.1 | 105.9 |
| Hail | 12.7 | 7.8 | 14.3 | 17.9 | 5.4 | 0.2 | 0 | 0 | 0 | 8.9 | 32.0 | 5.2 | 103.2 |
| Al Madinah | 12.3 | 1.0 | 9.2 | 10.0 | 5.0 | 0.4 | 0 | 0 | 0 | 0.7 | 5.3 | 2.0 | 47.4 |
| At Taif | 9.3 | 5.7 | 19.8 | 29.8 | 34.8 | 4.5 | 1.5 | 6.3 | 5.2 | 6.0 | 26.8 | 6.5 | 156.4 |
| Khamis Mushait | 8.1 | 16.8 | 52.9 | 34.2 | 34.7 | 7.3 | 23.0 | 18.1 | 7.5 | 0.4 | 18.1 | 12.0 | 270.9 |
| As Sulayil | 1.5 | 0 | 90.0 | 0 | 0 | 0 | 0 | 0 | 0 | 0 | 0 | 0.2 | 92.5 |
| Ar Riyadh | 18.9 | 4.5 | 29.4 | 36.3 | 16.5 | 0 | 1.0 | 0.1 | 0 | 0 | 3.9 | 7.7 | 117.1 |
| Al Gassim | 19.4 | 8.7 | 21.7 | 25.1 | 8.5 | 0 | 0 | 0 | 0 | 1.0 | 21.0 | 18.6 | 110.2 |
| Dhahran | 10.0 | 8.3 | 21.6 | 3.1 | 0.6 | 0 | 0 | 0 | 0 | T | 0.4 | 26.1 | 75.1 |
| Bahrain | 10.0 | 8.5 | 25.3 | 2.3 | T | 0 | 0 | 0 | 0 | 0 | 2.0 | 5.0 | 73.7 [1] |
| Kuwait | 41.0 | 15.0 | 25.0 | 22.7 | 0.3 | 0 | 0 | 0 | 0 | 0 | 8.5 | 19.7 | 129.0? |
| Bishah | 9.7 | T | 24.6 | 21.9 | 25.2 | 0 | 0.6 | 1.2 | 0 | 0 | 0 | 22.9 | 110.0 |
| Al Qaysumah | 14.8 | 7.4 | 24.1 | 7.7 | 15.1 | 0 | 0 | 0 | 0 | 0 | 0 | 31.8 | 101.6 |
| Aden [1] | ? | ? | ? | ? | ? | ? | ? | ? | ? | ? | ? | ? | 45.7? |
| Masqat [1] | ? | ? | ? | ? | ? | ? | ? | ? | ? | ? | ? | ? | 101.6? |
| Abqaiq [2] | 13.2 | 24.4 | 4.3 | 15.2 | 1.8 | 0 | 0 | 0 | 0 | 0 | 4.6 | 8.9 | 72.9 |
| Dhahran [3] | 15.5 | 10.9 | 10.7 | 8.9 | 1.8 | 0 | 0 | 0 | 0 | 0 | 9.1 | 14.5 | 71.4 |
| Ras Tannurah [4] | 23.6 | 13.2 | 9.7 | 10.7 | 3.0 | 0 | 0 | 0 | 0 | 0.5 | 7.9 | 18.8 | 88.1 |

T Trace, less than 0.05 millimeters.
? Insufficient data.
[1] Information from G. R. RUMNEY, 1968 (period of reference unknown).
[2] Information from ARAMCO data from 1960 to 1969.
[3] Information from ARAMCO data from 1935 to 1974.
[4] Information from ARAMCO data from 1945 to 1974.

ARAMCO has collected data on precipitations in the Gulf coastal region over long periods of time (40 years, 24 years). As a result, the value of average annual precipitation between As Safaniyah and Abqaiq along the eastern coast is considered a reliable figure of about 84 millimeters.

The Ministry of Agriculture and Water in Ar Riyadh operates a dense grid of weather stations in the Kingdom, collecting data on rainfall. They published an isohyetal map of mean annual rainfall over the years 1966–1974.

On the whole, the higher stations in the northern, the southwestern and the southeastern parts of the peninsula receive more rain than the central and coastal areas (Fig. 6). The 'Asir province is a continuation of the Yemen mountains. According to G. R. RUMNEY (1968, p. 436) the mountains of Yemen receive a wearly rainfall averaging over 40 inches (1,016 millimeters). These rains are caused by the monsoons of the Indian Ocean. Climatologically, the

'Asir province immediately north of Yemen belongs to the same southern fringe of the peninsula, and receives, therefore, a generous amount of rain. The same applies to the area south of Masqat. Most of the precipitation takes place during the winter months and spring (December–March/April).

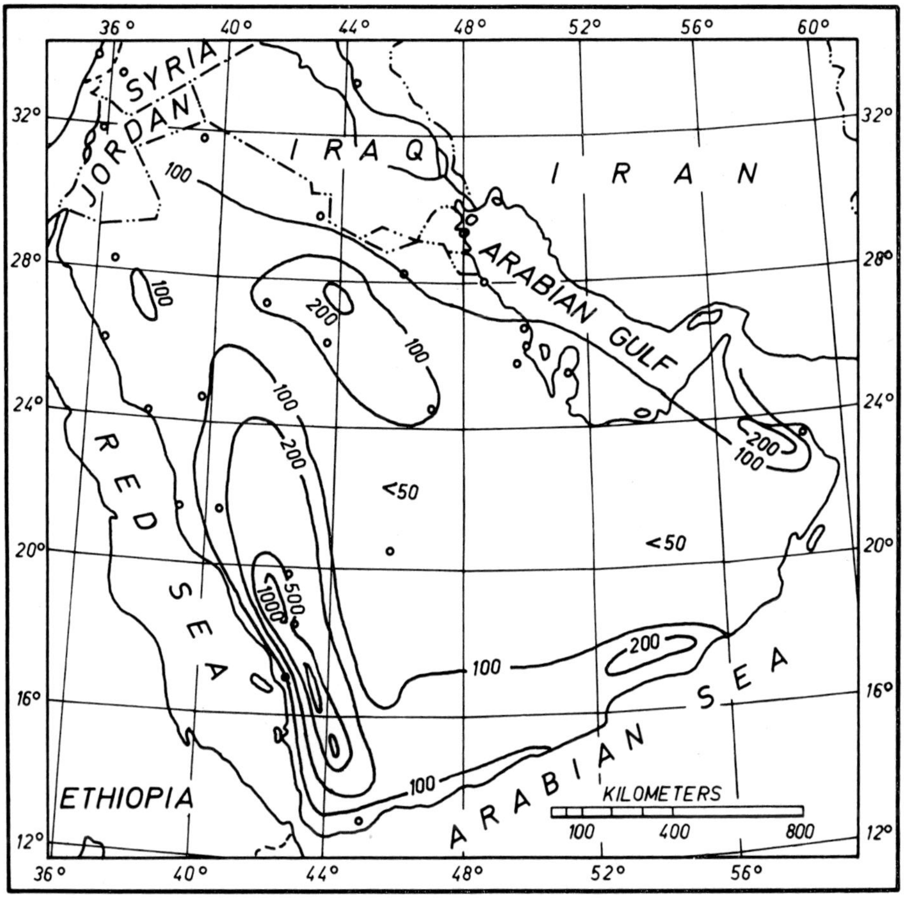


Fig. 6. Isohyetes of the average yearly precipitation in millimeters. From A. Baumgartner, E. Reichel (1975), completed with data from ARAMCO (Dhahran) and the Ministry of Agriculture and Water (Ar Riyadh)

## 1.3.4. Relative Humidities

There is, of course, a close relationship between relative humidity and evaporation rate. M. A. Kettani (1973) recently published values of these for the Dhahran International Airport, measured during the year February 1970–February 1971. The mean annual cloud cover over the peninsula is less than 20 percent except for the coastal area facing the Arabian Sea (South Yemen, Dhufar, Oman), where the annual cloud cover averages between 20 and 40 percent. It is, therefore, understandable that evaporation is far in excess of precipitation. M. A. KETTANI (1973, pp. 26–27) came to the following values for the Dhahran station:

*Potential evaporation rates:* "As expected, the peak months were June and July, but May and August were also high. A maximum weekly average of 26.4 millimeters per day was recorded for the week of July 7–13, and a minimum weekly average of 2.8 millimeters per day was recorded for the week of January 5–11. The maximum monthly evaporation was 691 millimeters (22.3 millimeters per day) in July, while the minimum monthly was 144 millimeters (4.7 millimeters per day) in January 1971. For the entire year the total evaporation was 4.9 meters, or a yearly rate 13.4 millimeters per year [1]. During the weeks of June 2–8 and June 23–29, two low rates are observed, 14.8 and 16.0 millimeters per day respectively."

Table 4. *Average mean monthly relative humidities calculated over the years 1966–1974*

| Station | Jan | Feb | Mar | Apr | May | Jun | Jul | Aug | Sep | Oct | Nov | Dec |
|---|---|---|---|---|---|---|---|---|---|---|---|---|
| Al Wajh | 53 | 57 | 56 | 60 | 67 | 73 | 75 | 73 | 74 | 65 | 56 | 54 |
| Yenbo' | 59 | 58 | ? | ? | ? | 60 | 63 | 63 | 65 | 65 | 63 | 67 |
| Jeddah | 61 | 60 | 57 | 58 | 58 | 62 | 58 | 60 | 68 | 68 | 63 | 61 |
| Rafha | 71 | 70 | 65 | 33 | ? | 16 | ? | 18 | 19 | ? | ? | ? |
| Jizan | 74 | 72 | 73 | 65 | 64 | 64 | 63 | 64 | 66 | 63 | 68 | 72 |
| Tabuk | 51 | 43 | 38 | 30 | 26 | 23 | 23 | 25 | 27 | 33 | 47 | 49 |
| Turayf | 78 | 72 | 63 | 42 | 30 | 27 | 28 | 30 | 33 | 37 | 56 | 75 |
| Hail | 57 | 48 | 34 | 34 | 25 | 16 | 14 | 15 | 17 | 26 | 53 | 57 |
| Al Madinah | 41 | 35 | 28 | 25 | 21 | 14 | 15 | 16 | 15 | 19 | 37 | 40 |
| At Taif | 59 | 49 | 44 | 40 | 34 | 21 | 23 | 24 | 26 | 34 | 54 | 57 |
| Khamis Mushait | 65 | 56 | 55 | 49 | 48 | 36 | 46 | 47 | 33 | 33 | 53 | 64 |
| As Sulayil | 47 | 36 | 37 | 25 | 18 | 15 | 17 | 16 | 19 | 25 | 34 | 46 |
| Ar Riyadh | 48 | 41 | 35 | 32 | 23 | 13 | 13 | 13 | 15 | 19 | 36 | 46 |
| Al Gassim | 52 | 44 | 37 | 36 | 29 | 17 | 16 | 18 | 18 | 22 | 43 | 51 |
| Dhahran | 77 | 72 | 66 | 57 | 43 | 37 | 40 | 54 | 58 | 68 | 66 | 72 |
| Bishah | 44 | ? | ? | 25 | 36 | 18 | 19 | 21 | 17 | 18 | 25 | 42 |
| Al Qaysumah | 68 | 45 | 30 | ? | 24 | 15 | 13 | 16 | 14 | ? | ? | 68? |
| Abqaiq [1] | 57.6 | 54.4 | 49.5 | 46.3 | 39.5 | 28.4 | 29.5 | 35.1 | 42.6 | 48.7 | 54.5 | 56.7 |
| Dhahran [2] | 67.5 | 64.6 | 59.7 | 55.3 | 47.6 | 41.3 | 43.9 | 50.2 | 53.4 | 58.2 | 62.7 | 66.8 |
| Ras Tannurah [3] | 73.3 | 71.3 | 69.7 | 67.2 | 64.0 | 59.8 | 58.8 | 62.9 | 66.4 | 69.7 | 73.0 | 71.9 |

? Insufficient data.
[1] Information from ARAMCO data from 1950 to 1970.
[2] Information from ARAMCO data from 1935 to 1974.
[3] Information from ARAMCO data from 1963 to 1972.

*Relative humidity* (p. 27): "Interestingly, the lowest rate occurs during July, with a monthly average of 28.4 percent; the low weekly average of 21.7 percent occurred during the June 30–July 6 week. December was the month of maximum humidity, with an average of 73.5 percent. A yearly average of 51.1 percent was obtained for 1970."

[1] This should, of course, read: 13.4 millimeters per day.

Mean monthly and annual relative humidities vary from place to place, and they also change from year to year. However, the variability is much less than for the precipitation values (see Table 4).

It is interesting to compare the mean monthly relative humidities for the inland stations with those along the Arabian Gulf and the Red Sea.

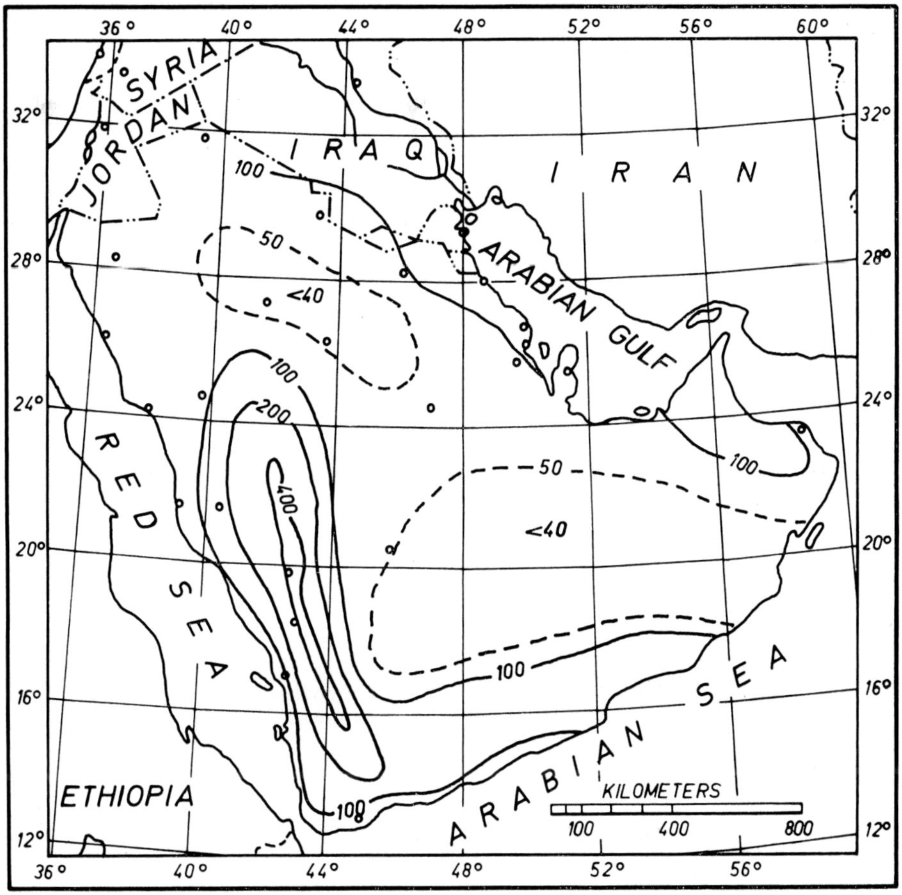


Fig. 7. The yearly average of the actual evaporation. From A. BAUMGARTNER, E. REICHEL (1975) completed with data from ARAMCO (Dhahran) and the Ministry of Agriculture and Water (Ar Riyadh)

The inland stations show the highest relative humidities during the winter months (December through February), ranging from 35 percent to 78 percent, and the lowest during the summer months (June through August), ranging from 13 percent to 47 percent.

The same pattern is found for the stations along the Arabian Gulf: high relative humidities during the winter, ranging from 64.6 percent to 73.3 percent, and low relative humidities for the summer, ranging from 37 percent to 62.9 percent.

The Red Sea stations reveal a completely different picture. Jeddah and Jizan show a more or less even relative humidity throughout the year, ranging from 58 percent to 74 percent. Al Wajh on the other hand, shows exactly the reverse of the inland and Gulf coastal stations: high relative humidities during the summer months (73–75 percent) and low relative humidities during the winter months (53–58 percent).

The actual evaporation depends on the depth of precipitation and temperature. The calculated yearly average is shown in Fig. 7.

### 1.3.5. Prevailing Wind Directions

The following conclusions could be drawn from Table 5. Variable wind directions were not taken into consideration for the following discussion. They could, however, be significant in indicating changes of seasons in some instances.

Eight stations recorded fairly constant wind directions throughout the year:

| | | | |
|---|---|---|---|
| Al Madinah | W | Dhahran | NW-NNW-N |
| At Taif | W | Jeddah | NNW-N |
| Al Wajh | WNW-NW | Al Qaysumah | NNW-N-NNE |
| Yenbo' | NW | Al Gassim | N-NE |

Six stations recorded fairly constant wind directions during different months:

| | |
|---|---|
| Jizan | May through October: W (summer and fall)<br>November through April: SW (winter and spring) |
| Tabuk | March and April: SW (spring)<br>May through October: NW (summer and fall) |
| Hail | June through September: N (summer)<br>October through February: S (fall and winter) |
| Khamis Mushait | September and October: E (fall)<br>November through August: SW (winter, spring, summer) |
| As Sulayil | July and August: NW and N (summer)<br>September through June: E (fall, winter, spring) |
| Ar Riyadh | June through October: N-NE (summer and fall)<br>December through April: SSE (winter and spring) |

Three stations recorded insufficient data: Rafha, Turayf and Bishah.

It appears impossible to assign the various wind directions to geographical areas.

Table 5. *Average monthly prevailing wind directions recorded over the years 1966–1974*

| Station | Jan | Feb | Mar | Apr | May | Jun | Jul | Aug | Sep | Oct | Nov | Dec |
|---|---|---|---|---|---|---|---|---|---|---|---|---|
| Al Wajh | NW | NW | NW | NW | WNW | WNW | WNW | WNW | NW | NW | NW<br>VAR | N<br>VAR |
| Yenbo' | NW | NW | NW | NW | ? | NW | NW | NW | NW | NW | NW | NW |
| Jeddah | NNW | NNW | NNW | NNW | NNW | NNW | NNW | NNW | N | N | N<br>VAR | N<br>VAR |
| Rafha | ? | ? | VAR | ? | ? | NW | NNW | NNW | VAR | SE | N<br>VAR | VAR |
| Jizan | SW | SW | SW | SW | W | W | WNW | W | W | W | SW | SW |
| Tabuk | VAR | VAR | SW | SW | NW | NNW | NW | NW | NW | NNW | VAR | N<br>VAR |
| Turayf | W | ? | VAR | ? | ? | W | WNW | WNW | ? | ? | VAR | VAR |
| Hail | S | S | VAR | VAR | VAR | N | NNW | NNW | N | S | S | S |
| Al Madinah | VAR | VAR | W | W | W | W | W | W | W | VAR | VAR | VAR |
| At Taif | VAR | W | W | W | W | W | W | W | W | W | VAR | W<br>VAR |
| Khamis Mushait | SW | SW | SW | SW | SW | VAR | VAR | SW | E | E | SW | SW |
| As Sulayil | E | E | E | E | E | E | NW | N | E | E | E | E |
| Ar Riyadh | SSE | SSE | SSE | SSE | VAR | N | NNW | N | NNE | NE | E<br>VAR | SSE<br>NE |
| Al Gassim | NE | NE | VAR | NE | NE | NE | N | NE | NE | VAR | VAR | |
| Dhahran | NW | NNW | N | N | N | N | N | NE<br>VAR | N<br>VAR | NW<br>VAR | NW | NW |
| Bishah | ? | ? | ? | VAR | VAR | VAR | NW | WNW | ENE | ENE | ENE | ENE |
| Al Qaysumah | ? | ? | NNW | NNW | NNE | NNW | NNW | VAR | N<br>VAR | NNW<br>VAR | N | N |

? Insufficient data.
VAR Variable wind directions.

## 1.3.6. Seasons

An attempt is made to identify the four seasons in the Arabian Peninsula. This is done by taking into account both mean monthly temperatures and mean monthly relative humidities, measured over many years.

In general, the peninsula experiences rapid changes from summer to winter temperatures, except for the stations along the Red Sea coast, where the difference is small and the relative humidity more evenly distributed throughout the year, at least in Jeddah and Jizan.

| Areas | Temperature ranges during the summer (June, July, August) | Temperature ranges during the winter (December, January, February) |
|---|---|---|
| Inland stations | 27.5–36.8 °C | 4.5–20 °C |
| Arabian Gulf | 30.6–36.8 °C | 11.0–22.2 °C |
| Red Sea stations | 27.9–33.5 °C | 18.6–26.9 °C |

These mean monthly temperature ranges show, at least for the inland stations and the Arabian Gulf area, that there is a summer and a winter season, during which the maximum winter value is about 7.5 °C lower than the minimum summer value.

The months of March, April and May show gradual increases of the mean monthly temperatures until the summer level is reached. These months have, therefore, been assigned to the spring.

The months of September, October and November show gradual decreases of the mean monthly temperatures until the winter level is reached. Consequently, these months have been assigned to the fall season.

The dates of these gradual increases and decreases of the mean monthly temperatures vary slightly from year to year, but over a large number of years, the averages of the monthly temperatures provide sufficient evidence for the distinction of the four seasons in the peninsula.

The identification of the four seasons is confirmed by the changes of the relative humidities. The inland stations and those along the Arabian Gulf show high relative humidities during the winter, and low ones during the summer season. The Red Sea stations do not conform to this pattern, i.e., Jeddah and Jizan show no definite difference in summer and winter relative humidities, whereas Al Wajh shows high values of relative humidity for the summer and low ones for the winter season.

| Areas | Relative humidity ranges | |
|---|---|---|
| | during the summer (June, July, August) | during the winter (December, January, February) |
| Inland stations | 13–47% | 35–78% |
| Arabian Gulf | 37–62.9% | 64.6–73.3% |
| Red Sea: | | |
| Jeddah and Jizan | 58–64% | 60–74% |
| Al Wajh | 73–75% | 53–57% |

## 1.3.7. Solar Radiation

The average annual percentage of sunshine recorded in the Arabian Peninsula ranges from 75 percent at the coast to more than 85 percent in the interior.

M. A. Kettani and E. Y. Lam (1974) published a map of the peninsula showing isolines of average daily solar radiation received on the ground during the year 1970, varying from 409 to 663 calories per square centimeter per day, the lowest value (409) in the northwest, and the highest (663) at As Sulayil at the western edge of Ar Rub' Al Khali in the southern half of the peninsula.

D. Uhlig (1971) reports for the Al Hasa area, about 50 kilometers south of Abqaiq, a solar radiation during June of 725 calories per square centimeter per day.

# 2. Regions of Investigation

## 2.1. Gulf Coastal Region and Its Hinterland

### 2.1.1. General Geology

(D. H. Johnson)

The area described here is a northwesterly trending belt about 50–100 kilometers wide between As Summan Plateau and the Arabian Gulf coast (see Fig. 8).

#### *2.1.1.1. Topography*

The area is generally smooth, rising gradually from the sea to an elevation of about 300 meters along the edge of As Summan Plateau. Over much of the area the transition from coastal zone to plateau is gradual and no real boundary is evident; elsewhere, as near Al Hofuf, there is a steep escarpment rising more than 100 metres. Local relief is generally low, with sand dunes or low rounded hills raising 5 or 10 meters above the general level, although in the Dhahran vicinity a number of steep-sided erosional buttes or jabals stand several tens of meters above the lowlands.

#### *2.1.1.2. Stratigraphy*

The surface rocks of this part of Saudi Arabia include consolidated sedimentary formations ranging from Paleocene to Middle Eocene age and Miocene to Pliocene age. Upper Eocene and Oligocene time is represented by a regional unconformity. Unconsolidated materials include gravel of Tertiary age and various sediments of Quaternary age, comprising beach gravel and sand, gravel and silt in basin deposits, sabkhah sediments, and various aeolian sand deposits. The following sections discuss the surface formations but subsurface geology is not treated (Fig. 8).

**Umm er Radhuma Formation.** The oldest rocks in this part of Saudi Arabia belong to the upper part of the Umm er Radhuma Formation, which is exposed only in one small outcrop about a kilometer north of Dhahran. The exposed thickness is about 3 meters, and the rocks continue to an undetermined depth; at the reference section, R. W. Powers and others (1966, p. D 85) give a thickness of 243 meters. In the outcrop mentioned above the rocks are yellowish-gray, very fine-grained to medium-grained crystalline dolomite, with chert nodules and quartz geodes up to 15 centimeters in diameter. J. W. Tleel (1973, p. 561) believes the rocks were originally "slightly pelletal aphanitic limestone" deposited in a shallow sea and later dolomitized.

"Ghost" fossils include indeterminate Foraminifera, echinoid spines, and pelecypod fragments. The Umm er Radhuma Formation has been dated (R. W. Powers et al., 1966, p. D 86) as Paleocene to early Eocene; the rocks discussed here must be of early Eocene age.

**Rus Formation.** The Rus Formation is exposed in the central portion of the Dammam Dome, where the type section occurs on Jabal Umm Ar Russ. Both J. W. Tleel (1973, p. 563) and R. A. Bramkamp (R. W. Powers et al., 1966, p. D 145) divided the formation into three units, though they differed in delimiting the two upper zones. As described by J. W. Tleel (op. cit.) the lower unit consists of about 21 meters of yellowish-gray microcrystalline to finely crystalline dolomite and dolomitic limestone with quartz geodes and light colored chert nodules up to 30 centimeters in diameter. The middle unit consists of 10 meters of massive yellowish-gray microcrystalline limestone that contains numerous 5–10 centimeter calcite geodes and scattered "wafer like" mud balls. The upper zone consists of about 25 meters of yellowish-gray, microcrystalline, partly dolomitic limestone and white chalky limestone with three thin layers of yellowish gypsiferous shale in the upper part. Near the top of the formation the white chalky limestone contains pebble- to boulder-size nodules of brown to black chert that remain on surfaces where the limestone has weathered away.

No specifically identifiable fossils have been found. J. W. Tleel (op. cit., p. 565) believes the depositional environment was similar to that of coastal sabkhahs with occasional shallow marine transgressions. Later dolomitization has destroyed much of the original texture. The age is Lower Eocene as the formation lies between Lower Eocene formations.

**Dammam Formation.** In the Gulf coastal area the Dammam Formation crops out around the circumference of the Dammam Dome, from whence it takes its name, and in four smaller areas about 10 to 25 kilometers to the west. It conformably overlies the Rus Formation and is bounded at the top by a regional unconformity. It has been subdivided (R. W. Powers et al., 1966, p. D 88) into five members, in ascending order: Midra Shale Member; Saila Shale Member; Alveolina Limestone Member; Khobar Member, subdivided into the Khobar marl and Khobar dolomite; and the Alat Member, subdivided into Alat marl and Alat limestone.

The Midra Shale Member consists of about 3 meters of yellowish-brown, fissile, thinly laminated shale, gray marl and impure limestone. It contains scattered fossil sharks teeth and was at one time informally known as the "shark tooth shale". J. W. Tleel (1973, p. 566) reports the occurrence of *Ostrea turkestanensis.* Gypsum occurs in cross-fiber lenses and seams, and there are nodules of goethite pseudomorphous after pyrite.

The Saila Shale Member has a basal 0.6 meter bed of yellowish-gray, impure, fossiliferous, calcarenitic limestone overlain by brownish-yellow sub-fissile clay shale. The basal limestone contains abundant *Alveolina* cf. *A. decipiens* and *Nummulites globulus,* and common *Coskinolina ballasilliei* and *Dictyoconoides* sp. (J. W. Tleel, 1973, p. 566).

The Alveolina Limestone Member consists of about 1 meter of yellowish-gray, microcrystalline, partly recrystallized, dolomitized limestone, occasionally silicified. It contains common *Alveolina elliptica* (SOWERBY) var. *flosculina* SILVESTRI, and internal molds of *Lucina pharaonis* (J. W. TLEEL, op.cit., p. 567).

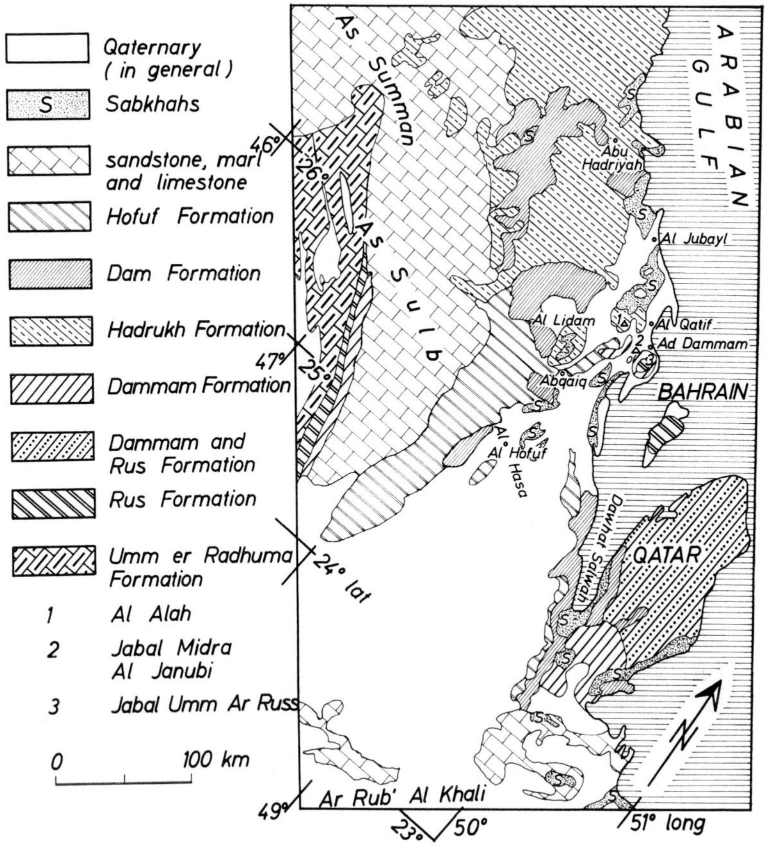


Fig. 8. Generalized geologic map of the Gulf Coastal region and its hinterland

The Khobar Member is subdivided into a lower unit, the Khobar marl, consisting of 1.5 meters of light gray to tan dolomitic marl, and an upper unit, the Khobar dolomite, which despite its name, consists predominantly of limestone, with 4 meters of light gray, partly recrystallized, non-porous, nummulitic limestone, above which is 1 meter of yellowish-brown, soft, marly limestone, which is overlain by 3 meters of yellowish gray, massive, hard, nummulitic, calcarenitic limestone. According to J. W. TLEEL (ibid), the fossils are *Nummulites somaliensis*.

The Alat Member also is subdivided, into a lower 6 meters of light-colored dolomitic marl, locally argillaceous, and an upper 9 meters of cream to tan, chalky, porous, commonly dolomitic limestone with numerous molds of mollusks, and other indeterminate fossils.

J. W. Tleel (ibid.) believes the Dammam Formation was laid down in a turbid, open-marine environment, with clearing to permit deposition of limestone containing *Alveolina* and *Nummulites*. The pyrite in the Midra Shale Member suggests a restricted rather than open-marine environment during Midra time. The fossils indicate that the Alveolina Limestone Member and overlying members are of Middle Eocene age, while the lower two members are of Lower Eocene age.

**Hadrukh Formation.** The Hadrukh Formation extends in a band along the Gulf coast from the northern border of Saudi Arabia as far south as lat 26° 30′ N, and in scattered patches beyond Abqaiq to about 25° 52′ N. Over most of this area the formation is non-marine, but a small area extending perhaps 60 kilometers from Al Qatif–Al Alah has marine layers near the top of the formation.

The thickness of the Hadrukh Formation ranges from about 20 meters in the Abqaiq vicinity to as much as 120 meters around Abu Hadriyah (lat 27° 20′ N, long 48° 38′ E) and to the northwest. The rocks consist of marly sandstone, sandy marl, sand, clay and sandy limestone, with abundant chert in some layers and minor gypsum. Colors are variegated; mainly green and gray, with some red, brown, white and pink. Within the marine layers are poorly preserved mollusks.

**Dam Formation.** Near the Gulf coast, the Dam Formation occurs largely in scattered patches from about lat 27° 30′ N, southeasterly to Qatar and beyond. In the type locality at Jabal Al Lidam (lat 26° 22′ N, long 49° 28′ E) the Dam is 90 meters thick and consists of pink, white and gray marl, and red, green and olive clay with interbedded sandstone, limestone and coquina.

At the Dammam Dome the Dam unconformably overlies Rus and Dammam rocks. Only part of the formation occurs there, as the upper part has been eroded away. At Jabal Midra Al Janubi on the west side of the dome, J. W. Tleel (1973, p. 568) found about 2 meters of yellowish-gray, microcrystalline, sandy limestone overlain by two stromatolitic limestone beds 1 meter thick separated by 0.6 meter of microcrystalline limestone. The upper stromatolitic bed is overlain by 31 meters of interbedded molluscan, pelletal and foraminiferal calcarenites, calcirudites and calcarenitic microcrystalline limestone with some argillaceous and algal limestones. This is followed by 6 meters of calcirudite with pebbles and boulders of cryptocrystalline limestone and calcarenite, and then 9 meters of cliff-forming, yellowish-gray, massive, cryptocrystalline limestone with possible algal growths. At the apex of the dome, the Dam Formation on the northern five peaks consists of a massive reef complex containing algal growths, siderastrian-type corals, pelecypods and gastropods in growth positions. The four southern peaks show an off-reef facies represented by 11 meters of interbedded reef limestone (coral-algal), calcarenitic limestone (mollusks), calcarenitic limestone (peneroplid—miliolid), crystalline limestone (miliolid) and crossbedded calcarenite (pellets and oolites).

J. W. TLEEL (1973, p. 568) reports the following fossils from the Dam Formation: *Echinocymis* sp., *Peneroplis farsensis, Sorites orbiculus, Borelis* (= *Neoalveolina*) *melo, Taberina malabarica* and *Archaias angulatus.* The age of the formation is Middle Miocene.

J. W. TLEEL (ibid., p. 570) believes the sea gradually covered the dome, producing first tidal flats around a central island, and later when the sea covered the entire dome, a pinnacle-type reef on the highest parts. In the surrounding shallow-marine waters, calcarenites were formed in high energy areas while carbonate muds were deposited in quiet lagoons. The interbedded off-reef deposits at Jabal Umm Ar Russ indicate fluctuating sea level and shifting environments.

**Hofuf Formation.** Although the Hofuf Formation covers large areas to the west of Al Hasa oases, along the coast there are only scattered outcrops from about the latitude of Al Hofuf (lat 25° 22′ N) to the Qatar Peninsula and beyond. The Hofuf formation is quite lenticular and heterogeneous, and both the thickness and lithology vary from place to place. At the type locality the thickness is about 95 meters. The lower part is 19 meters of red and white conglomerate consisting of boulders and pebbles of limestone in a quartz sand matrix, and greenish-gray to red sandy marl at the base. Above this is light grey, partly impure, sandy limestone 18 meters thick, followed by 49 meters of alternating red and light gray argillaceous sandstone and then 9 meters of gray marly conglomerate with limestone boulders in a quartz sand matrix. There are occasional fresh water fossils such as *Lymnaea* and *Chara;* the age may be either late Miocene or Pliocene. In many places there is developed on Hofuf rocks a calcareous duricrust; this is discussed by R. W. CHAPMAN, sec. 2.1.4.

**Tertiary gravel.** Although Tertiary gravels consisting of well-rounded white quartz pebbles occur in the Eastern Province of Saudi Arabia, e.g., lat 24° 15′ N, long 48° 30′ W, none is known in the coastal area.

**Quaternary beach gravel and sand.** From Al Jubayl southeastward to the southern end of Dawhat Salwah, a series of discontinuous raised beaches occur near the shore 1 to 4 meters above mean high tide. They are typically in the form of low terraces or mesas covered by sand, shells, shell fragments and coquina. A few scattered small mesas about 15 kilometers south of Al Jubayl and several kilometers from the shore seem to have a partially similar origin; they have numerous chert nodules beside the sand and shell material.

**Quaternary gravels.** Several low basins are floored with thin but extensive layers of gravel representing concentrations of coarse debris from the surrounding area, residual material derived from older sediment or rock by removal of intergranular fines, and lag concentrations from earlier gravel deposits. They may consist of mixed limestone and quartz pebbles along with quartz and other pebbles derived from the basement complex.

**Sabkhah deposits.** Numerous coastal and inland sabkhahs are filled with silt, clay, and muddy sand along with gypsum and anhydrite, and frequently a salty surface covering. They are discussed in section 2.1.5.

**Aeolian sands.** Large areas, particularly in the southeastern coastal area are covered by aeolian sand deposits. These include areas of *dikakah*, an irregular surface of bush and grass covered sand; areas of barchan dunes and various undulating sand sheets, and toward Ar Rub' Al Khali, sand mountains 50 to 300 meters high, and areas of *'uruq*, or parallel, longitudinal dunes separated by broad sand valleys.

#### *2.1.1.3. Structure*

The coastal area is part of the Arabian platform, an area that has been stable from the Paleozoic to the present time. Formations are thin but extensive, and dips are generally less than 1°. G. F. BROWN (1972) indicates a very gentle slope northward and eastward toward the Arabian Gulf. A few minor structures break the pattern. The most prominent is the Dammam Dome which is believed to be the result of deep seated salt intrusion because of its strong negative gravity anomaly and oval shape. Its major axis trends about N 35° W, and according to J. W. TLEEL (1973, p. 574), average dips on the flanks are from 1° to 1.5°. At the crest of the dome, Rus beds dip north at 2° 30′ and overlying Dam beds dip south at 1° 60′, producing an angular unconformity of 4° 20′. A few very minor anticlines and synclines are superimposed on the dome, but J. W. TLEEL (ibid.) observed no faulting. He estimates a minimum structural relief of 100 meters.

#### *2.1.1.4. Economic geology*

This area includes some of Saudi Arabia's major oil and gas deposits, but these will not be discussed here as adequate treatment would require a much longer discussion. Several formations, notably the Umm er Radhuma, and the Khobar and Alat Members of the Dammam are important aquifers. A little salt is produced from some of the sabkhahs, limestone is extensively quarried for road metal, aggregate, and to a small extent for building stone; and marl and marly shale are excavated for use as impermeable underlayment for garden areas around building sites. Otherwise there are no economic products, although someday sabkhah brines may find some use.

### 2.1.2. Sea Level Fluctuations During the Quaternary Period

(H. FELBER, H. HÖTZL, V. MAURIN, H. MOSER, W. RAUERT, and J. G. ZÖTL)

The eustatic fluctuations of sea level during the Pleistocene and Holocene resulted in remarkable shifts of the shoreline along the relatively flat Arabian Gulf. The youngest, i.e. postglacial, rise of the sea level, the so-called

Flandrian transgression, is mentioned by a number of authors (J. J. HOUBOLT, 1957; D. A. HOLM, 1960; J. C. M. TAYLOR and L. V. ILLING, 1969; G. EVANS, V. SCHMIDT, P. BUSH, and H. NELSON, 1969; M. SARNTHEIN, 1971 and 1972; P. KASSLER, 1973), while descriptions of Pleistocene fluctuations are rare and mostly indirect (G. M. LEES, 1928; N. L. FALCON, 1947; D. A. HOLM, 1960; K. W. GLENNIE, 1970; R. W. POWERS et al., 1966; P. KASSLER, 1973).

To acquire a better knowledge of the Quaternary sea level fluctuations along the east coastal zone of Saudi Arabia, three areas have so far been investigated: the coastal area south of Al Jubayl (Fig. 8), a shoreline of about 10 kilometers extension north of Ras Tannurah, and the coastal area of Salwah near the border with Qatar. The coastal region south of Dhahran was also briefly visited. In contrast to the area of the Red Sea and the eastern border of the Arabian Gulf, the coastal area between Kuwait and the peninsula of Qatar was of relatively low tectonic activity during the Upper Quaternary apart from local periclinal folding.

Dunes and dikakahs in many places reach the shore line which often cuts Miocene and Pliocene strata.

### *2.1.2.1. Prewürmglacial sea level fluctuations*

Along the recent Gulf coastal area of Saudi Arabia, only in the Bay of Salwah does the hinterland rise in a scarp up to 130 meters, where a plain extends for more than 150 kilometers towards the west. This plain is mainly a large gravel fan having its roots in the Wadi As Sah'ba and sloping towards the east with a gradient of about 1 permil. D. A. HOLM (1960) interpreted this fan as delta deposits. Except for some small gravel deposits, details cannot be recognized, because the dunes of Al Jafurah cover the plain.

The western border of this plain, morphologically impressive, is a scarp, 80 meters to 120 meters high, the eastern escarpment of As Summan Plateau. Ranging from Haradh over Al Hofuf and Abqaiq towards the north it is particularly well developed in the area of Al Hasa, where numerous inselbergs give evidence of a larger extent of As Summan Plateau eastward in the past.

Marine beach sediments found adjacent to this scarp in the area of Al Hasa, the delta sediments of Haradh, the dissection of the scarp, and the inselbergs with wave-cut notches and evident forms of marine abrasion leave no doubt that this scarp is a line of cliffs caused by sea transgression during the Late Pliocene. This age is assumed from the facts that, on the one hand, the rocks of As Summan Plateau (see also Shedgum Plateau) consist of superimposed layers from Miocene to Lower Pliocene, and on the other hand, the gravels of the delta fan of Haradh, which is more than one million years old, i.e., from Late Pliocene to Eearliest Pleistocene, cover the transgression plain mentioned above (see 2.1.3.). The marine geomorphogeny of the cliff was recognized by D. A. HOLM (1960). He concluded from this that the sea level of the Earliest Pleistocene was about 150 meters higher than now. He consequently came to the conclusion that the numerous sabkhahs which now have reached an altitude of 150 meters above sea level were formerly coastal sabkhahs. K. W. GLENNIE (1970) made the same conclusion concerning inland sabkhahs south of the Trucial Coast (on the other hand see 2.1.5.). Descriptions of higher sea levels during the Upper Pliocene and the Lower Pleistocene are to be found in many coastal areas (F. E. ZEUNER, 1952; P. WOLDSTEDT, 1961; etc.).

Table 6. *Steps of the slope in the hinterland of Salwah (meters above sea level)*

| Step No. | Coastal area of Khashm Umm Huwayd | | Coastal area west of Salwah | |
|---|---|---|---|---|
| I | 1.5–2.5 m, | coastal sabkhah | 2–2.5 m, | coastal sabkhah |
| II | | | 5–7 m, | step on claystone and marl; scattered oyster shells |
| III | 10 m, | tilted strata of beach sediments (sandy limestone) with abrasion gravels | 10 m, | step on claystone and marl |
| IV | 15 m, | step on marl | 15–18 m, | step on claystone and gypsum; internal moulds of shells and sea-urchins (*Fibularia sp.*) [1] |
| V | 25 m, | step on coarse crystalline gypsum of the Dam Formation, more than 100 m broad | 25 m, | step on claystone and sandstone |
| VI | about 35 m, | step on sandy limestone | 38–40 m, | step on claystone and sandstone covered by gravels |
| VII | about 50 m, | step on gypsum and partly oolitic sandstone | 45–50 m, | step with cover of gravels and pieces of oyster shells |
| VIII | | | about 70 m, | steps on crossbedded sandstone of the Hofuf Formation, cover of delta sediments |
| IX | about 120 m, | uppermost plain on delta sediments of Pliocene/ Pleistocene age | about 110 m, | uppermost plain on delta sediments of Pliocene/ Pleistocene age (0.7–2.5 m thickness of the gravel cover) |

[1] Determination by Prof. Dr. H. FLÜGEL, University of Graz.

R. W. CHAPMAN (1971, 1974) describes in some contrast to above the morphogeny and development of the steep scarp of As Summan Plateau (Shedgum Plateau, 2.1.4.) and its canyons, details of which are discussed in chapters 2.1.3. and 2.1.4.

After the transgression during the Upper Pliocene and the development of the cliff-line of the eastern scarp of As Summan Plateau, the regression of the sea during the Latest Pliocene and the Earliest Pleistocene was not continuous but took place in several phases. During the Pleistocene this

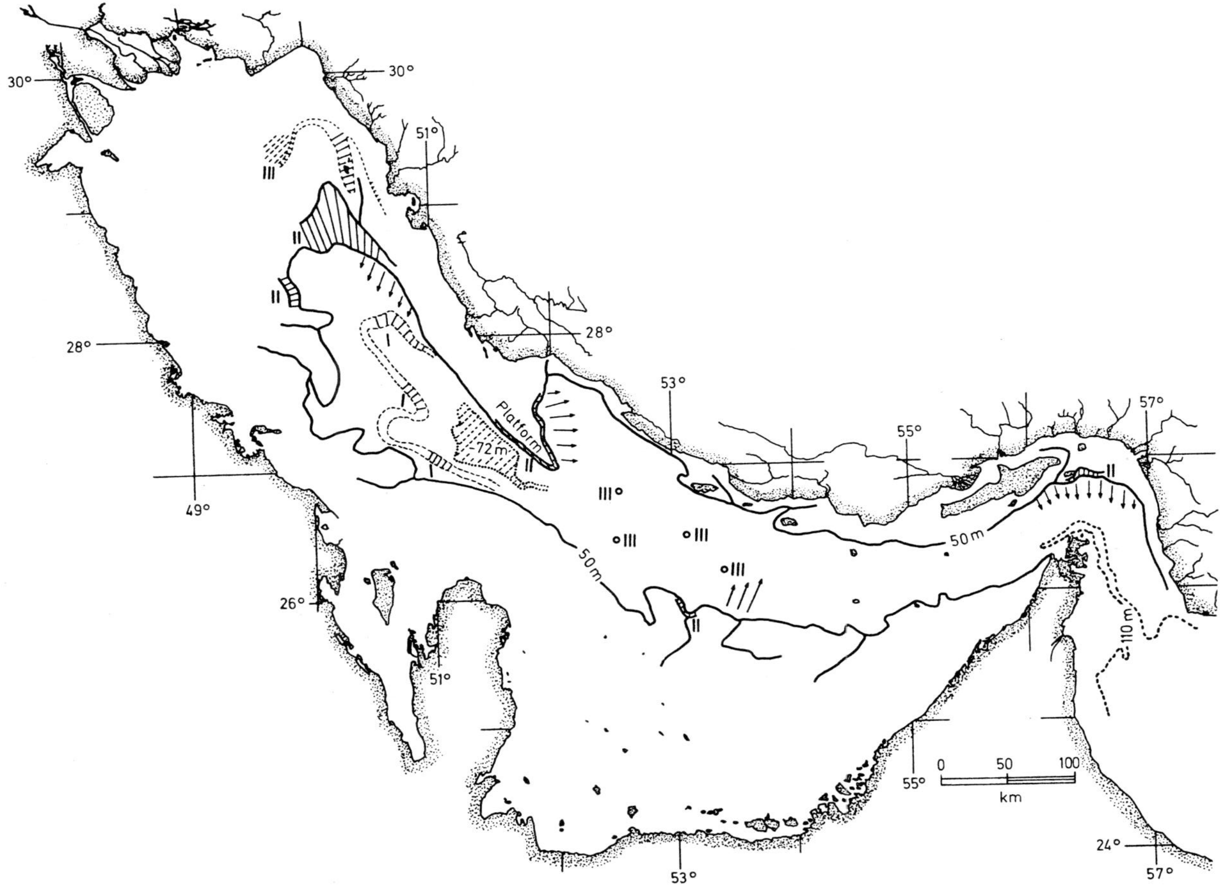


Fig. 9. Palaeogeographic map showing the Arabian Gulf at various times during the post-glacial transgression. *I*, *II*, and *III* correspond to the shoreline positions during three periods of approximate stillstand at minus 62, 50, and 30 meters sea level. From M. Sarnthein, 1972

regression was influenced by eustatic fluctuations corresponding to periods of different climate. In general the shorelines of the high sea levels of the interglacials lie lower the more recent is the corresponding interglacial (R. W. FAIRBRIDGE, 1961; P. WOLDSTEDT, 1961).

Traces of marine shorelines and abrasion terraces of the Pleistocene are to be found in various parts of the coastal zone and its hinterland of Saudi Arabia between Kuwait and Abu Dhabi. In the course of this work terraces at a higher altitude were studied only in the hinterland of the Bay of Salwah where the slope rises steeply up to an altitude of 130 meters. The field reconnaissance was concentrated in the area of Khashm Umm Huwayd, about 80 kilometers north of Salwah, and the area situated directly west of Salwah. In spite of the great distance between the two areas, the profiles and terraces are in good correspondence to each other as can be seen from Table 6.

Steps No. IV–VIII are covered with duricrust, but not the lower ones. It was not possible to identify the origins of the oyster shells at step no. VII. It was also impossible to prove any connection between the higher "terraces" (steps no. V–IX) and the sea levels of the Pleistocene. D. A. HOLM (1960) describes possible beach gravels at an altitude of 110 meters near Nibah, 30 kilometers south of the Bay of Salwah.

Although P. WOLDSTEDT (1958, Vol. II, p. 366) finds that the altitude of Pleistocene marine shorelines of the Arabian Gulf corresponds well to those of the classic Pleistocene marine niveaus in Mediterranean coastal areas, the Prewürm sea level fluctuations of the Arabian Gulf cannot be dated in detail because of great distances as well as probable tectonic movements. The Bay of Salwah must have its origin in a tectonic subsidence of Pliocene/Pleistocene age because it interrupts the sediments of the delta plain which now have an altitude of 130 meters in the hinterland of the bay and the sediments of the same delta now isolated at a corresponding altitude in the peninsula of Qatar (P. KASSLER, 1973). Only very detailed sedimentological and morphological studies of the area east of As Summan Plateau between Kuwait and Ar Rub' Al Khali may produce a better knowledge of the Prewürm sea level fluctuations in the Arabian Gulf and of the morphogeny of its marine shore terraces at higher altitudes.

#### *2.1.2.2. Sea level fluctuations during the Würm time*

The marine regression during Würm time resulted in a sea level more than 100 meters below the present water level. In this period the Arabian Gulf was dry and the shoreline was in the Gulf of Oman with an estuary in the area of Hormuz (M. SARNTHEIN, 1972; lowest water level of the Late Pleistocene minus 105 to minus 120 meters; Fig. 9).

The marine transgression during the interstadial period of the Middle Würm (40,000–26,000 years B.P.)—discussed with much fervour in former years—became evident in various parts of the world owing to modern investigations (H. FAURE, P. ELOUARD, 1967; J. D. MILLIMAN, K. O. EMERY, 1968). The maximum water level lay a few meters higher as compared to the present sea level.

During the writers' 1973 field campaign, oyster shells were collected below the 10 meter terrace of Salwah (Fig.10 ). An age of > 38,800 years B.P. was revealed by $^{14}C$ measurements (W. RAUERT; Table 7, No. 5).

The shell deposits are at an altitude of 5 to 7 meters above sea level. The 10 meter niveau showed no beach pebbles. In contrast, in the area of Khashm Umm Huwayd the writers found typical, slightly-weathered, cylindrical pebbles on the 10 meter step (Table 6) which is there an old strand

Fig. 10. 10-meters terrace north of Salwah. The eastward dipping strata of claystone and marl are cut by the flat terrace. (Photo: V. MAURIN, 1973)

plain covering a width of up to 50 meters. A sandwall 3 to 5 meters high and now at a distance of 150 meters from the sea accompanies the old beach, probably a remnant of a submarine bank of the Middle Würm. The present shoreline is 1.5 kilometers eastward; the area between the former and the present shore is now mainly a coastal sabkhah.

These conditions are partly comparable with the results of investigations carried out by P. KASSLER (1973). He reports on samples of marine limestone at a depth of 22 meters collected northeast of Qatar with a $^{14}C$ age of 26,700 years B.P., and a fossil beach at the Halul Island dated 32,700 to 44,300 years B.P.

Description of various shelf regions indicate that the postglacial transgression began about 15,000 years B.P. (J. E. MILLIMAN, K. O. EMERY, 1968). The rise of sea level occurred in phases interrupted by brief phases of standstill and partly by regression. For the Arabian Gulf this is illustrated by the topography of its floor (J. J. HOUBOLT, 1957; P. KASSLER, 1973; Fig. 9) and by its autochthonous to parautochthonous relict sediments (M. SARNTHEIN, 1971, 1972).

The first rise of the post Würm sea level reached an altitude of — 65 meters presumably 11,000–12,000 years B.P. An enrichment of coarse quartz sand and aragonite mud on submarine platforms, now at depths of 64–61 meters, 53–40 meters and 30 meters corresponds with phases and interruptions respectively of the rise of sea level during the Late Würm (Fig. 9).

Table 7. *Carbon-14-measurements of samples collected 1973 at the shoreline of the Arabian Gulf and Al Hasa*

| | Locality | Geographic position | Lab. | Reg. No. | Material | Age (years B.P.) |
|---|---|---|---|---|---|---|
| 1 | Shoreline, wave-cut bench | lat 26° 53′ N<br>long 49° 56′ E | IRM [1] | Lab. 3649 | shells of *Cardies* and *Pectes* | 4,670 ± 190 |
| 2 | shell bank | lat 26° 50′ N<br>long 50° 00′ E | IRM | Lab. 3650 | shells of *Cardies* and *Pectes* | 3,380 ± 180 |
| 3 | wave-cut terrace | lat 26° 30′ N<br>long 50° 00′ E | IRKW [2] | VRI-383 | Oyster shells | 1,090 ± 80 |
| 4 | shell bank | lat 26° 30′ N<br>long 50° 00′ E | IRKW | VRI-406 | cemented shells of *Cardies* and *Pectes* | 3,990 ± 90 |
| 5 | shells 1.5 km west of the shoreline | lat 24° 45′ N<br>long 50° 45′ E | IRM | Lab. 3652 | Oyster shells | > 38,800 |
| 6 | small hill in Al Hasa | lat 25° 30′ N<br>long 49° 37′ E | IRM | Lab. 3653 | Snail shells | 14,280 ± 430 |
| 7 | Al Hasa 2 m below surface | lat 25° 30′ N<br>long 49° 40′ E | IRM | Lab. 3654 | Snail shells | 2,180 ± 210 |
| 8 | small hill in Al Hasa | lat 25° 30′ N<br>long 49° 37′ E | IRKW | VRI-405 | Peat and charcoal | 8,290 ± 120 |

[1] Institut für Radiohydrometrie, München.
[2] Institut für Radiumforschung und Kernphysik, Wien.

### *2.1.2.3. Holocene sea level fluctuations*

A well-developed abrasion terrace exists near Al Jubayl and the adjoining coastal area. Two to three meters higher than the recent sea water level, it is very similar to the Tapes or Nizza terrace of the Mediterranean (see Fig. 11). Measurements of $^{14}C$ on oyster shells spread across the surface of the dunes covering this coastal region south of Al Jubayl show a young age (Table 7, No. 3). In some places the wind has transported these shells far into the hinterland.

Nevertheless, an exact dating of this terrace was possible in the area north of Ras Tannurah. A shell bank was found there below this terrace, from which the very well-cemented shells (*Cardies* and *Pectes*) were collected with hammer. Measurement of $^{14}C$ revealed an age of about 4,000 years B.P. (Table 7, No. 1, 2, 4). These measurement confirm the existence of the Holocene transgression and a higher sea level during the Middle Holocene (Table 6)

Fig. 11. Well developed coastal terraces and wave-cut of the "Neolithic pluvial", 3 meters above sea level. Arabian Gulf south of Al Jubayl. (Photo: J. G. Zötl, 1973)

in the east coastal zone of Saudi Arabia. These measurements, together with three $^{14}C$ measurements of shells of *Cerithies* from sand barriers in Qatar (3,930 ± 130, 4,200 ± 200, and 4,340 ± 180 years B.P.; J. C. Taylor, L. V. Illing, 1969) leave little doubt that the lowest step of the slope in the hinterland of Salwah (Table 6) also belongs to this system, which comprises the whole coastal zone for the area from Ras Al Misha'ab to Salwah. D. A. Holm (1960) presumed this transgression but was not able to date it exactly.

Information about the duration of the Holocene sea water transgression in the Arabian Gulf was obtained from sabkhah investigations in Abu Dhabi by G. Evans et al. (1969). Thirty-six $^{14}C$ measurements showed that this transgression began about 7,000 years B.P., reaching its greatest extent from 6,000 to 4,000 years B.P. Along the Trucial Coast the sea level during this period was about two meters above the present level. About 3,750 years B.P., the sea level first fell back rapidly by one meter and then it had a successive regression down to the present level 1,000 years B.P.

### 2.1.3. Geologic History of the Al Hasa Area Since the Pliocene

(H. Hötzl, V. Maurin, J. G. Zötl)

The oasis of Al Hasa, with its center in the town of Al Hofuf, is about 70 kilometers towards the interior of the country from the Gulf coast near Al 'Uqayr, at elevations of 130 to 160 meters above sea level. The plain between the oasis and the coast, sloping with a very low gradient towards the east, is covered by the sand dunes of Al Jafurah and is partly a dikakah. Westwards the steep scarp of the Shedgum Plateau rises to an altitude of 290 meters.

Al Hasa is one of the largest oases of the world. Situated between the rock desert of As Summan Plateau in the west and the sand dunes covering the adjoining plain in the east, the oasis owes its existence to large karst springs at the eastern scarp of As Summan Plateau.

Geologically a part of the Arabian Shelf, the Mesozoic and Cenozoic series of strata dip slightly towards the east. In the area of Al Hasa the outcrops consist only of Tertiary and Quaternary sedimentary rocks. The morphogeny and hydrogeology of this area is strongly influenced by the Ghawar structure laid out by Cretaceous and Lower Tertiary tectonics. This north-south-trending anticline west of Al Hasa is 20 to 40 kilometers wide and contains the world's largest oil field.

The field studies carried out in 1973 and 1975 provide a basis to investigate the geological history and morphogeny of this area since the Pliocene, including not only results from the area of Al Hasa but also from the surrounding region from Abqaiq in the north to Haradh in the south. Studies of D. A. Holm (1960), R. W. Powers et al. (1966), and R. W. Chapman (1971, 1974) will be compared and discussed.

#### *2.1.3.1. Terrestrial sedimentation of the Lower Pliocene*

As Summan Plateau between Abqaiq and Haradh consists of a sedimentary complex of Miocene and Pliocene strata (R. W. Powers et al., 1966), of which the upper layers are very well exposed in the scarp towards the east near Al Hasa. This upper series of strata, the Hofuf Formation of the Upper Miocene and Lower Pliocene, consist of terrestrial sediments with almost no fossils at all. They lie on the marine strata of the Dam Formation, limestones, marls and clays of the Middle Miocene. These marine strata form the base of the scarp of As Summan Plateau north and south of Al Hofuf.

The Mio-Pliocene age of the Hofuf Formation was a conclusion drawn only from the stratigraphic situation until the late nineteen sixties. A fossil mastodon (Pontian, upper half of Lower Pliocene) in similar sediments of Jabal Barakah in the coastal area of Abu Dhabi (K. W. Glennie, B. D. Evamy, 1968) permits a more accurate dating.

The different strata of the sedimentary complex from the Upper Miocene to the Lower Pliocene are the consequence of the changing climatic conditions during these periods. In the area of Al Hofuf a thin layer of marl is overlain by conglomerates with a thickness of 17 meters. The gravels of this Miocene stratum, partly from Mesozoic rocks and partly from rocks of the Arabian Shield, were transported under humid climatic conditions by rivers and are a product of fluviatile sedimentation. The humid period was followed by the development of fresh water lakes and the deposition of sandy limestone, as shown in the profile of Al Hofuf, to a thickness of 18 meters.

The strata overlying the limestone are white and reddish sandstones, witnesses of an arid climate. Field studies conducted northwest of Al Hofuf showed a different character of these sandstones. Partly, they originated in clayey and silty sedimentation into shallow closed pans (small layers of gypsum support this theory); partly, as eolian deposits, at some places still showing the structure of dunes but elsewhere unstratified or having a narrow network of cemented small pipes of roots. According to K. W. GLENNIE, B. D. EVAMY (1968) sandstones of this latter type originated from dikakah sediments. Strata overlying these sandstones are marly limestones and sand (about 9 meters), also deposited in a more humid climate. The uppermost stratum is freshwater limestone (1–2 meters), occurring in many parts of As Summan Plateau (R. W. CHAPMAN, 1971).

#### *2.1.3.2. Pliocene-Pleistocene marine transgression and regression*

Presumably caused by tectonics in connection with the folding of the Zagros Mountains, marine transgression from the east took place during the Middle and Upper Pliocene. The consequence thereof was the erosion of parts of the terrestrial sediments of the Upper Miocene and Lower Pliocene and the creation of the steep cliff of the recent eastern border of As Summan Plateau discussed in the following chapters.

The abrasion plain of this transgression lies in the hinterland of Salwah at altitudes of 100 to 130 meters, ascending with a gradient of 0.5–0.8 per mill up to 150 meters along the foot of the cliff. The plain itself is covered by delta sediments and/or the dunes of Al Jafurah.

The marine origin of this plain, dipping towards the east, as well as of the eastern scarp (cliff) of As Summan Plateau was already presumed by D. A. HOLM (1960). This theory is confirmed not only by the morphology of the eastern escarpment of As Summan Plateau but also by the existence of marine sediments in front of the cliff, as well as the impressive breakers caves of Jabal Al Qarah, and the delta of Wadi As Sah'ba (Pre- or Lower Pleistocene; see 2.1.3.3. and 2.1.3.4.). The fossil cliff, now the eastern escarpment of As Summan Plateau with its abrupt rise of an average of 80 meters extends more than 150 kilometers from Abqaiq to Haradh. The numerous outliers and their series of strata dipping slightly towards the east, corresponding to the cliff, preclude any development of the escarpment along a fault; the same applies to the morphology of the cliff itself.

In the area of Al Hasa the cliff and its morphology are especially well developed. Here, the top plain of the Shedgum Plateau (i.e., the eastern part of As Summan Plateau) averages between 270 and 280 meters above sea level. The steep escarpment shows a number of terraces; the smaller ones are caused by the different resistances to erosion of the lithologic sequence. Larger terraces are to be found at altitudes of 240–250 meters and 200–220 meters above sea level; they are developed at the slope of the plateau as well as at the outliers (e.g., Barqa Ar Rukban, Jabal Burayqa, Jabal Sha'bah, Jabal Al Qarah). At the eastern outliers also abrasion niveaus of lower altitude are to be found. There is no doubt that all these large terraces extending over long distances are not only caused by the lithology but are also reflecting interruptions of the Upper Pliocene uplift of As Summan Plateau.

The abrasion plain in front of the cliff exposes, over large areas, gray marls probably belonging to the uppermost strata of the Dam Formation. Beginning with the last phases of the marine regression the cliff line was dissected and transformed by the runoff from the plateau. Eroding short and deep gullies, the runoff deposited numerous alluvial fans over the abrasion plain from the Lower Pleistocene until recent times. These fans can be found mainly in the area between the margin of the plateau and the nearest outliers. Towards the east thin sheets of sand and gravels with a high percentage of quartz pebbles cover the marls of the Dam Formation. The quartz pebbles are derived from the sandstones and conglomerates of the Hofuf Formation destroyed by the marine abrasion.

The marine transgression is proved by beach sediments on the abrasion plain near the cliff. Although field reconnaissance in this area was limited, algal limestone locally combined with corals and bryozoa was detected in three places. One of these places lies south of Jabal Sha'bah about 11 kilometers northeast of Al Hofuf at an altitude of 135 meters (about 500 meters west of the newly drilled well HAW 1 of the Ministry of Agriculture and Water). It is a small remnant of algal limestone about 1.5 meters higher than the surrounding plain. The bedding is not well defined. A thin-section shows a network of a blue-green algae and well rounded, transparent, white to reddish quartz grains. These algae are typical of beach sediments because they can exist only in the foreshore. Many signs indicate a phase of marine regression.

The second local occurrence of algal limestone is in the area of Al Hofuf on the road to Dhahran about 2 kilometers northeast of the train station. The limestone is similar to that mentioned above but has many more quartz grains cemented by calcite.

The third remnant is south of the village Ash Sha'bah about 7 kilometers northeast of Al Hofuf (about 3 kilometers east of the administration building of HIDA). The limestone is similar to that mentioned above and also contains corals and bryozoa besides the algae. This algal limestone is presumably underlain by quartz sand, which has a thickness of 1.80 meters in a pit near the last mentioned limestone. The quartz grains are in size and shape and colour similar to those enclosed in the crust of algae. Because of the well-rounded shape, the good sorting and the absence of stratification a marine origin has to be presumed. The strata underlying this sand belong to the Miocene Dam Formation. It is supposed that the quartz sand was produced by waves of the sea transgression, unlike the algal limestone which indicates a later phase of regression.

Similar but more differentiated series of strata were described by G. Einsele, D. Herm and H. V. Schwarz (1974), illustrating a transgressive-

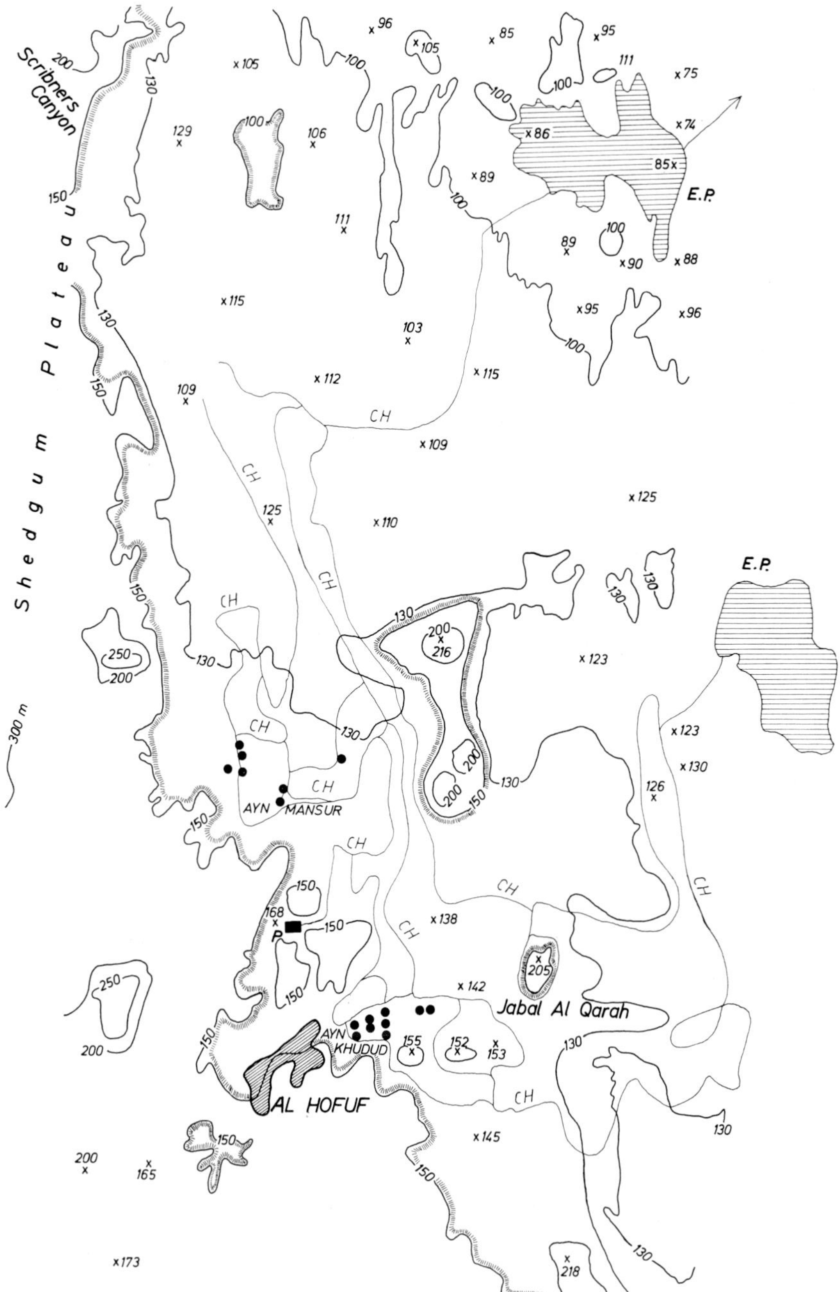


Fig. 12. Morphographic map of Al Hasa area; altitudes (× 109 = meters above sea level), main springs (black circles), main channels (*CH*), evaporation pans (*E.P.*), and pumping station (*P*)

regressive cycle of Pleistocene beach sediments in Mauretania. Recent sediments of this kind may be found in the coastal area of the Arabian Gulf.

One must emphasize that an exact survey of these marine sediments combined with paleontologic and sedimentological studies should result in a detailed knowledge about the marine transgression and regression during the Plio-Pleistocene. It is remarkable that this marine transgression largely co-

Fig. 13. Pillar-like erosional remnants and cliff at the lower niveau at the eastern side of Jabal Al Qarah. (Photo: J. G. Zötl, 1973)

incides in its regional extension with the marine transgression of the Middle Miocene and the marine sediments of the Dam Formation (R. W. Powers et al., 1966) respectively. The very locality of these two coast lines seems to be connected with the position of Al Ghawar anticline which runs parallel to the escarpment. The uplift of this anticline, mainly in the Cretaceous and Lower Tertiary periods, must have continued until the Lower Pleistocene.

Some other aspects of the morphogeny of the eastern escarpment of As Summan Plateau are discussed by R. W. Chapman (1971, 1974) and in his description of the Shedgum Plateau in 2.1.4. For the morphogeny it is decisive that along the 150 kilometers of the escarpment between Abquaiq and Haradh there exists not a single valley system of remarkable extent. The wadis eroded into the scarp are usually less than 2 kilometers long and have no connection with a larger river or wadi system in the hinterland. This also seems to be genuine evidence of the marine origin of the escarpment.

### 2.1.3.3. *Breakers terraces and caves of Jabal Al Qarah*

Jabal Al Qarah, about 10 kilometers east of Al Hofuf, is one of the completely isolated erosional remnants in front of the cliff of As Summan Plateau; its top is some 70 meters higher than the surrounding oasis (top 205.5 meters above sea level; road along the eastern slope of the jabal, 132.5 meters above

Fig. 14. View over the palm gardens of Al Hasa Oases. At the lowest steps around Jabal Al Qarah the development of gorges and collapses is a significant feature. (Photo: J. G. Zötl, 1973)

sea level, see Fig. 12). The jabal consists of marl and marly sandstone of the Hofuf Formation, and its sides are just as steep as the cliff of the Shedgum Plateau (i.e., the part of As Summan Plateau northwest of Al Hofuf). Jabal Al Qarah covers an area of 1.7 square kilometers.

Concerning the geological features, within the sandstone, layers of silt and clay are interbedded, and their yellow and reddish colours are typical. The base of the jabal consists of grey marls probably belonging to the Dam Formation.

Morphologically, Jabal Al Qarah shows an alternation of step-like small plains and steep rises. The largest plain is the slightly eastward-sloping plateau at an altitude of 190 meters above sea level, bordered by small hills rising to 205.5 meters above sea level at its northern and southern edges. At the western, southern, and especially the eastern sides of the jabal, distinctly visible niveaus are developed at altitudes of about 170 meters,

155 meters, and 145 meters above sea level. At the lower level, cliffs and pillar-like erosional remnants have been formed (Fig. 13).

There is no doubt that the above-mentioned niveaus were developed primarily by marine abrasion. For any later shaping of the topography two facts exerted a decisive influence: the protection of the terraces by duricrust, and the vertical erosion by linear surface runoff. At the lowest steps the development of gorges and collapses is a significant feature (Fig. 14).

Measurements of the joint system and the evaluation of aerial photos show a NW-NNW-trending joint system of Jabal Al Qarah. This had an influence on the development of the cliffs. Wave-cut gorges follow the direction of the joint system. The largest of these gorges—in some places 10 meters wide, and reaching almost 300 meters into the jabal—can be found in the area of the village of At Tuhamiyah, where it cuts into the southeast flank of the jabal. In its inner part, where the walls are still 20–30 meters high, collapsed walls formed barriers so that behind them young fanglomerates could pile up to a thickness of 5–8 meters.

At the western side of Jabal Al Qarah the ruptures meet the cliff line in an acute angle. There, the cliff face is strongly influenced by the lithologic change of strata and the selective marine abrasion. This can be very well observed south of the village of Al Qarah. An abrasion niveau 6 meters above the road is made up by grey marls (Dam Formation?), and its terrace is now used as threshing floor. The cliff bordering the abrasion niveau is 5–6 meters high and consists of sandstone and clayey sandstone. Wave plantation and breakers exerted an effective influence depending on the alternation of beds. Beside hollows, sea caves extending more than 10 meters into the jabal were developed, today partly resting on natural pillars and partly collapsed. In some places blocks, as remnants of wall collapses, protected the inner parts of caves, e.g., the block in the central part of Fig. 15.

Behind this lowest cliff step at the western side of Jabal Al Qarah another breakers terrace is developed at an altitude of 150–155 meters. However, this terrace is split up into many large blocks. The reason for this phenomenon is either the collapse of a cliff overhang or a rockslide. Such a rockslide, partly submarine, was detected at the eastern side of the jabal. At present, there exists a maze-like cave system directly behind the above-mentioned threshing floor, in parts episodically flooded by runoffs from the western side of the jabal. The cave, originally a sea cave, was later modified by fluviatile erosion as is demonstrated by young runoff channels (Fig. 15). Furthermore, salt weathering was particularly effective in clayey sandstones. In this case the water-damming marls also play an important role. Finally, some parts of this cave have been artificially widened; they were used for storage or as places of shelter. Some other remnants of breakers caves are still used for pottery factories or as sheep and goat stables.

Morphogenically, the cave "Ghar An Nashab" (cave of the archer) located at the eastern side of Jabal Al Qarah is highly remarkable. With its various branches the entire system is more than 1.5 kilometers long, developed along mainly rectangularly crossing joints. Vertical shafts extend in many cases up to the land surface, which makes it possible to inspect large parts of the

Plate I

**Ghar An Nashab I**

(The cave of the archer)

Jabal Al Qarah, Al Hasa, Saudi Arabia

Measured: H. Hötzl and V. Maurin

Drawn: V. Maurin 1976

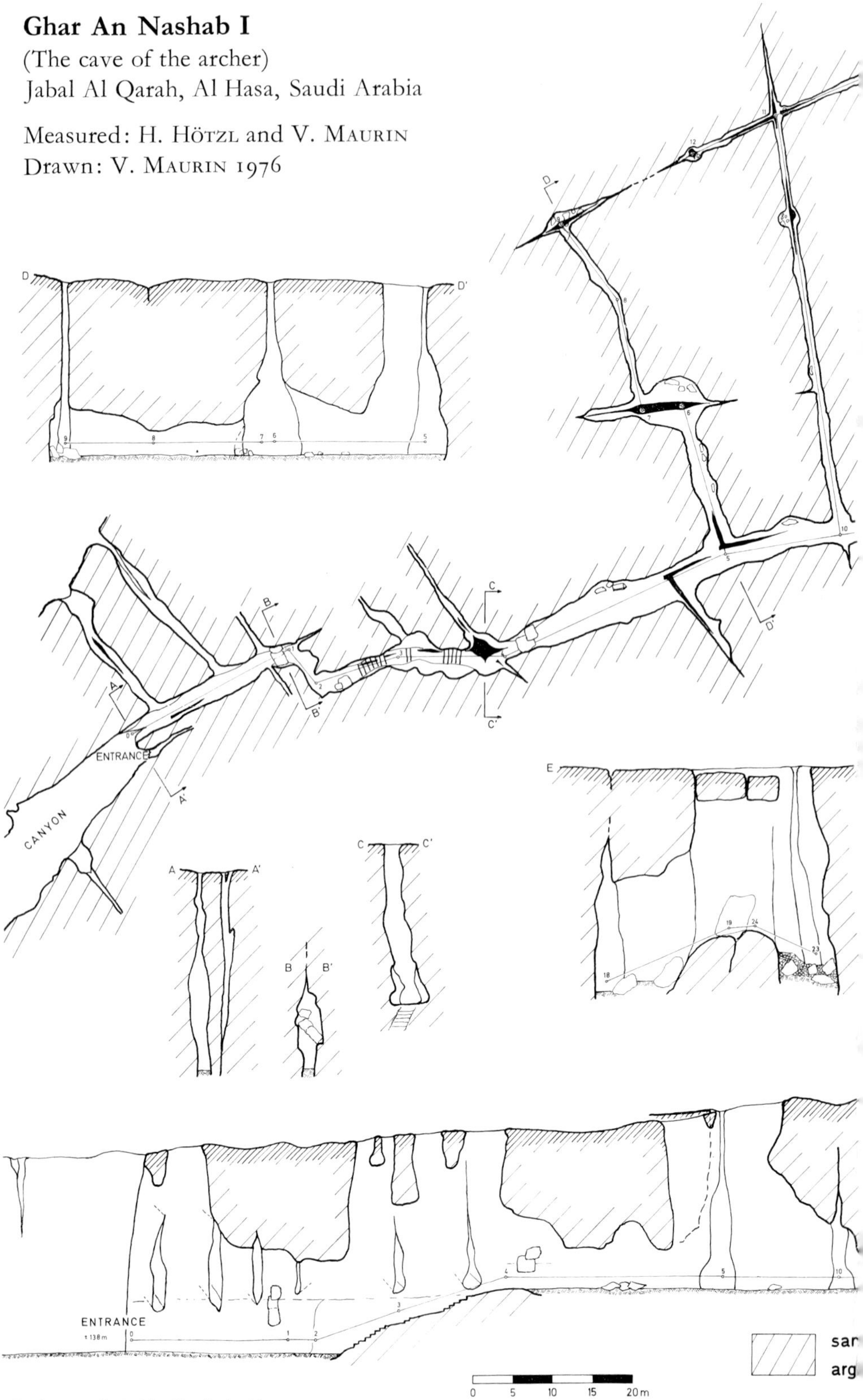


Quaternary Period in Saudi Arabia
Springer-Verlag Wien New York
Druck: Adolf Holzhausens Nfg., A-1070 Wien

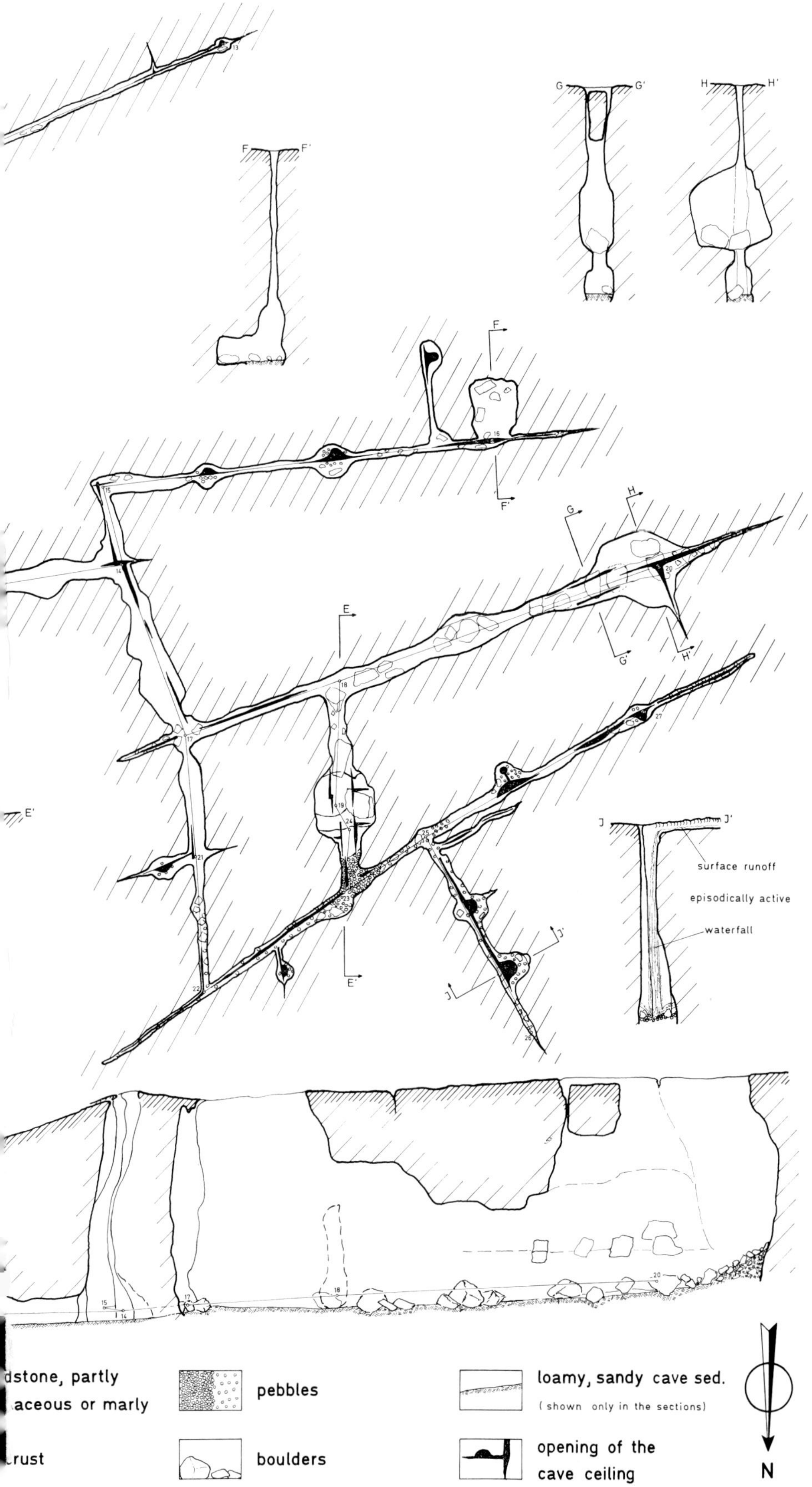

F F'
G G'
H H'
E E'
J J'
13
14
15
16
17
18
19
20
21
22
23
24
25
26
27
surface runoff
episodically active
waterfall
dstone, partly
aceous or marly
rust
pebbles
boulders
loamy, sandy cave sed.
( shown only in the sections)
opening of the
cave ceiling
N

cave without artificial light (Fig. 17). Beside numerous clefts two cave systems, reaching about 200 meters into the jabal, are of major interest. The southern one of these caves—we called it Ghar An Nashab I—is accessible by car. During the field reconnaissance of 1975 a geodetic levelling was carried out together with a polygon measurement of 584.1 meters of the cave's total

Fig. 15. Opening of a sea cave at the western side of Jabal Al Qarah. Developed by selective marine abrasion influenced by the lithologic change of strata, today the cave is partly resting on natural pillars and partly collapsed. Height of the block on pillars ca. 8 meters. (Photo: J. G. Zötl, 1973)

length of 746 meters (see Plate I). For lack of time it was not possible to make the same geodetic survey in the northern, more or less parallel, system (Ghar An Nashab II).

Detailed investigations conducted in the cave system as well as on the land surface enabled us to reconstruct the morphogeny of this cave system. Primarily, the system was developed at the same time as the breakers terrace at the present altitude of 145 meters at the eastern side of Jabal Al Qarah. The marly and clayey components of the terrace were partly layered below sea level and had plastic properties, no doubt. This was the reason for an enormous sandstone rockslide. This sandstone acted as a capping bed reaching a thickness of about 30 meters and had a compact consistency. The joint system running north-northwest, i.e., parallel to the cliff, favoured such a rockslide. This decomposition as well as a subsidence of a few meters caused by the sliding can be observed on the land surface of the block field.

At the same time breakers and tides, as well as infiltrating precipitation, influenced the development of the cave system. During the marine regression and after reaching a lower erosional base level, the forming of the topography of the 145-meter terrace and the headward erosion into the cave floor began. The longitudinal section on Plate I reveals a depth of 7 meters for this headward erosion under polygon point 2. The cave system II shows for longer distances the so-called "keyhole"profile, which is typical of the second phase of erosion. In cave system II this development is still at work since the episodic surface runoff from the top plateau makes use of this system. Canyon-like channels (Fig. 16), eroded shafts, and the accumulation of well-rounded pebbles (accumulated in protected places to a thickness of 2 meters) give evidence of the present activities.

The plateau at the top of Jabal Al Qarah displays a dendritic drainage pattern. Some of the surface channels have depths of 4 meters and floors of a few meters breadth. All the recent episodic surface runoff disappears into shafts. The upper and flatter parts of the runoff channels are covered with duricrust, but not the steep flanks in the lower parts. This confirms our opinion of a mainly pre-Pleistocene development of the duricrust, an age also assumed by R. W. Powers et al. (1966).

Under present climatic conditions salt weathering plays an important role in the widening of the cave's cross-section. Primarily, it works on the fine clastic layers in the lower parts of the cave systems. For this type of weathering the low permeability of the clayey and marly material has also to be taken into account. Other factors recently working on the further cave development are rockfall and collapses. Physical and chemical weathering lead to breaking up of the rock surface. The vertical joint system is very favourable for rockfall and subsidence of larger blocks. Examples can be seen in Profiles E–E′ and G–G′ on Plate 1. More comprehensive examples of this phenomenon can be found in Ghar Al Nashab II. Some collapses opened shafts reaching up to the land surface, one of them some 70 meters in from the entrance has a diameter of 8 meters. System II also reveals by the displacement of the cross joints the mutual shifting of gigantic sandstone blocks as a consequence of submarine rockslide.

Investigations made in the large joint and cave systems confirm the relationship between their origin and the existence of the Upper Pliocene cliff development at Jabal Al Qarah. Origin only from a more underground erosion during Pleistocene pluvials (R. W. Chapman, 1971; W. A. Goellner, 1972) seems impossible considering the very small recharge area of the jabal's plateau. This proof of a marine origin does not preclude periods of stronger development during more humid phases of the Quaternary.

Numerous similar vertical caves exist along the cliff of As Summan Plateau (R. W. Chapman, 1971). They were likewise formed by breakers and salt weathering. A very good example of a cliff landslide and cliff falls can be seen in a small wadi southwest of the Coast Guard station of Al Jadidah.

Finally, the problem of existing waterways reaching below the surrounding land surface should be mentioned. The strata at the orographic base of Jabal Al Qarah are probably not completely impermeable. Floods after episodic heavy rainfalls reach only the innerpart of the cave Ghar An Nashab I but not the entrance, it seems. Consequently, seepage occurs in spite of clayey and marly strata—probably through the north-northwest joint system. With regard to the problem of the amount of recent recharge of springs and groundwater in the oasis of Al Hasa such possibilities of infiltration are of principal

Fig. 16. Young runoff channel modifying the original sea cave of "Ghar An Nashab II". (Photo: V. Maurin, 1975)

Fig. 17. Vertical shafts developed along joint systems in the cave extend in many cases up to the land surface. Entrance to "Ghar An Nashab I". (Photo: H. Hötzl, 1975)

interest. In the course of a field reconnaissance carried out along the eastern edge of the Shedgum Plateau, episodically water-bearing shafts without any connection to caves in the cliffs were detected. Likewise, the surface of Barqa Ar Rukban, an inselberg about 17 kilometers north-northwest of Al Hofuf, has an open vertical shaft where the surface runoff of a depression disappears after heavy rainfalls. These observations lead to the assumption that the large As Summan Plateau belongs partly at least to the recharge area for the component of younger waters which get mixed with old waters discharging from aquifers at a greater depth.

#### *2.1.3.4. The Pliocene/Pleistocene Delta of Wadi As Sah'ba*

South of the oasis of Al Hasa, the plain between As Summan Plateau and the coastal area from Salwah to Sabkhat Matti is covered by an enormous gravel fan. This fan begins at the cross channel of Wadi As Sah'ba through As Summan Plateau near Haradh and ranges from Al Hofuf, in the north, to the southwestern part of Qatar, in the east, down to Ar Rub' Al Khali, in the south. The radius of this arch is more than 150 kilometers. The main extension runs east-southeast where one finger reaches down to Sabkhat Matti 250 kilometers from Haradh. The geometric form of this gravel fan is a flat semi-cone with its apex near Haradh at an altitude of 300 meters above sea level. The gradient to its border is roughly one permil, as can be seen in Table 8.

Table 8. *Distance from the beginning of the gravel fan of Haradh (300 meters above sea level) to its border and the gradient of the fan*

| Border Area | Altitude | Distance from Haradh in km | Average gradient in ‰ |
|---|---|---|---|
| Al Hofuf | 145 m | 145 | 1.06 |
| Khashm Umm Huwayd, Bay of Salwah | 110 m | 185 | 1.02 |
| Jal Abu Arzilah, south of Salwah | 108 m | 175 | 1.10 |
| Khashm An Nuksh southwest Qatar | 95 (?) m | 205 | 1.00 |
| Bunayyan, 120 km north of Al 'Ubaylah | 118 m | 175 | 1.04 |

The fan's bedding on the transgression plain of the Upper Pliocene, its correspondence to a correlating high base level of erosion, its morphologic form, and the type and structure of its sediments lead to the conclusion that this must be of deltaic origin. Already D. A. Holm (1960) and R. W. Powers et al. (1966) have assumed this morphogeny. D. A. Holm came to this theory because of the general shape, the forset beds at several places along the margins, and the divergent fanning of gravel train. Topographically, these trains are just a few meters higher than their surroundings, yet despite the cover formed by dunes (Al Jafurah) they can be found over a distance of 50 kilometers and more. Showing a divergence towards the margin of the delta, they were the proper river channels comparable to the Mississippi "passes". Deposits of coarse material occur in these channels while fine-grained deposits lie between the channels. The channels within the delta of Wadi As Sah'ba also are filled up with coarse gravels and pebbles while between the channels sand, partly with small gravel lenses, and silt can be found. The size of grains becomes noticeably smaller from Haradh towards the margin of the fan.

The process of exhumation of the channel's gravel fill began, after the regression of the sea, by deflating the fine material between the channels. So, in these areas, the intercalated gravels remained just at the surface giving the impression of a uniform gravel cover, partly caused by a secondary scattering. Now only the small ridges formed by the relief inversion indicate in contrast to the deflation depressions the original facies differentiation within this delta fan. A very good example of the effectiveness of wind deflation is a ridge consisting of partly cemented gravels about 50 kilometers southeast of Al Hofuf extending more than 40 kilometers towards the northeast (M. STEINEKE et al., 1958). As can be seen in the areas which are not covered by dunes, this ridge is 1.5 to 2 kilometers wide and about 20 meters higher than the deflation pans which accompany the ridge. The crest of this ridge is 10 to 30 meters wide; its slopes gently merge into flat pans of deflation. However, in details the structure of this ridge is much more complicated. Pits and quarries show that the ridge consists of marl and limestone (Hofuf Formation?) covered by duricrust, on the crest. Scattered gravels were the reason for mapping this ridge as gravel trains (M. STEINEKE et al., 1958). However, quarries providing materials for road construction and the building industry show that on the crest of the ridge only a channel 12 to 15 meters wide and about 8 meters deep, cutting into the limestone, is filled up with gravels. The excavation of this material provides an excellent view into the channel with vertical sides of limestone covered with duricrust. This channel, cut into older rocks, shows that the delta deposits do not originate in one uniform foresetting sedimentation but in a sequence of a marine regression, interrupted by a standstill with fluviatile erosion, followed again by a new sedimentation process.

A similar formation is supposed for the enormous Ad Dibdibah fan, the delta of Wadi Ar Rimah into which the channel of Wadi Al Batin was eroded. The delta enlarges towards Kuwait and the Arabian Gulf, and the river system of Tigris and Euphrates. In both deltas, that of Wadi As Sah'ba as well as that of Wadi Ar Rimah, the thickness of the sediments is small (mostly less than 10 meters) in spite of a large extent of more than 200 kilometers, although the material is coarse. The small thickness allows assuming a flat and stable subsoil; the coarse gravels depend on the river flow and its gradient and on the climatic conditions prevailing in the recharge area. These features cannot be compared to the large recent deltas of the Mississippi, Niger or Nile, which consist of thick layers of fine clastic sediments. A certain similarity to the flat delta of the Mekong River and its "sand channels" seems to exist with regard to the thickness of sediments. This delta is formed on a flat descending platform below shallow sea water (J. P. MORGAN, 1971).

There are no fossils to date these deltas. However, the general stratification as well as the origin of various gravels do permit dating the origin of these deltas from the Upper Pliocene to the Lowest Pleistocene. There is no doubt that the delta of Wadi As Sah'ba belongs to an old river system crossing the Hofuf Formation (Mio-Pliocene) west of Haradh. Furthermore, the obvious orientation and superposition of the delta on the marine transgression plain likewise formed the cliff at the eastern scarp of As Summan Plateau in sediments of the Lower Pliocene. Therefore, the delta must have its origin in the post Lower Pliocene.

The other limit of origin of the delta follows from the basalt gravels in the delta sediments. The basalt eruptions in the area of the Shield took place from the Miocene until the Holocene.

In the course of the field trip of 1975 into the area of the eastern part of the shield, basalt samples were collected in two different places. The younger of these basalts (Potassium-Argon age of about a million years; see chapter 2.5) displays hardly any erosion at all and lies flatly on a granitic plain which shows almost no denudation effect since this last volcanic event. It is impossible that the coarse basaltic components of the delta gravels of Wadi As Sah'ba have their origin in this younger basalt. In contrast to this an older basalt in the same region (age determination of 3 million years) is strongly eroded, showing a lateritic weathering. This intensive weathering and erosion may have provided the material for the delta sediments.

From these observations the delta accumulation of Wadi As Sah'ba took place before 1 million years ago, i.e., during the uppermost Pliocene (Astian) and/or the Lower Pleistocene (Calabrian). Concerning the climatic conditions, the lateritic weathering of the older basalt permits the presumption that there must have been a humid period with high temperatures between 3 and 1 million years ago, also proving a discharge of the amount of water necessary for the transportation of the coarse gravels (reaching a diameter of 15 centimeters) found in the delta sediments down to a distance of 450 kilometers towards the east.

### *2.1.3.5. Quaternary erosion and sedimentation*

The topography of the area of Al Hasa has probably not changed much since the Plio-Pleistocene. The striking morphologic features such as the cliff and its outliers, the plain towards the Gulf and the delta of Wadi As Sah'ba essentually developed in the Upper Pliocene. The Quaternary mainly produced an erosive activity of the wind as well as eolian transport and its ensuing accumulation.

This eolian activity, through deflation of fine material from the delta sediments, proved to be very effective. As mentioned before (2.1.3.4.) the consequence thereof was a relief inversion by the exhumation of small gravel ridges (former river channels). The difference in altitude between the tops of these young ridges and the lower deflation plains along them is about 30 meters and may well be the total amount of deflation. It was the following enormous sand transportation from the northwest, the main wind direction, which covered almost the whole area with the transverse dunes of Al Jafurah.

At As Summan Plateau, west of the cliff, deflation seems to have been much less effective as a consequence of a relatively thick and hard duricrust covering the land surface. On the other hand, wind accumulation and dunes are very rare at the most exposed parts of this plateau, which is not protected by any mountains.

Only the narrow short gullies which are steeply cut into the front of the cliff are witnesses of fluviatile erosion of the Quaternary (Fig. 18). Not quite clear is the development of the recent channel of Wadi As Sah'ba. South of Haradh, in the area of the King Feisal Settlement Project, the wadi is

Fig. 18. One of the narrow gullies of young age in a tributary of Scribners Canyon. Depth of the gully ca. 4 meters. (Photo: J. G. Zötl, 1973)

2–2.5 kilometers wide and incised into the delta sediments as well as the underlying strata of the Hofuf Formation. The surface of the delta sediments lies about 25 meters above the recent wadi floor, which reaches an altitude of 270 meters above sea level near the experimental farm. Numerous water well drillings show that the bottom of the wadi channel can be found at a depth of 30 meters, i.e., at 240 meters above sea level, and that the channel is filled with sand and fine gravels, and marly limestone, gypsum and various salt crusts up to the recent wadi floor (information from the King Feisal Settlement Project, Haradh, not yet published).

Proceeding from this general situation of the recent Wadi As Sah'ba it seems possible that its erosion occurred as a consequence of the lower base level of erosion during the glacial periods of the Pleistocene. On the other hand, the study of the geological map of R. A. Bramkamp, L. F. Ramirez (1959 b) leads to the conclusion that this channel has its base level of erosion in the area of Al Ghafah, i.e., about 100 meters above the recent sea level, or the base level of erosion of the Lowest Pleistocene. Because of the lack of field studies in the area of Al Ghafah this problem is still unsolved.

There is no doubt that the valley fill of the recent Wadi As Sah'ba near Haradh is of Quaternary age. The upper layers of the wadi fill, sand and silt with various crusts of gypsum, can be found up to 150 kilometers downstream from Haradh (R. A. Bramkamp, L. F. Ramirez, 1959 b). Further downstream the wadi floor is covered by dunes or gravels, the latter can be ascribed to the removal of the fine material through wind activity.

Sand and silt, partly together with fine-grained gravels and/or crusts of gypsum, also occur overlying the delta sediments in a small area from Wadi As Sah'ba to Al Hofuf (M. Steineke et al. 1958). Situated in front of the first dunes of Al Jafurah this material is presumed to have been transported by the episodic runoff leading off the cliff.

The Quaternary sediments in the area of Al Hofuf are of various natures, including sand and fine-grained material from the surface runoff, freshwater limestone, clayey marl with gastropods, and sabkhah plains. The dunes have temporarily dammed up the runoff from the huge karstwater springs, thus creating freshwater lakes and swamps; some small ones persisted until recently. Neolithic spears and fish-hooks prove that a good deal of fishing was done during the Middle Holocene (H. A. McClure, 1971). White calcareous mud of 2.5 meters thickness was found overlying Tertiary clay and marl. How-

---

Fig. 19. Longitudinal section of the underground cave of the spring "'Ayn Khudud" (Al Hasa). From an unpublished survey of the German divers group "KITAG" (Brandemer, Kistenmacher, 1968). 'Ayn Khudud is the largest spring of all Al Hasa oases (see Fig. 12). After "WAKUTI" (vol. 2, 1964) measurements of water level fluctuations were carried out before the installation of the modern irrigation system and construction of catchment buildings. The amount of the fluctuation was 28 centimeters (144.15 meters up to 144.43 meters above sea level); the largest discharge was 1.718 cubic meters/second. Symbols: *1* limestone and marl of the Dam Formation, *2* blocks from collapses, *3* sand and pebbles, *4* fine sand and silt, *5* land surface of the oasis, *6* road, *7* water level, *8* flow direction of the water

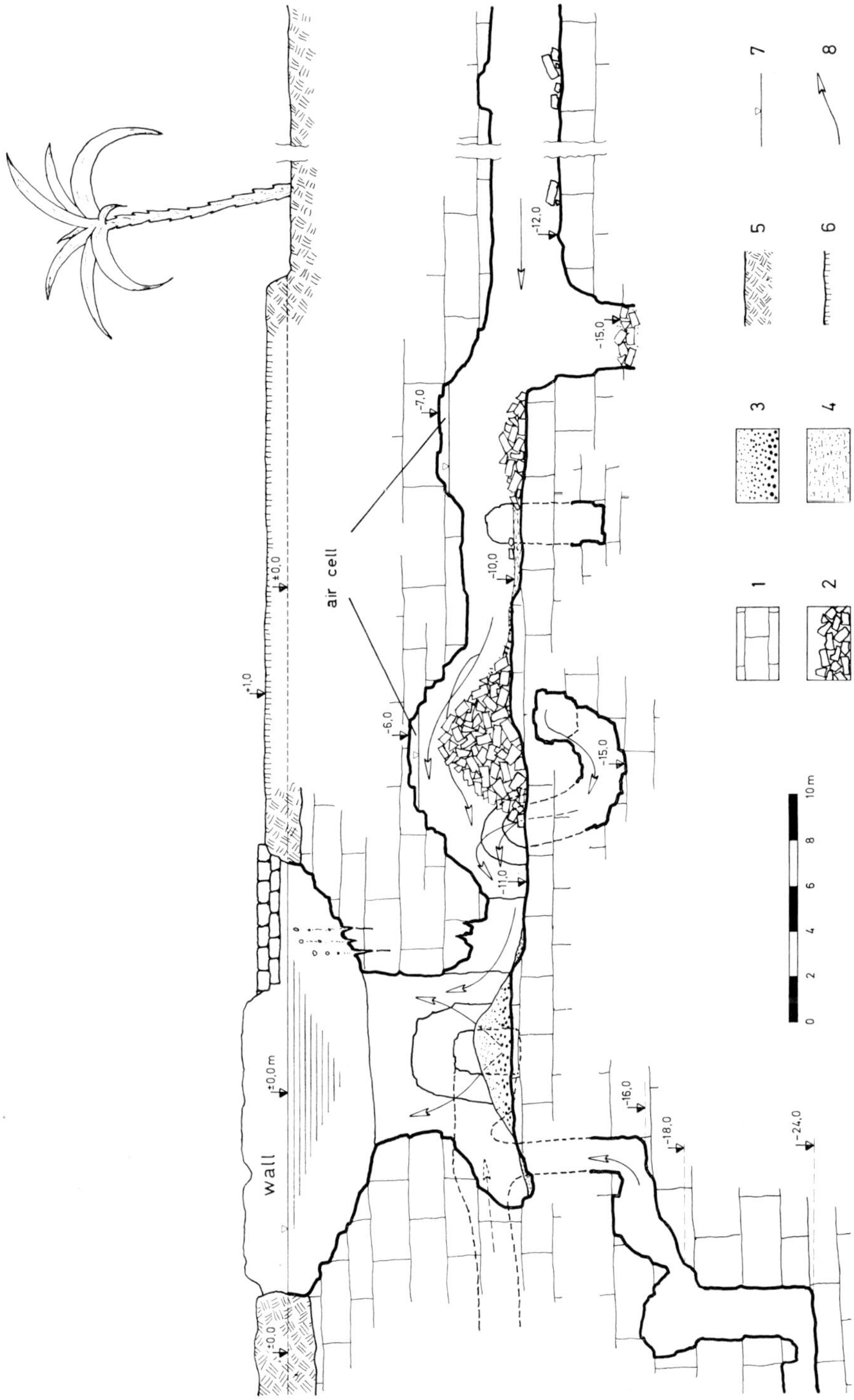
air cell
wall
±0,0 m
±0,0
±0,0
-1,0
-6,0
-7,0
-10,0
-11,0
-12,0
-15,0
-15,0
-16,0
-18,0
-24,0
1
2
3
4
5
6
7
8
0
2
4
6
8
10 m

Fig. 19

ever, traces of plants and shells of ostracoda could not be identified. In the course of the construction of the drainage channel northeast of Al Hawtah, numerous gastropods, mainly *Melanopsis tuberculata*, were found. They show a good state of preservation of nacre. $^{14}$C-measurement revealed an age of 2,180 ± 210 B.P. (Table 7, No. 7). Further shells of gastropoda with an age of 14,280 ± 430 (Table 7, No. 6) were collected on a small hill east of Al Mrah. Settlement in the Early Holocene, proved by artefacts and implements, was dated by peat and charcoal (Table 7, No. 8).

In the plain extending east of Al Hasa, silt, clay and salty clay were deposited in large sabkhahs. These sabkhahs allow the conclusion that the oasis must have had its drainage towards the Gulf until recently. Two drainage patterns have to be distinguished, one of them coming from the northern part of the oasis, the area of Al Muhtaraqah-Al Qarn, the other from the southern area of Al Qarah-At Taraf, both of them running parallel towards the northeast.

Recently the surface runoff has been interrupted by large dune areas (Al Jafurah). However, in the area of Al 'Uqayr a surface runoff exists again from the local coastal sabkhah. It is assumed that this surface runoff is fed by an underground seepage from the evaporation pans of Al Hasa. This assumption is confirmed by the fact that this runoff in the area of Al 'Uqayr has increased since the construction of the irrigation system of Al Hasa.

### *2.1.3.6. Climatic and hydrologic conditions*

The series of strata of the Miocene and Lower Pliocene (Hofuf Formation) testify that the arid climate of this era was interrupted by humid and semi-arid periods. For the Upper Pliocene there is a lack of these changing terrestrial strata which are necessary to give exact evidence of the climate and its changes.

The marine transgression during the Upper Pliocene was primarily combined with denudation, but a few remnants of beach sediments with algal and coral limestones allow presuming just the same temperature for the sea water as in recent times. The Upper Pliocene and Lower Pleistocene must have had temporarily a much wetter climate with a greater depth of precipitation in the highlands because of the runoff necessary to transport the coarse gravel of the delta of Wadi As Sah'ba.

An unsolved problem is the fact that the cliff line of about 150 kilometers extent was not dissected by canyons of appreciable length eroded headward into the plateau during these humid periods. Several circumstances—probably of common efficiency—may be the reason: 1. more arid conditions in this gulf area than in the mountains in the west; 2. north-south running drainage systems along the western flank of the uplifting Al Ghawar anticline; 3. a mainly underground drainage, this especially if the land surface consists of well-jointed sandstone covering a karstified limestone. Presently, the karst springs of the Al Hasa oasis have a total discharge of approximately 13 cubic meters per second. As can be seen from investigations by Philip Holzmann

Ltd. (KITAG—divers group) these springs are the overflow of horizontal cave systems through vertical shafts. The horizontal cave systems show several stages with karst tubes of considerable length and diameter (see Figs. 19, 20). Fig. 19 shows a cross-section through the cave system of 'Ayn Khudud with horizontal tubes, at depths of 8 to 10 meters (passed by divers over a length of 60 meters) and 23 meters, connected by vertical shafts. Similar situations are found at other springs. This karst system is developed in limestone of the Dam Formation. The limestone was deposited during the Middle Miocene in the course of a short marine transgression reaching towards the eastern flank of Al Ghawar anticline. A profile of a borehole near Al Midah, 11 kilometers northeast of Al Hofuf (HAW 1, drilled for the Ministry of Agriculture and Water 1975) shows for its uppermost 50 meters the following sequence of strata, probably belonging only to the Dam Formation:

| Meters, altitude above sea level | Lithological description of the strata |
|---|---|
| 135 | Land surface |
| down to 128 | Light coloured marl with fibrous gypsum |
| down to 125 | Marly limestone with fibrous gypsum, cream coloured; small layers of blue and reddish silt |
| down to 122 | White marl, soft; partly gypsiferous, small layers of pink limestone |
| down to 119 | Soft marly limestone; pink limestone, some gypsum |
| down to 116 | Soft marl and white marly limestone, moderately hard |
| down to 113 | Light coloured and grey marl |
| down to 107 | Cream coloured marl; some limestone |
| down to 104 | Marl and calcareous sandstone |
| down to 98 | Marly and cream coloured limestone |
| down to 95 | White limestone, very fine grained |
| down to 80 | White marl and chalky limestone |

The compact grey limestone and marl quarried by the cement factory south of Scribners Canyon at an altitude of 148 meters also belongs to the Dam Formation. These are the same limestones to be found at shallow depth in the main spring area of Al Hofuf.

The first karstification of the limestone of the Dam Formation may have occurred in an early stage, e.g., after the marine regression of the Middle Miocene. In this case the limestone had to be uplifted above the sea level. A climate and vegetation supporting this karstification can be concluded from the basal conglomerate of the Hofuf Formation including crystalline pebbles. The karstified limestone was later covered by terrestrial sediments of the Hofuf Formation (approximately 100 meters thickness), and afterwards partly exhumed and reactivated by the marine transgression of the Upper Pliocene. On the other hand it is possible that the first karstification took place after the marine regression of the Early Pleistocene following the lower base level of erosion.

The water discharging recently from the large karst springs derives from aquifers of the Umm er Radhuma Formation as can be seen from the hydrochemical and isotopic measurements (see 2.1.6. and 2.1.7.). The Umm er Radhuma limestones crop out about 100 to 150 kilometers west of the spring area at altitudes of 300 to 450 meters above sea level; they were karstified between the Lower and Upper Tertiary. The water of these aquifers rises through tectonic structures into higher horizons of younger stratigraphic complexes. The chemical composition of this water shows that it has hardly any karst corrosive efficiency. How far this efficiency can be reactivated through mixing corrosion caused by waters from higher horizons or infiltration has to be investigated.

Beside the surface and the subterranean drainage system, duricrust also is an important material from which to study the climate of the Upper Pliocene. This duricrust, mentioned for As Summan Plateau as early as in the nineteen thirties (R. P. MILLER, 1937), has been intensively studied in the cliff area of the Shedgum Plateau near Al Hasa by R. W. CHAPMAN (1971, 1974; see 2.1.4.). The basis of his conception of a fluviatile development of the topography of this escarpment is that the duricrust has its origin in the various interpluvial phases of the Quaternary Period.

For this topic it must be considered that the young Quaternary sediments such as calcareous mud and fine clastic sediments, as well as the marine sediments of the Upper Pliocene and the above mentioned delta accumulations are free of duricrust. Contrary to this, various delta channels were eroded during and after the Plio-Pleistocene regression of the sea, into a limestone which was covered by duricrust (see 2.1.3.2.).

R. W. CHAPMAN (1974) describes a different maturity of the duricrust covering the various erosional terraces of Scribners Canyon north of Al Hofuf. R. W. CHAPMAN explains their differences in maturity and thickness by the different ages of the terraces. In contrast, the duricrust at the surface of Jabal Al Qarah as well as at some comparative sections of the cliff itself seems to be relatively uniform. There the thickness of the crust becomes smaller in proportion to the gradient of the slope. However, it increases once more on the flat erosional steps. Wherever steps are exclusively caused by the change of strata, a lithological influence on the development and thickness of the duricrust, too, seems to become effective. Another factor is the different nodular weathering of the duricrust under the arid climatic conditions of the Quaternary.

The climatic conditions necessary for the development of duricrust have been minutely investigated, especially by F. KABELAC (1962) and A. GOUDIE (1973). From these studies, a semi-arid climate with an annual precipitation of 200 to 500 millimeters is estimated for the period of the Upper Pliocene (Astian) and/or Lower Pleistocene (Calabrian). With regard to the transport of the coarse delta gravels (diameters up to 15 centimeters) this amount of precipitation seems to be too low. Towards the west, with an increasing altitude, also the depth of precipitation was probably considerably greater. Consequently just a few very large river systems, having their recharge area in the crystalline rocks of the shield, disposed of such runoff quantities to transport the load of rock material accumulated in the deltas of Wadi As Sah'ba and Wadi Al Batin.

In the eastern part of the Arabian peninsula a larger independent discharge system could not develop during the Quaternary because of the low precipitation rate and the abundant karstification. In flat limestone areas the

Fig. 20. Karst springs of Al Hasa before the installation of the modern irrigation system. (Photo: J. G. Zötl, 1970)

infiltration percentage of rainwater was high, and both weathering and karstification were very active. High temperatures and in some parts shallow groundwater encouraged evaporation and the development of duricrust.

## 2.1.4. Geomorphology of the Eastern Margin of the Shedgum Plateau

(R. W. Chapman)

### *2.1.4.1. Introduction*

The Shedgum Plateau is that portion of As Summan Plateau that lies southwest of Abqaiq and north of Al Hofuf in the Eastern Province of Saudi Arabia. The Shedgum Plateau is exceedingly flat and its eastern edge is a prominent, irregular escarpment that overlooks the Arabian Gulf coastal region of the Arabian Peninsula. This escarpment has been deeply eroded and it now displays an intricate array of landforms covered by a protective blanket of calcareous duricrust [1].

### 2.1.4.2. *Factors controlling the geomorphogeny*

Before discussing the landforms and the duricrust, it is advisable to consider briefly the two factors that were most instrumental in their development: 1. the nature of the Pliocene, Pleistocene, and Holocene climates in eastern Arabia, and 2. the characteristics of the bedrock of the Shedgum Plateau.

Although little information is available directly from Arabia, it seems reasonable to assume that the ancient climates of the Eastern Province were rather similar to those in northeast Africa. The temperature of the Pliocene was probably about the same as that of today. In the Pliocene periodically there was considerable moisture, distributed evenly over the year. The Pleistocene climate probably fluctuated widely; relatively cool, moist intervals or pluvials alternated with relatively long, warm, semiarid or arid intervals. The Holocene was a time of fluctuating climate.

Today, the Eastern Province is arid with daytime temperatures frequently reaching 50 °C. Winters are short and cool with temperatures rarely descending below 5 °C. Few areas of the province receive more than about 11 centimeters of rain per year. In the rainy season, from November through May, sporadic, violent downpours occur, and the resulting accelerated runoff brings about brief but intensive erosion.

The Shedgum Plateau is underlain by flat-lying sedimentary rocks of continental origin, known collectively as the Hofuf Formation. This formation is either late Miocene or Pliocene in age, and in the Shedgum area, at the type locality, it is 95 meters thick. The Hofuf Formation consists of red to reddish-brown, gray to rarely off-white conglomerate, sandstone, sandy limestone, sandy marl, and sandy shale. The upper beds of fresh-water sandy limestone cap the plateau and many of the erosional outliers.

Four characteristics of the Hofuf Formation have influenced its weathering and erosion, and have controlled the shape and arrangement of the landforms. First, the beds are all horizontal, and because of slight differences in resistance, tend to weather at different rates. Some beds break down rapidly while others stand as bold cliffs, and as a result structural benches are common.

Second, most of the beds are calcareous and at least partially soluble. This has facilitated weathering and the formation of sinkholes, caves, and duricrust.

Third, the formation is transected by sets of vertical joints. The most prominent set strikes northwest, and another less prominent set strikes north-

[1] The writer (R. W. Chapman, 1971, 1974) has studied the landforms and duricrust in considerable detail in the Shedgum area, a rectangle bounded approximately as follows: on the north by lat 25° 47′ N, on the south by lat 25° 33′ N, on the east by long 49° 35′ E, and on the west by long 49° 25′ E. The present paper reviews the results of those studies. Most of the material presented here, including Fig. 21, has been taken from the two Geological Society of America publications cited above. The writer is grateful to the Geological Society of America for giving permission to do this.

east. The joints have performed the following functions: 1. have divided the sedimentary beds into blocks, thereby hastening their destruction; 2. have been the prime cause of the formation of caves; 3. have determined the courses of the wadis; 4. have been instrumental in the development of pedestal rocks; and 5. have been remotely involved in the formation of erosional outliers.

Fourth, the duricrust, developed on the Hofuf Formation, has preserved many of the landforms that otherwise would have been eroded away.

### *2.1.4.3. Landforms*

#### *General description of the plateau region*

In the Shedgum area the plateau is exceedingly flat and is capped by resistant beds of fresh-water sandy limestone. The surface is preserved by a protective layer of calcareous duricrust. The plateau presents a steep, roughly north-south front extending from about lat 25° 47′ N to lat 25° 34′ N. The front is an irregular, curving escarpment facing east. In the north the rim of this escarpment has an elevation of about 260 meters, but southward it rises to about 290 meters above sea level. At its midpoint the escarpment face is about 90 meters high, but northward it decreases to about 60 meters, and at its extreme southern end it is only 45 meters. The whole escarpment is indented by innumerable small wadis, or dry watercourses, 1 kilometer or so in length, and by one larger wadi in the north known locally as Scribners Canyon.

From the foot of the escarpment, a broad, flat piedmont plain slopes gently eastward for many kilometers toward the Arabian Gulf. The western part of this plain, up against the escarpment, is veneered with gravel and sand. Eastward the veneer thickens, and the surface merges imperceptibly with the Arabian Gulf coastal plain.

Near the mouth of Scribners Canyon, numerous erosional outliers rise above the piedmont plain. These range from 30 to 60 meters high, and the elongate ones trend north to northwest more or less parallel to the wadis in the escarpment front. Other outliers, more distant from the front, occur to the east and south of the area. The broad piedmont plain and the distant outliers indicate that over the years the escarpment has retreated westward for many kilometers.

#### *Geomorphogeny of the plateau surface*

It is believed that the Shedgum Plateau began to rise in the Pliocene and continued to move upward throughout the Pleistocene, but during that long time its surface apparently changed very little. The flatness of the plateau, the absence of well-defined stream channels over its interior, the presence of broad, shallow sinkholes, and the permeability of the underlying rocks indicate that infiltration was far more important than runoff during the wet Pleistocene pluvials. During the warm semiarid intervals, carbonate enrichment developed on the surface a massive duricrust which became thicker with time. This crust, now as much as 3 meters thick, has helped to preserve the surface. Today the plateau is well armored by a lag gravel of duricrust concretions that prevents wind abrasion and deflation. The infrequent rains sink quickly into the permeable rocks and collect below the water table.

*Geomorphogeny along the plateau front*

The plateau front has six distinctive geomorphic features: 1. short, steep wadis which transect it; 2. terraces along the wadis; 3. short stretches of vertical escarpment between the wadis; 4. deep vertical crevasses or caves; 5. pedestal rocks; and 6. erosional outliers. The wadis and terraces are especially important and are considered in later sections.

Along the plateau front, between the wadis, a vertical or steep escarpment descends to the flat desert floor. Generally the two surfaces meet at a sharp angle, but in a few places a gravity slope has developed on fine-grained detritus at the base. Here, the softer limestone beds in the escarpment are crumbling, thereby undermining any massive, more resistant beds that lie above. Thus the escarpment face remains steep or essentially vertical and retreats slowly toward the west. Evidence presented elsewhere in this volume indicates that the escarpment is probably a marine cliff formed during a higher stand of the Arabian Gulf.

Large crevasses or caves are ubiquitous features along the plateau front and around the erosional outliers. They are especially well developed at Jabal Al Qarah, about 10 kilometers east of Al Hofuf. The caves are vertical openings, 15 meters or more high and 3 meters or more wide, which penetrate into the rock in some cases as much as 100 meters. They are believed to have been formed by wave action and intense weathering along vertical joints (sec. 2.1.3.3.).

Pedestal rocks in the Shedgum area consist of a column of white limestone topped with duricrust, and field relations show that they were formed mainly by weathering along joint cracks. The general profiles were developed in a more humid climate but are changing rapidly today. The pedestals are younger than the duricrust that tops them, but older than later, thinner crust that developed along their sides.

The erosional outliers in front of the plateau show the same degradational features as the escarpment, that is, steep to vertical cliffs, small wadis, caves, and pedestal rocks, and it is clear that the same geomorphic processes are responsible for them. Their development was controlled by high rock solubility, horizontal beds, pronounced vertical joints, and a duricrust cover.

*Wadis and gullies*

The most notable features of the Shedgum area are the innumerable dry valleys or wadis that indent the eastern edge of the plateau. They are of two types: 1. comparatively large well-developed wadis, averaging 1 to 2 kilometers in length, which antedate the duricrust; 2. small, narrow gullies which post-date most of the duricrust.

Scribners Canyon, its main tributaries, and the valleys along the plateau front to the south belong to the first group. Scribners Canyon differs from the other wadis in having: 1. greater length (6 kilometers) and width, 2. several clearly defined tributaries, and 3. numerous erosional outliers that border it on the east and northeast. It appears to have retained most of its original length, whereas the other wadis appear to be vestiges of former, longer wadis. It is believed that northwest-southeast joints established the courses of the streams that cut the wadis.

Fig. 21. View looking northeast across Scribners Canyon in the Shedgum area, showing landforms and duricrust. Man is standing on a crusted terrace surface which slopes about 3° towards the canyon. A deep gully is incised into the terrace. High flat spur on the right is a remnant of an older terrace. On the horizon are flat-topped outliers detached from the plateau. (This photograph has been reprinted from the Bulletin of the Geological Society of America, Vol. **82**, 1971)

Typical wadis are short, deep, narrow, and steep-walled. Their heads, cut in bedrock, are very steep, whereas their lower courses have gentler gradients. Remnants of terraces may be seen along their sides. The heads of most wadis end abruptly against the resistant limestone on the plateau surface. However, the head of Scribners Canyon is rather broad and grades up gently to the plateau. At the head of many wadis the surface limestone has been undercut, forming bowl-shaped depressions or amphitheaters.

The wadis were cut by ancient streams flowing off the plateau at a time when the front stood farther to the east. The flow started along well-developed, northwesterly-trending joints which were widened quickly. The short courses of the streams and their steep gradients caused rapid downcutting and the formation of deep, narrow gorges.

The gullies, or valleys of the second group, are younger and narrower than the wadis. The longer ones run the full length of some of the older tributary wadis, and their winding channels have cut through the terraces and into the softer limestone beneath. Normally, they are about 15 meters deep and very narrow with precipitous walls (Fig. 18). The shorter gullies, 60 to 80 meters long, either are tributaries of the longer ones or flow directly into Scribners Canyon. Since the gullies are incised in the terraces and duricrust, they are clearly younger, and they disclose a period of renewed stream activity probably brought on jointly by continued regional uplift and increased rainfall.

*Terraces*

Four well-defined terraces are still preserved in Scribners Canyon. Because of their duricrust cover, these surfaces can be traced and their slopes determined with accuracy. Three of them extend along the full length of the canyon, mostly on the southwest side, and up into many of the tributaries (Fig. 21). All four surfaces slope gently downstream (southeastward), and, at the same time, show a slightly steeper inclination from the canyon walls inward. The three older terraces converge on the plateau surface at the head of the canyon at an elevation of approximately 260 meters, and they diverge noticeably and uniformly downcanyon. The youngest terrace, however, converges downcanyon.

These surfaces are true bedrock terraces, not alluvial terraces; everywhere in the canyon they are dipping whereas the underlying strata are horizontal. It is believed that they were formed in wetter climates by active streams striving to develop low-gradient floors during interludes of stability between plateau uplifts. Degradation was the principal process involved.

### *2.1.4.4. Calcareous duricrust*

*Characteristics*

Throughout the Shedgum area, the plateau surface, the tops of erosional outliers, the sloping divides and walls of wadis and gullies, and all terrace surfaces are covered by duricrust which has developed on the calcareous beds of the Hofuf Formation. In Scribners Canyon, where the topography is especially rugged, the crust even clings to vertical gully walls. It is especially important to note that the duricrust in Scribners Canyon rests on erosion surfaces of different ages, and that the crusts on these various surfaces have different characteristics. The upper part of the duricrust has weathered to a rubble of concretions which forms a thin lag gravel or desert pavement over the whole region and imparts an unusual aspect to the scenery (Fig. 21).

The duricrust is typically gray or reddish-brown and concretionary, and it has been divided into three types: very immature, immature, and mature. The very immature crust rests on the youngest topographic surfaces and is the youngest, thinnest, and least thoroughly developed. The immature crust rests on somewhat older surfaces, and is older, thicker, and more thoroughly developed. The mature crust rests on the oldest topographic surface and is the oldest, thickest, and most thoroughly developed.

Three general conclusions can be drawn from petrographic studies. First, the crusting was a concretionary process involving two simultaneous actions: 1. the recrystallization, in situ, of original calcite in the parent limestone, and the introduction of additional calcite probably either from above or below; and 2. the replacement of detrital quartz and feldspar by calcite. Second, the upper part of each of the two nonmature crusts shows greater rock alteration than the lower part. Third, the older crusts are thicker and more intensely altered than the younger crusts. The oldest crust in the region, that on the plateau surface, is relatively uniform petrographically throughout its observed thickness, and it is believed to be essentially mature.

Chemical data indicate that the crusting process resulted in the buildup of Ca in the upper part of the duricrust profiles. The concentration was accomplished by highly alkaline ground water moving alternately downward and upward in response to changes in rainfall and evaporation. The older a particular crust is, the richer and more uniform it is with respect to Ca. The data also show that Si decreased generally with advancing crustal maturity. Much of this decrease is, no doubt, a percentage change reflecting the addition of Ca to the profiles, but some must represent the actual removal of Si from the profiles.

### *Origin*

Three facts have been established: 1. the calcareous duricrust in the Shedgum area rests on topographic surfaces of different ages, and the age of any particular portion of the crust corresponds approximately to the age of the surface on which it rests; 2. the older portions of the crust are thicker than the younger portions, and they have been more thoroughly altered physically and chemically; and 3. the period of time during which the crust as a whole must have developed, that is, late Pliocene, Pleistocene, and perhaps early Holocene, was one of wide climatic variations and gradually increasing aridity. It is concluded from these facts that the duricrust as a whole evolved progressively, though intermittently, over a long period of time, and that the characteristics of any particular portion of it are partly a function of the climate under which it formed.

The events that gave rise to the topography and duricrust began with the initial uplift of the Shedgum Plateau, probably in late Pliocene time. This uplift was apparently intermittent, and continued throughout the Pleistocene and perhaps into the very early Holocene. Repeated fluctuations of rainfall and probably temperature accompanied the uplift, especially in the Pleistocene.

During times of cool temperature and considerable rainfall there were physico-chemical weathering, active wadi cutting, and terrace formation. In the semiarid periods, on the other hand, the calcerous duricrust is thought to have formed.

The crusting involved essentially 1. the recrystallization of original calcite in the parent sandy limestone, 2. the introduction of additional calcite either from above or below, and 3. the replacement of quartz and feldspar by calcite and removal of the released silica. All of these changes were promoted by ground water moving through the rock, probably both downward and upward. It is believed that the depth to which these processes extended was governed principally by the amount of rainfall, and that the intensity of their effects, that is, the degree of rock alteration, was directly proportional to the duration of the processes. In other words, the thickness of a particular portion of the duricrust is a function of the climate, whereas the intensity of its alteration is largely a function of its age.

### 2.1.5. Sabkhahs of Eastern Saudi Arabia

(D. H. Johnson, M. R. Kamal, G. O. Pierson, and J. B. Ramsay)

#### *2.1.5.1. Introduction*

Although there has been much study devoted to sabkhahs in the past dozen years or so, most of the published work deals with coastal sabkhahs of Abu Dhabi and other parts of the former Trucial Coast of the Arabian Peninsula (e.g., P. Bush, 1973; P. R. Bush, 1970; G. P. Butler, 1969, 1971; G. P. Butler et al., 1965; R. Curtis et al., 1963; G. Evans, C. G. St. C. Kendall, P. A. D'E. Skipwith, 1964; G. Evans, D. J. J. Kinsman, D. J. Shearman, 1964; G. Evans, V. Schmidt, P. Bush, H. Nelson, 1969; G. Evans, D. J. Shearman, 1964; K. W. Glennie, 1970; K. J. Hsü, J. Schneider, 1973; L. V. Illing et al., 1965; C. G. St. C. Kendall, P. A. D'E Skipwith, 1968, 1969; D. J. J. Kinsman, 1964, 1967, 1969; E. G. Purdy, 1965; D. J. Shearman, 1963, 1966; W. Sudgen, 1963; J. C. M. Taylor, L. V. Illing, 1969). Virtually nothing has been published on the sabkhahs of Saudi Arabia except for very brief mentions by R. W. Powers et al. (1966) and K. W. Glennie (1970), and their depiction on maps by the Saudi Arabian Directorate General of Mineral Resources and the United States Geological Survey. Preliminary results of some of the writers' work has been presented orally at several Arab scientific meetings by M. R. Kamal.

The writers began studying sabkhahs in the vicinity of Dhahran in 1973. Except for a few reconnaissance visits to other sabkhahs, work has been restricted to part of the Sabkhat Ar Riyas, concentrating on the water chemistry.

In the following sections are presented some reconnaissance observations on various sabkhahs and then some preliminary results from Sabkhat Ar Riyas.

It seems advisable here to define the term sabkhah (spelled in some publications sabka, sabkha, sebka, sebkha, sebkhah, sebkra, subkha, etc.) as there has been confusion in various usages of the term. D. J. J. Kinsman (1969, p. 833), D. G. A. Whitten and J. R. V. Brooks (1972, p. 397), and K. W. Glennie (1970, p. 60) give good explanations of sabkhahs as the term is used in Saudi Arabia, in contrast to brief, inaccurate definitions appearing in some geologic dictionaries (e.g., M. GARY et al., 1972, p. 639; P. W. THRUSH, 1968, p. 977); the term is not included in some geologic dictionaries (e.g., S. AL SINAWI and N. AL ANSARI, 1970; J. CHALLINOR, 1973).

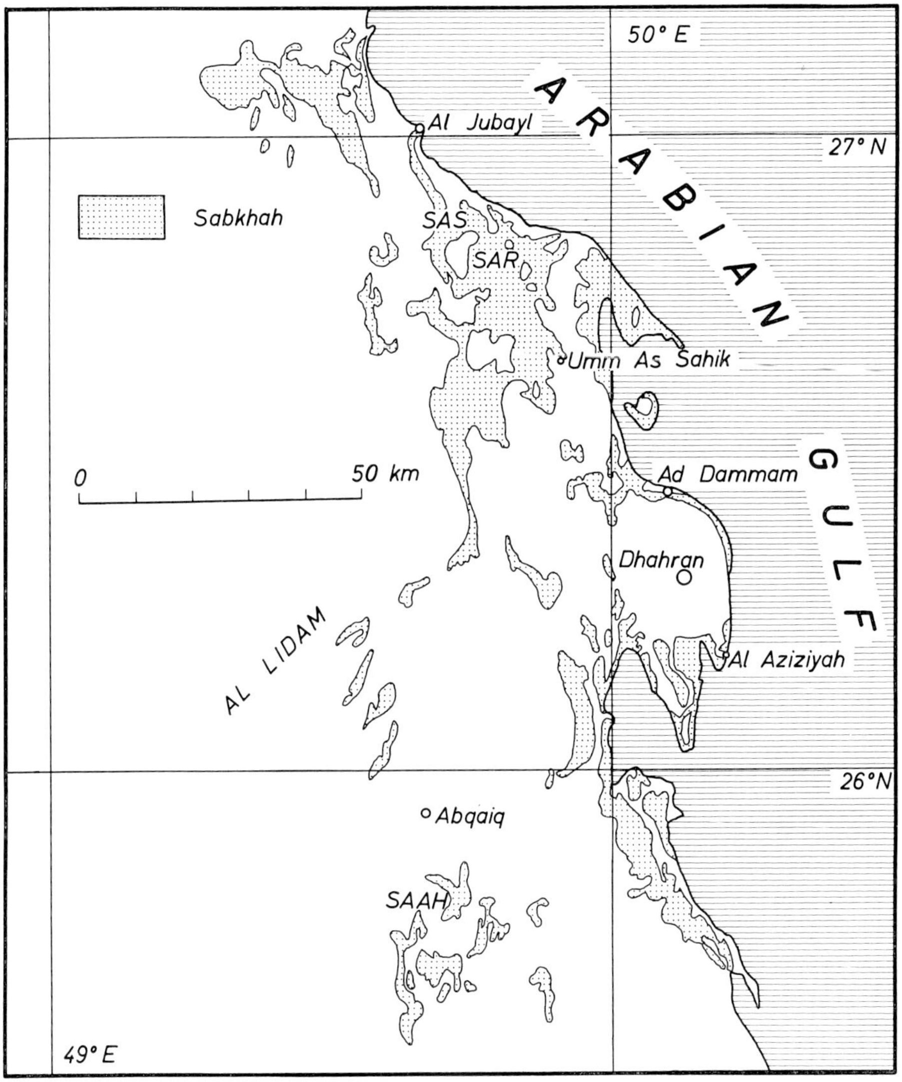


Fig. 22. Sabkhah areas in the coastal region and its hinterland between lat 25° 45′ and 27° 10′ N. *SAR* Sabkhat Ar Riyas, *SAS* Sabkhat As Summ, *SAAH* Sabkhat Aba Al Hamam (from M. STEINEKE et al., 1958)

Sabkhahs are saline flats underlain by clay, silt and sand, and often encrusted with salt. They are equilibrium surfaces whose level is largely controlled by the local groundwater table. It is necessary to distinguish between *coastal sabkhahs* and *inland sabkhahs.*

*Coastal sabkhahs* are supratidal surfaces produced by depositional offlap of marine sediments with, in some cases, the addition of aeolian sediments. The associated brines are derived primarily from seawater.

*Inland sabkhahs* lie away from the coast, and represent areas of equilibrium between aeolian sedimentation and deflation, controlled by the local water table. The water table necessarily is higher than bedrock, and the base level of deflation lies just above the capillary fringe in the sediments above the water table. Evaporation through the surface causes the formation of brines and the precipitation of evaporite minerals.

### *2.1.5.2. Reconnaissance observations*

Saudi Arabia has large numbers of sabkhahs, both coastal and inland. In the short time the writers have been studying these sabkhahs, they have been able to make reconnaissance visits to only a few of them, and to observe only the Sabkhat Ar Riyas in somewhat more detail. The reconnaissance observations are reported here, and the work on Sabkhat Ar Riyas in a later section. Fig. 22 shows the sabkhah areas in the coastal region between lat 25° 45′ N and lat 27° 10′ N and in its hinterland.

*Coastal sabkhah south of Al Aziziyah.* The surface of this sabkhah was just a few tens of centimeters above sea level, with several scattered tidal rills. The surface was hard enough to support a Land Rover in some parts and in others was so soft that a person walking would sink in nearly to knee depth. The surface was everywhere rather damp, and the water table only a fraction of a meter deep. In many places, even above any likely level the sea would reach even at storm tides, the surface showed numerous *Turritella*-like shells, showing that the surface had been under water in past time. In several places hard limestone with fossil shells cropped out at the surface. It is not known whether this represents recently indurated sediments or Tertiary bedrock. The sabkhah is bordered and in places overlain by sand dunes.

*Sabkhat As Summ, about 17 kilometers south of Al Jubayl.* This inland sabkhah had a dry surface covered in places by a few millimeters of wind-blown sand and elsewhere by silty material with scattered flakes of gypsum, and in places scattered *Turritella*-like shells. The water table was less than a meter below the surface, and the walls of a hole dug to the water table showed fairly distinct banding of sand and silt. Standing 1.5 to 2 meters above the surface are several small flat-topped hills, on which are found a variety of marine shells, including gastropods, pelecypods and sea urchins. The hills are obviously remnants of a higher surface, but it is not known whether that surface was also sabkhah.

*Sabkhat Aba Al Hamam.* This sabkhah, about 15 kilometers south of Abqaiq covers an area of several square kilometers and lies about 100 meters above sea level. In one part, covering an area of two or three hectares, the water table lies at the surface and brine pools are formed and halite crystallizes out. In places the pools are more than two meters deep, but it is not known whether or not this represents the original depth, because for a number of years salt has been harvested from these pools by various means including the use of a back-hoe. Daily production appears to be of the order of a few tons. The surface around the mined pools consists of salt with considerable aeolian sand.

Immediately south of the mine area is an expanse of several hectares characterized by polygonally patterned ground. The polygons are from $^1/_3$ to 1 meter in average size, and many have an irregularly hexagonal shape. The polygons are separated by fractures 2 to 5 centimeters wide, and along the fractures the edges of the polygons have built up to heights of 5 to 10 centimeters above the centers of the polygons. The surface within the polygons consists of salt with a small to moderate amount of included aeolian sand. Many of the polygons form arches so that an open space of several centimeters exists between the cover and a rather silty floor. When the polygon surface is broken, the air in the hollow below can be felt to be much warmer than the air above, even on a hot day. Apparently an original crust formed and tension fractures developed as it dried. The central portions of the polygons are impervious, so continued rise of water by capillarity occurs at the boundary fractures, and as evaporation proceeds, the borders are built up. Wind-driven sand drifts across the area and the polygonal ground sometimes changes with seasons. After a year with less than normal rain, the polygonal area largely disappeared, but the following year, after rather heavy rainfall, the polygons developed again.

*Inland sabkhahs in the northeastern Rub' Al Khali.* Flying visits were made by D. H. Johnson to a drilling site (20 minutes on the ground) and to a seismograph party's camp (3 hours on the ground) in Ar Rub' Al Khali, and in those brief periods made the following observations.

The sabkhahs are a few square kilometers in extent, and are surrounded by great sand dunes that rise more than 100 meters above the sabkhah surface. The sabkhah surface seems to be largely gypsiferous silt and sand with traces of salt. In places outcrops of sandy limestone rise slightly above the surface. Bulldozer pits dug for drilling water indicate that the water table is about 1.5 to 2 meters below the surface. The bulldozers had brought up quantities of poorly formed "sand crystals", crystals of gypsum that enclose so much sand that they appear to be crystallized sand. In some of the loosened material near the surface sand crystals appeared to be in the process of growing. Where wind had blown loose sand away, rough crystal outlines could be seen in the remaining sand, but the forms were so fragile that they broke when an attempt was made to pick them up. Apparently the amount of interstitial gypsum was too little to impart any strength to the structures.

### *2.1.5.3. Observations on Sabkhat Ar Riyas*

Part of Sabkhaht Ar Riyas was chosen for more detailed study because the sabkhah is fairly large, easily accessible by road, and because reconnaissance measurements had shown appreciable concentrations of potassium and magnesium in the brines. Studies to date have dealt largely with brine sampling and analysis, and preliminary, qualitative geologic observations in some parts.

Sabkhat Ar Riyas begins about 30 kilometers north-northwest of Dhahran, Saudi Arabia, and extends about 70 kilometers north-northwesterly to the coastal town of Al Jubayl. Southerly from Al Jubayl for about 12 kilometers, it is only 2 to 3 kilometers wide, but then widens abruptly to 8 kilometers, gradually increasing to about 12 kilometers, then pinching down to 4 kilometers or less near its southern end. About 4 kilometers north of the village of Umm As Sahik the Arabian American Oil Company (ARAMCO) was building a pipeline across the sabkhah at the time the writers began their studies (spring 1973).

Parallel to the pipeline, about 100 meters to the south, a series of 21 sample stations was established in a straight line at 304.8 meters (1,000 feet) intervals, from the Dhahran-Al Jubayl highway to a small jabal (hill) that rises about 5 meters above the sabkhah. About 2 kilometers farther south another line of 7 stations was established between two small jabals. At each station, holes were hand dug by shovel to below the water table to reveal the nature of the sabkhah sediments and to determine the depth of the water table. The holes were continued by posthole augers below the water table to emplace plastic sewer pipes, 10 centimeters in diameter. At each station, it was planned to set one pipe to a depth of 1.2 meters and another to 2.2 meters below the surface, but in a few places it was not possible to set the deeper pipes with the tools available. Each pipe was covered with a plastic cap to prevent evaporation and dilution by rainfall, but was provided with a pair of tiny vent holes so that air pressure would remain equal inside and outside the pipes.

The surface of the sabkhah is generally even. Leveling along the 7 kilometers line of sample stations showed a maximum relief of only 1.4 meters and local variations of a few centimeters. A few small jabals rise to heights of 5 to 10 meters above the sabkhah surface. These jabals are covered with wind blown sand stabilized by scattered low bushes, and on some of them are outcrops of marly limestone of the Hadrukh or Dam Formations of Miocene age. Sand deposited in wind shadows behind rocks and bushes on these jabals and elsewhere indicates that the prevailing wind is from about N 25° W.

The texture of the sabkhah surface is quite varied. During dry summer months it ranged from a very smooth crust, to a faint polygonal pattern in which gypsiferous sandy rims, 1 to 2 millimeters high and 5 to 10 millimeters wide, bordered flat irregular polygons 10 to 15 centimeters across, to a "bubbly" gypsiferous crust 5 to 10 millimeters thick covering myriad flat empty bubbles 10 to 15 centimeters in diameter and 1 to 4 centimeters high. In places, where standing water has evaporated or a very wet surface has dried, a thin discontinuous crust of halite may exist briefly, but is soon covered and/or dispersed by blowing sand. Adhesion ripples similar to those described by K. W. Glennie (1970, p. 71) have been seen in places. The sabkhah surface is strongly hygroscopic in some places; during several humid days in August 1973 the surface was so muddy that walking was difficult, even though it had been dry previously, no rain had fallen, and the water table was about a meter below the surface.

When observations were first begun (July and August 1973) the maximum depth to the water table from the sabkhah surface was 1.15 meters; the minimum, 0.30 meter. Total relief on the water table over a distance of 7 kilometers was 0.75 meter. Over the next three months, the water table was observed to subside 10 to 15 centimeters. The subsidence was not uniform at all stations, so the inference is that evaporation is not uniform over all the sabkhah surface and that horizontal permeability at shallow depths is also not uniform.

In December 1973 a shallow pond was observed forming on the sabkhah a few hundred meters south of the line of sample stations. The source of the water was several tiny creeks that result from overflow in the Umm As Sahik irrigation system; rain supplied a minor amount of water to the pond. By mid-January 1974, the pond had spread until it was about 2 kilometers long by $^3/_4$ kilometer broad. Its northern edge had reached part of the sample line, and at stations 4, 6, and 7 the water was several centimeters deep. A hole dug near the edge of the pond showed that the water table was still about 30 centimeters below the surface, so there were about 30 centimeters of relatively dry sediments between the water table and the bottom of the pond. This indicates a very low vertical permeability in that area, as the water had not been able to sink in and fill the sediments in several weeks. The pond began shrinking in February 1974 and by the end of April it was gone. Similar ponds formed in the winters of 1975 and 1976.

Sabkhah sediments were observed in the holes dug for setting sample pipes, but only partial data could be obtained, and much more sampling and observation is needed. Holes were dug with shovels down to the water table, and the walls of the holes could be observed fairly well, but below the water table, post-hole augers were used, sample recovery was poor, and sampled material was badly disturbed.

The sediments are distinctly layered. At the immediate surface is usually a one-centimeter crust of fine-grained sand lightly cemented with halite (NaCl) and gypsum ($CaSO_4 \cdot 2\,H_2O$). In one place possible small amounts of sylvite (KCl) and epsomite ($MgSO_4 \cdot 7\,H_2O$) were tentatively identified in the surface crust.

In many places below the crust there are several alternating 5 to 10 centimeters layers of poorly sorted fine- to medium-grained sand and grey mud; the mud contains considerable amounts of calcite ($CaCO_3$) and dolomite ($CaMg[CO_3]_2$). Below these layers are usually somewhat thicker layers of sand of varied sizes. The contacts between the various sand layers are in some places gradational and elsewhere very sharp; the sizes range from fine-grained to very-coarse-grained, and the sorting from poor to fair. The sand consists largely of quartz with a few grains of white, gray and black chert. The grains are well-rounded and most have frosted surfaces. In some holes there are a few 1 to 3 centimeter layers of brown to gray mud between the sand layers.

In a number of holes, nodular, snow-white, sugary anhydrite ($CaSO_4$)

is found in 5 to 10 centimeters layers about 10 to 30 centimeters below the surface, and sometimes a few centimeters above the water table. The layers of anhydrite range in firmness from a thick plastic mud to rock hard.

In almost every hole prismatic gypsum crystals 1/2 to 5 and occasionally 10 centimeters long are found, especially clustered in a layer just above and extending several centimeters below the water table. Except for the smallest sizes, all these crystals contain abundant inclusions of sand or mud. The crystals above the water table usually have tan sand inclusions; those below have gray sand or mud. In a few places flaky white 1 to 2 millimeters gypsum crystals are abundant in the surface itself.

In several holes the sediments are lightly to moderately cemented at about 30 to 50 centimeters below the surface, and in the majority of holes there is moderate to fairly firm cementation just above and/or below the water table. Gypsum is the principal cement but carbonates are sometimes also involved.

At station 4 large halite cubes up to 5 centimeters across were found at a depth of 1.3 meters below the surface (1.0 meter below the water table). About 200 meters north of Station 7 a little salt is being extracted commercially by scraping off the surface layers of the sabkhah to expose the water table, allowing halite to crystallize by evaporation, and then scraping up these crystals into sacks.

Around about the western quarter of the sampled stations, the sand tends to be somewhat coarser grained and the mud layers are minor or absent. Lenticularity is pronounced in the various layers, and holes only two or three meters apart show marked differences in the layers present. Because of the lensing there are considerable differences in permeability between adjacent holes and at different depths. This variable permeability appears all along the line of sample stations.

In a number of places small streaks and spots of dark reddish brown material appear. This is believed to be organic, perhaps algal, in nature. In one hole a few marine shells (gastropods) were found at a depth of one meter. Although a strong odor of hydrogen sulphide was noted in several holes, no pyrite ($FeS_2$) or other sulphide minerals were seen.

The thickness of sediments along the sampled line is not known. The sampling holes were not much more than two meters deep, and none reached bed rock, even at Station 21 near the jabal, although the bottom of the sabkhah must rise to where the jabal emerges from the sabkhah. A few measurements were made with a small portable refraction seismograph, but were not conclusive. There is a suggestion that the floor of the sabkhah slopes off the jabal and then flattens out at a depth of perhaps 5 to 8 meters, with some irregularity; the irregularity is apparently erosional rather than structural.

The jabal just west of Station 21 rises about 5 meters above the sabkhah surface, and the pipeline and its accompanying road were cut through the jabal, exposing buff to pink limestones and marls. The beds are essentially flat, but small folds of about 25 centimeters amplitude may be observed. The marly beds contain lenses one to two meters across and up to 10 centimeters thick of columnar calcite or aragonite ($CaCO_3$). Some of the limestone beds contain so many pelecypod shells that they almost might be considered coquina. The top and eroded flanks of the jabal are covered with myriad one-centimeter pelecypod fossils that have weathered out of the limestone. These rocks are probably from the lower part of the Dam Formation of Miocene age.

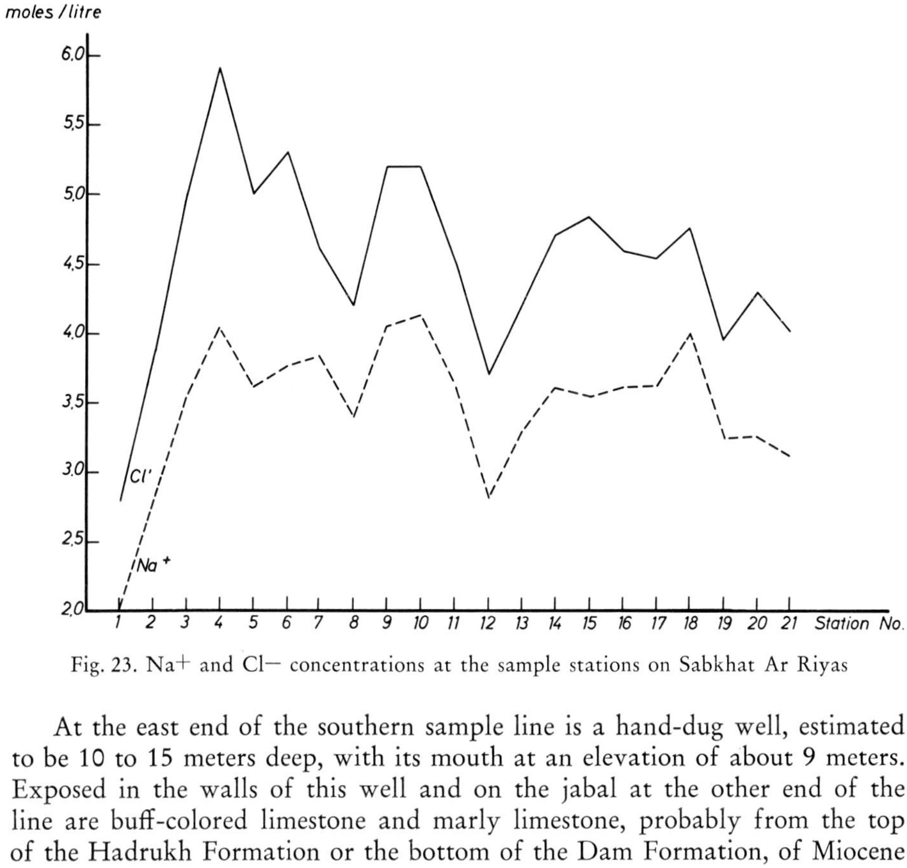


Fig. 23. Na+ and Cl− concentrations at the sample stations on Sabkhat Ar Riyas

At the east end of the southern sample line is a hand-dug well, estimated to be 10 to 15 meters deep, with its mouth at an elevation of about 9 meters. Exposed in the walls of this well and on the jabal at the other end of the line are buff-colored limestone and marly limestone, probably from the top of the Hadrukh Formation or the bottom of the Dam Formation, of Miocene age.

### *2.1.5.4. Brine studies at Sabkhat Ar Riyas*

The plastic pipes were set at each station on the sabkhah so that samples of brine could be taken over a period of time from identical sites. Numerous samples have been obtained, but only part of the planned assays have been made, and analysis of the results is only partially complete. Some preliminary observations and conclusions are given below.

There is considerable variation of brine concentration along the line of sample stations as indicated in Fig. 23, which shows the $Na^+$ and $Cl^-$ concentrations at the different stations. The large fluctuations are not apparently related to surface terrain but might be due to influx of water from springs below the water table.

Samples were taken of the water in the pond that forms on the sabkhah surface during the winter and of the tiny creeks that feed it. Table 9 gives typical assays. It is apparent that most of the salinity of the pond is due to solution of salts from the sabkhah surface, and that mixing of the pond water is incomplete.

Table 9. *Chloride ion concentration of surface waters in the area of Sabkhat Ar Riyas*

| Sample Location | $Cl^-$, mg/ml |
|---|---|
| Near Station 3 | 38.6 |
| Near Station 6 | 74.6 |
| Edge of Pond | 42.1 |
| Stream leading to sabkhah | 3.0 |
| Small pond 75 m upstream from main pond | 2.6 |

Table 10. *Typical ion concentrations (in mg/ml), ion ratios, and pH of Sabkhat Ar Riyas brines and Arabian Gulf water*

| Location | $Na^+$ | $Mg^{++}$ | $K^+$ | $Ca^{++}$ | $Cl^-$ | $SO_4^{--}$ | $HCO_3^-$ | $\frac{Na^+}{Mg^{++}}$ | $\frac{Na^+}{K^+}$ | pH |
|---|---|---|---|---|---|---|---|---|---|---|
| Station 1 | 46.7 | 5.25 | 2.81 | 2.25 | 99.3 | 5.58 | 0.095 | 8.9 | 16.6 | 7.1 |
| Station 4 | 93.3 | 9.17 | 3.22 | 1.50 | 209.2 | 4.45 | 0.082 | 10.2 | 29.0 | 6.9 |
| Station 7 | 88.3 | 6.83 | 2.28 | 1.63 | 163.1 | 5.87 | 0.054 | 12.9 | 38.7 | 7.0 |
| Station 10 | 95.0 | 8.00 | 2.90 | 1.60 | 184.4 | 5.46 | 0.039 | 11.9 | 32.8 | 6.5 |
| Station 13 | 76.7 | 7.17 | 2.50 | 1.48 | 148.9 | 7.12 | 0.048 | 10.7 | 30.7 | 6.6 |
| Station 16 | 83.3 | 8.33 | 3.08 | 1.99 | 162.9 | 4.66 | 0.044 | 10.0 | 27.0 | 6.4 |
| Station 19 | 75.0 | 6.58 | 2.06 | 2.54 | 139.6 | 3.55 | 0.032 | 11.4 | 36.4 | 5.7 |
| Salt Mine | 85.0 | 12.50 | 5.53 | 1.74 | 189.7 | 5.10 | 0.089 | 6.8 | 15.4 | 6.8 |
| Gulf Water | 15.3 | 2.3 | 1.0 | 0.81 | 34.5 | 4.71 | 0.110 | 6.7 | 15.3 | 8.1 |

Typical concentrations for the major ions, as well as some ion ratios and pH values are presented in Table 10. The observed concentrations are approximately six times those of sea water for the sodium and chloride ions, with somewhat lower concentrations near the ends of the line. Magnesium and potassium ions are somewhat less abundant, being three to four times sea water concentrations. The use of the term sea water in this discussion is not intended to imply that sea water is the sole source of these salts, particularly in view of the annual runoff from the nearby wells which passes through this area. A comparison of the ion ratios [Br/Cl] and [Br/K] for the brine samples and the water from the wells in the Al Qatif region (as reported by C. Job, 2.1.6.1.) is consistent with the source of the brines being the local ground water. However data are not available to compare these ratios for Arabian Gulf water.

Since the brines are usually in contact with either gypsum or anhydrite, it is expected that they will be saturated with respect to calcium sulfate. The calcium and sulfate concentrations shown in Table 10 clearly do not show a constant ion product, particularly in the center of the line where calcium is nearly constant while sulfate shows considerable variation. These variations can be explained if the variation of calcium-sulfate-solubility with changes in total salt concentration is considered.

W. M. MADGIN and D. A. SWALES (1956) have studied the solubility of gypsum and anhydrite in sodium chloride solutions. Fig. 24 presents their results. The points off the line in this graph represent the observed ion products for the sabkhah brines. These points all lie parallel to the line rather

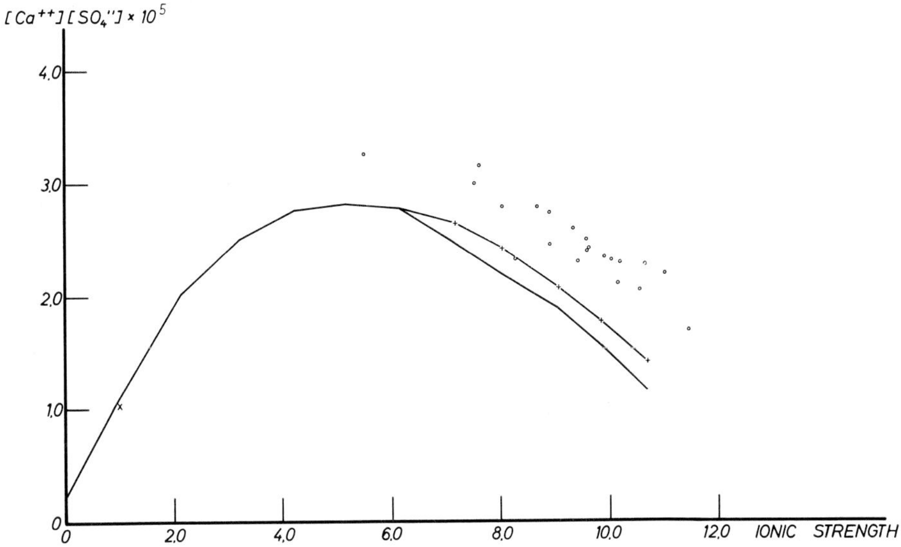


Fig. 24. Graph representing the observed ion products of Sabkhat Ar Riyas brines. Solubilities of gypsum and anhydrite in NaCl solutions (after W. M. MADGIN, D. A. SWALES, 1956)

than on it. This deviation may be due to a lowering of the activity of sulfate ion by a specific ion-pairing with magnesium which is present in the brines. Most of this salt-controlled variation in the ion product seems to be expressed as a variation in the sulfate concentration. The calcium concentration seems to be controlled by an independent process, possibly solubility of calcite or dolomite. Due to difficulties in the carbonate analysis, this hypothesis has not as yet been verified.

## 2.1.6. Hydrochemical Investigations in the Areas of Al Qatif and Al Hasa With Some Remarks on Water Samples From Wadi Al Miyah and Wadi As Sah'ba Near Haradh

(C. JOB)

### *2.1.6.1. Al Qatif oases* [1]

*Introduction*

The Al Qatif oases are palm gardens about 10 kilometers long and 2 to 3 kilometers wide along the coast of the Tarut Bay of the Arabian Gulf

[1] This article is written with special thanks to Mr. M. FAHMY, technologist at the University of Petroleum and Minerals in Dhahran, who accompanied me many times to the oases and contributed a great deal to the successful progress of the field work.

between Ad Dammam in the south and the small peninsula of Ras Tannurah in the north. There are numerous villages along the roads crossing the oases. Sabkhahs and sand areas divide the main oasis of Al Qatif from smaller palm gardens around the villages of Safwa and Umm As Sahik in the north and northwest, as well as the village of Al Ajam in the west and the town of Ad Dammam in the south. Also the island of Tarut is partly cultivated.

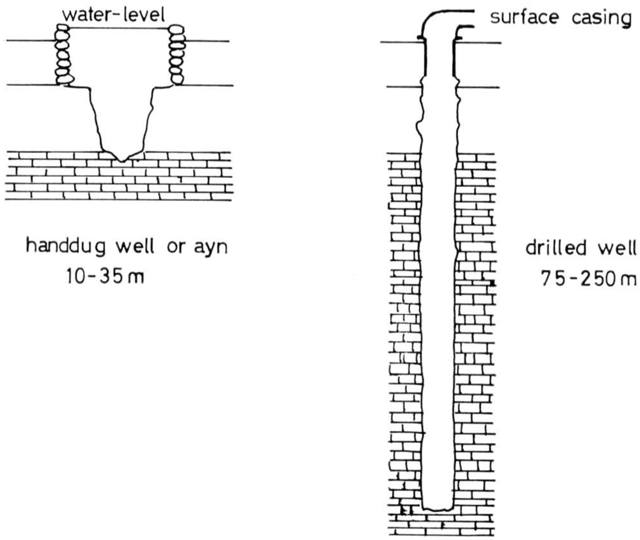


Fig. 25. Standard design of wells in the coastal belt oases. From: Water and Agricultural Development Studies for Area IV. Final Report, ITALCONSULT, Rome 1969

Altogether these oases cover an area of about 75 square kilometers and are the largest cultivated region along the eastern coastal area of Saudi Arabia. A wide arc of sabkhahs and dunes divides the oases from their hinterland.

## *Hydrological conditions*

The Qatif oases live on the fresh water outflow from a number of karst springs. The older ones, the so-called 'ayns, are hand-dug wells reaching a depth of 10 to 35 meters where the water discharges from joints of the Alat aquifer, a dolomitic limestone of the Eocene Dammam Formation.

During the last two decades numerous boreholes have been drilled down to the lower Eocene Khobar dolomite (70 to 140 meters deep) and into the Paleocene limestones and dolomites of the Umm er Radhuma Formation (230 to 600 meters deep). With the exception of some test wells, the casing of these boreholes reaches just a few meters below the land surface; consequently, water from all the aquifers encountered during drilling is mixed in the discharge. Fig. 25 shows cross-sections of an 'ayn and a borehole.

*Previous reports*

According to a contract with the Ministry of Agriculture and Water, the wells of the coastal belt region were surveyed by ITALCONSULT between February 1967 and May 1968 [1]. In the course of these investigations the discharge of the wells (springs), the temperature and the electrolytic conduc-

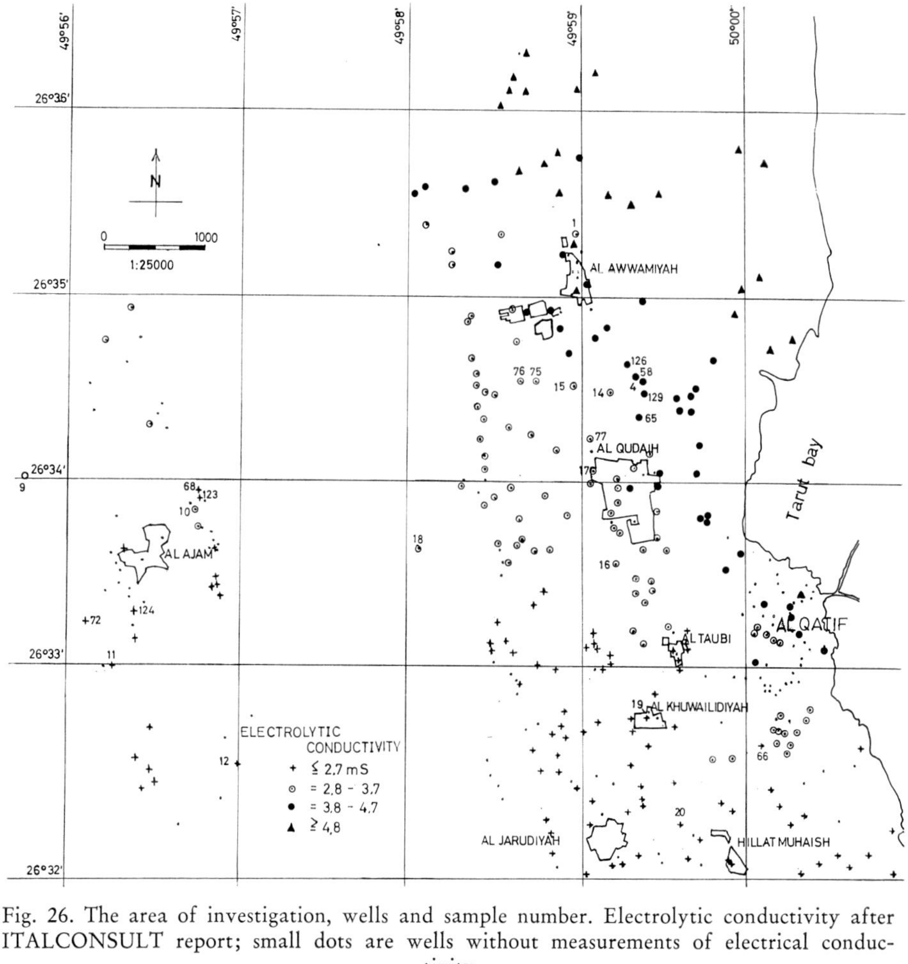


Fig. 26. The area of investigation, wells and sample number. Electrolytic conductivity after ITALCONSULT report; small dots are wells without measurements of electrical conductivity

tivity of the water were measured. In some wells their static water level was also checked. To acquire a more comprehensive knowledge of the hydrological conditions, ITALCONSULT drilled a few fully-cased test wells for pumping tests and chemical investigations of the water. All the data con-

[1] The report comprises five volumes "Water and Agricultural Development of Area IV"; however, only the volumes "Water Point Inventory" and "Discharge Tests" were available and could be evaluated.

cerning the Qatif area were interpreted and compared to the results of our measurements.

*General hydrological conditions*

The area of the central and northern part of the main Qatif oasis and the region of Al Ajam shown in Fig. 26 covers about 18.5 square kilometers.

Fig. 27. 'Ayn Al Labaniyah. (Photo: J. G. ZÖTL, 1975)

In this area 372 wells were investigated of which 121 were ruined or without appreciable discharge.

Discharge data, water temperature and electrical conductivity are to be found in the report of ITALCONSULT for 251 wells and springs. Thirty-two 'ayns had a total discharge of about 36,000 cubic meters per day and 219 boreholes some 168,500 cubic meters a day. This makes about 205,000 cubic meters water per day for the 18.5 square kilometers cultivated area. Compared to precipitation this quantity amounts to about 4,000 millimeters depth of rain per year, which actually ranges between 10 and 140 millimeters (Dhahran 1955–1960). For the entire cultivated area (75 square kilometers) a discharge of 800,000 cubic meters per day can be assumed. Considering the submarine springs in the Tarut Bay and the spring water evaporating from sabkhahs, the amount of karst water discharging in this coastal area can be estimated up to $1.2 \times 10^6$ cubic meters per day or about 14 cubic meters per second. The overflow from the boreholes ranges between 1 and 20 litres per second. The largest 'ayns, e.g., 'Ayn Al Labaniyah (Fig. 27), discharge from 30 to 50 litres per second. Four of these large 'ayns lie in the region of Al Awwamiyah-Al Qudaih, one northeast of Al Jarudiyah, all of them more or less along the N-S-striking Qatif anticline which is, in its lower geological strata, a very productive oil reservoir.

Data of water temperatures collected in the course of investigations carried out by ITALCONSULT were checked as to their frequency distribution in the area of Fig. 26. For the drilled wells the result was an approximately

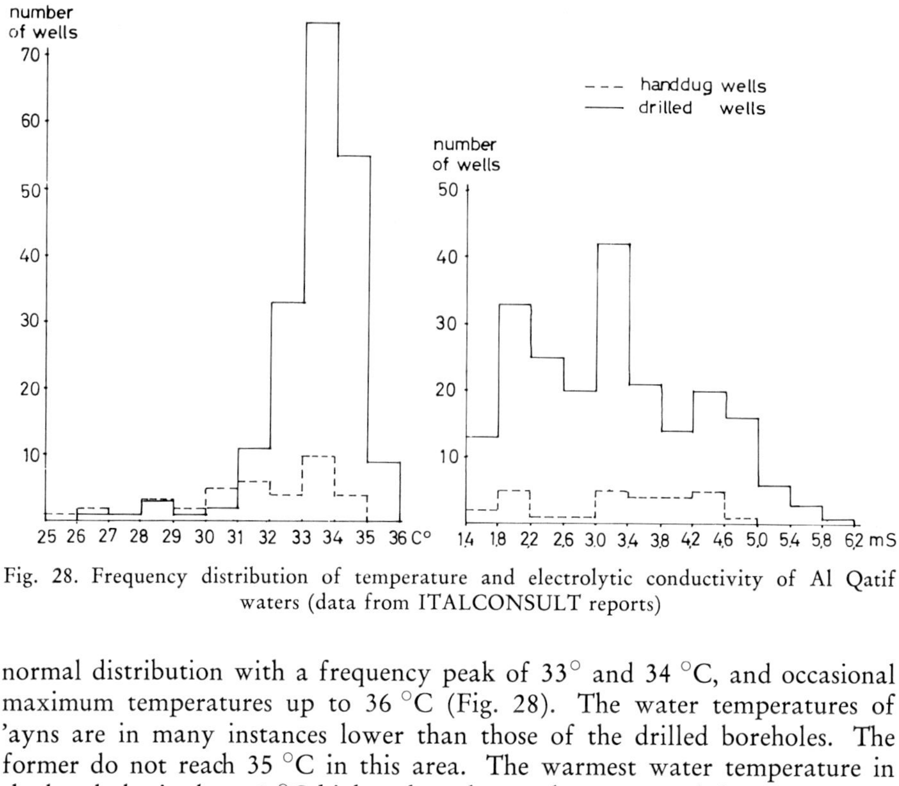


Fig. 28. Frequency distribution of temperature and electrolytic conductivity of Al Qatif waters (data from ITALCONSULT reports)

normal distribution with a frequency peak of 33° and 34 °C, and occasional maximum temperatures up to 36 °C (Fig. 28). The water temperatures of 'ayns are in many instances lower than those of the drilled boreholes. The former do not reach 35 °C in this area. The warmest water temperature in the boreholes is about 9 °C higher than the yearly average of the air temperature of Dhahran (about 27 °C). Calculating a geothermal gradient of 33 meters per 1 °C the primary aquifer lies at a minimum depth of 300 meters, i.e., in the Umm er Radhuma formation (230 to 600 meters deep). Cased boreholes drilled down to this aquifer discharge water with temperatures of 36° to 38 °C (e.g., boreholes Q 1 U/A, Q 4 U/A, Q 15 U/C) [1]. However, also in drilling into the overlying aquifer the so-called Khobar-strata (70 to 140 meters deep) water temperatures of 37.6° and 37.9 °C were measured (boreholes Q 2 K/A and Q 6 K/A). From these results it can be concluded that the water in the Khobar aquifer comes directly from the Umm er Radhuma Formation—at least in some places. The water from another

[1] Borehole numbers refer to ITALCONSULT identification.

test drilling into the Khobar aquifer showed only a temperature of 31.2 °C which is in the range of the waters from the upper Alat limestone feeding the hand-dug 'ayns. Concerning the geothermal gradient, the water discharging from the Alat aquifer, too, is too warm as this stratum reaches down to a depth of only 70 meters; thus the Alat aquifer is probably also alimented by the Umm er Radhuma Formation. It can be assumed that both the Alat and the Khobar aquifers—which have too small a quantity of recharge by infiltration—receive water from deeper layers rising up along faults through

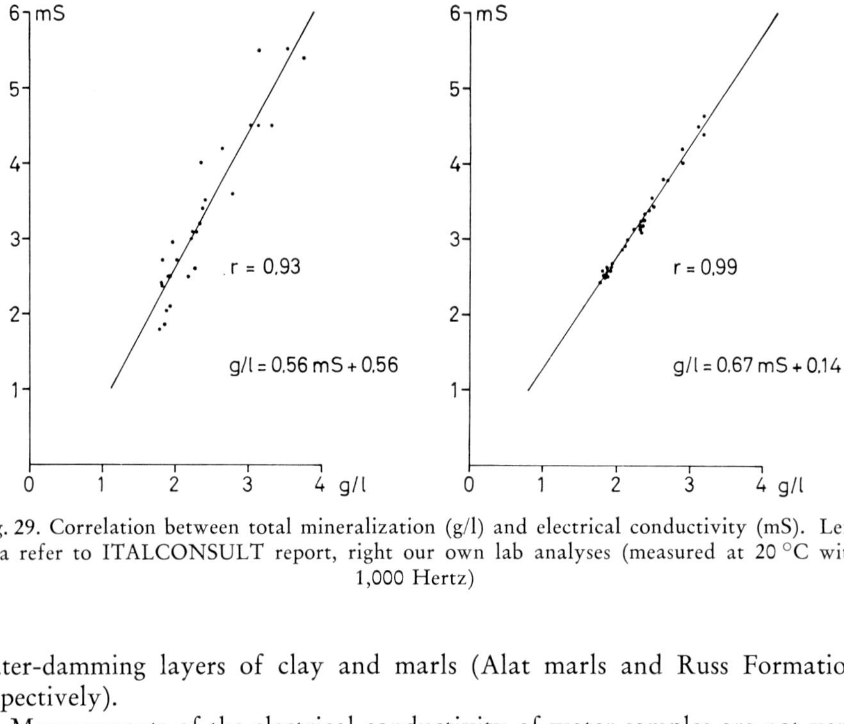


Fig. 29. Correlation between total mineralization (g/l) and electrical conductivity (mS). Left data refer to ITALCONSULT report, right our own lab analyses (measured at 20 °C with 1,000 Hertz)

water-damming layers of clay and marls (Alat marls and Russ Formation respectively).

Measurements of the electrical conductivity of water samples are not very exact when carried out in the field considering the large differences in temperature and technical problems. Consequently, these measurements only represent a rough estimate of the total concentration of the solution (Fig. 29). Water from boreholes without a casing show three classes of frequency distribution: 1.8 up to 2.2 mS, 3.0 to 3.4 mS, and 4.2 to 4.6 mS (Fig. 28). Not as clear, yet more or less similar, is the distribution of the electrical conductivity of 'ayn waters.

In all the wells electrical conductivity increases from southwest towards northeast (Fig. 26). This regional trend is independent of the strata from which the water discharges. Fig. 30 gives a comparison of electrical conductivity of waters from Alat-Khobar wells (uncased down to a depth of 140 meters) to the values of the occasionally nearest Alat-well (mostly 'ayns reaching down to 15 to 25 meters) and the Alat-Khobar-Umm er Radhuma well respectively (uncased down to a depth of more than 200 meters). The values lie along a line with a gradient of approximately 45°. This indicates that the conductivity of the water does not depend on the depths of the wells but on their regional position. Fig. 30 reveals once more that the electrical conductivity is not uniformly distributed over the sampling area. This phenomenon cannot be explained through a different number of wells in various regions, as a selection of wells of equal distribution within this region engenders three classes of total mineralization too. Thus, the increasing salt

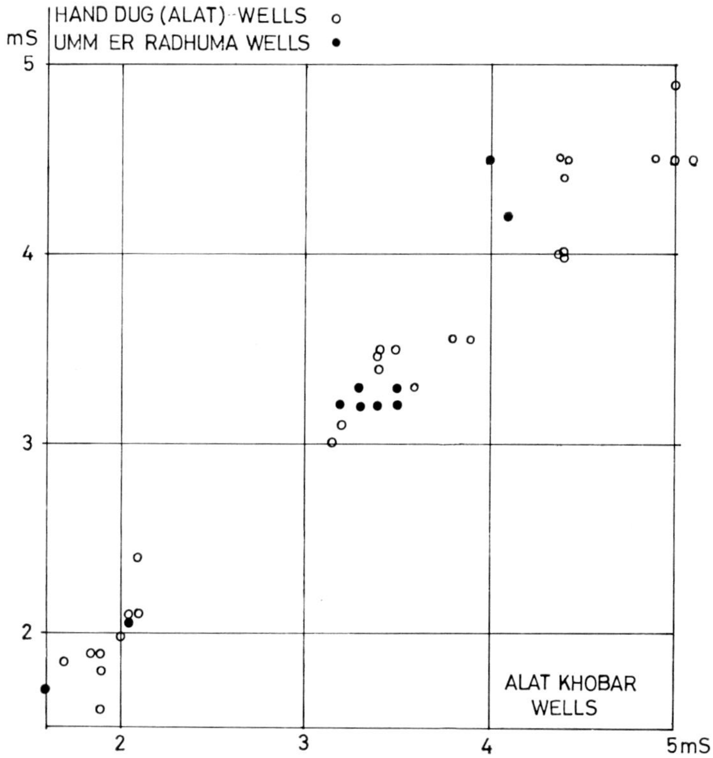


Fig. 30. Neighbouring wells of different depth show similar electrolytic conductivities. There exist three groups of different values (see text)

content from southwest towards northeast takes place in the form of three steps.

Water discharging from wells with full casings drilled into the Umm er Radhuma Formation (Q 1 U/A, Q 4 U/A) near Al Qasamah about 10 kilometers south of Al Qatif disclosed electrical conductivities of 4.4 and 3.9 mS respectively. The water of Q 1 U/A showed strong variation of concentration with time. Electrical conductivity decreased from 4.4 mS via 3.8 mS down to 2.9 mS. Remarkable variations were also detected in the water from a test drilling into the Umm er Radhuma Formation west of Umm As Sahik, about 15 kilometers northwest of Al Qatif (Q 15 U/C). During the drilling at depths from 263 down to 600 meters large changes in electrical conductivity between 3.2 and 8.8. mS were measured. These changes indicate that the Umm er Radhuma Formation contains water with very different salt contents.

The frequency distribution of the piezometric level of uncased drilled wells shows in general two horizons (Fig. 31): 6 to 7 meters at the lower-situated wells (about 3 to 6 meters above sea level), which in most cases have a free outflow and 8.5 up to 10.5 meters at the higher situated wells (8 to

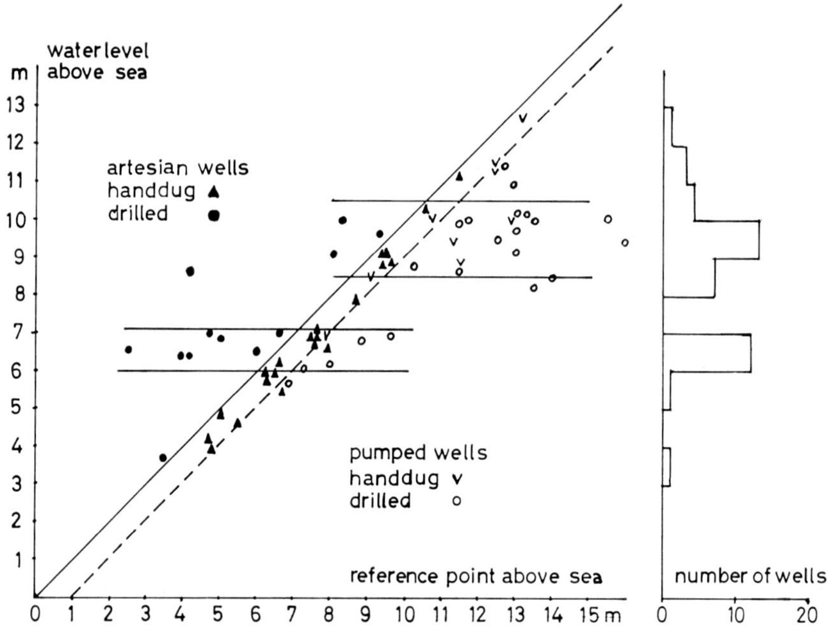


Fig. 31. Piezometric levels of uncased drilled and handdug wells. Right: Frequency distribution of piezometric levels of drilled and pumped handdug wells

16 meters above sea level), where the water level is mostly below the land surface. The decrease of the piezometric level of the lower situated wells indicates pressure release owing to many freely discharging wells (springs).

In fully cased test wells, the water of the Umm er Radhuma Formation showed a piezometric level of 13 to 15.6 meters above sea level, the water of the Khobar aquifer 8 to 11.5 meters and the Alat waters a level of 8.5 to 9.1 meters. This proves that there is a difference of pressure head between the waters of the Umm er Radhuma Formation and the overlying Khobar and Alat aquifers. These differences may be explained by the fact that the water confined in the Umm er Radhuma passes overlying strata of low permeability at some places, but later reaches the more permeable deposits of the Khobar and Alat aquifers. The high daily differences in the pressure head caused by pumping (Fig. 32) indicates that the water passing the less permeable strata overlying the Umm er Radhuma is restricted.

A study by T. DINÇER et al. (1974) interprets the differences of the piezometric level within the coastalbelt wells as a consequence of water penetrating from the Umm er Radhuma strata into the overlying aquifers. The authors hold the opinion that all the coastal-belt aquifers are connected with each other and act as one hydrological unit. This general picture is supported by investigations carried out by A. I. NAIMI (1965), which give a much more detailed account of the situation. A. I. NAIMI evaluated numerous data from drillings of the Arabian-American Oil Company (ARAMCO). This study discloses that the salt content in all the aquifers of eastern Saudi Arabia in-

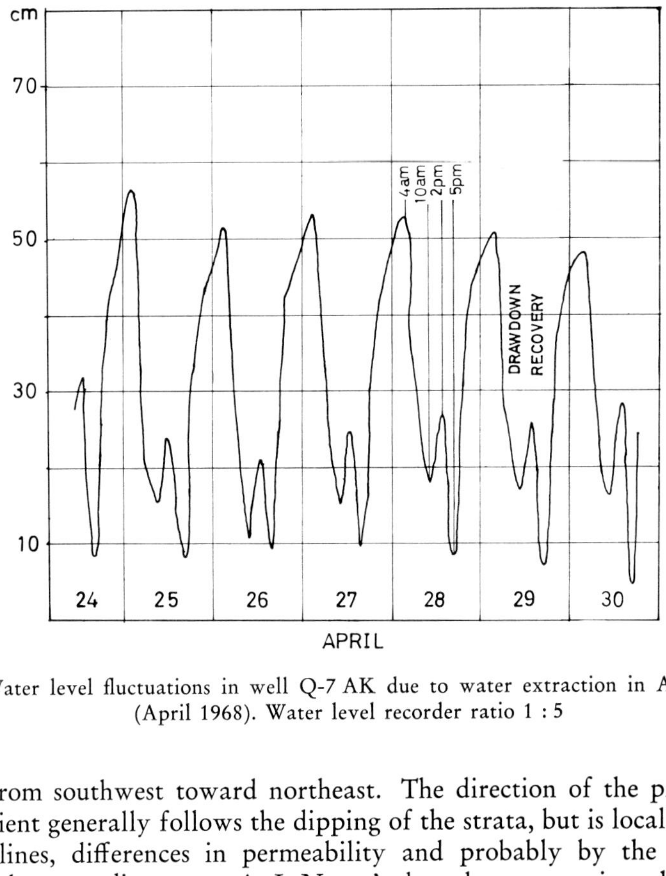


Fig. 32. Water level fluctuations in well Q-7 AK due to water extraction in Al Qatif oasis (April 1968). Water level recorder ratio 1 : 5

creases from southwest toward northeast. The direction of the pressure and salt gradient generally follows the dipping of the strata, but is locally modified by anticlines, differences in permeability and probably by the infiltration from the large wadi systems. A. I. NAIMI's data do not permit making a clear differentiation of the gradient of the piezometric level and the salt content in the coastal oases region. Only the Wasia aquifer at a depth of more than 1,000 meters shows a remarkably higher pressure head and a stronger salinity. Following the vertical direction towards the hydroisohypses given for the Umm er Radhuma aquifer a tongue of less mineralized water comes from the south, water with a higher salt content comes from the northwest. The influx of waters with different salt contents from different directions in Al Qatif area favors the appearance of mixed waters.

*Recent hydrochemical investigations*

During the field campaigns of 1973–1975 water samples were collected for chemical analysis and complemented by local measurements from 32 wells of the coastalbelt region. Of these, 17 samples were taken in the area of Al Qatif (between the villages of Al Awwamiyah and Al Qudaih), 9 in the surroundings of Al Ajam, 4 near Umm As Sahik and 2 at the island of Tarut. A number of samples were collected from a drainage channel, the runoff of joint drainage waters between Al Awwamiyah and Al Qudaih flowing towards the Tarut Bay (samples No. 60, 62, 132–140). From some of the wells samples were collected every year (Table 11 and Fig. 26). Some of the sampling places are not in the area shown in Fig. 26: these are Umm As Sahik samples No. 6, 7, 8, and 142 taken about 10 kilometers north of Al Ajam, the drilled well No. 74 situated a few kilometers south of Al Ajam and the two wells of the island of Tarut (No. 55, 66).

Table 11. *List of water samples collected in the Al Qatif Area from 1973—1975. For statistic calculations and hydrochemical diagrams only the italic numbers were used*

| Al Qatif | | | Al Ajam | | | Umm As Sahik | | Tarut |
|---|---|---|---|---|---|---|---|---|
| 1973 No. | 1974 No. | 1975 No. | 1973 No. | 1974 No. | 1975 No. | 1973 No. | 1975 No. | 1974 No. |
| *1* | — | — | *9* | 73 | 125 | 6 | — | *55* |
| 2 | 57 | *126* | *10* | — | — | 7 | — | *56* |
| 3 | *58* | 127 | *11* | 70 | 141 | 8 | — | |
| *4* | 59 | 128 | *12* | — | — | — | 142 | |
| 5 | 64 | *129* | — | *68* | — | | | |
| 13 | *65* | 130 | — | 69 | *123* | | | |
| *14* | 63 | 133 | — | *71* | *124* | | | |
| *15* | 61 | 131 | — | *72* | — | | | |
| *16* | — | — | — | *74* | — | | | |
| *17* | 78 | 143 | | | | | | |
| *18* | — | — | | | | | | |
| *19* | — | — | | | | | | |
| *20* | — | — | | | | | | |
| — | *66* | — | | | | | | |
| — | *75* | — | | | | | | |
| — | *76* | — | | | | | | |
| — | *77* | — | | | | | | |

*Methods of analyses*

Quantitative analyses were carried out by means of the flame photometer ($Na^+$, $K^+$), and by complexometric ($Ca^{2+}$, $Mg^{2+}$, $SO_4^{2-}$), titrimetric ($Cl^-$, $Br^-$, $HCO_3^-$), colorimetric ($NO_3^-$, $H_3BO_3$, $H_2SiO_3$), and electrometric methods ($H^+$, $F^+$). Qualitative measurements for $NH_4^+$, $Fe^{2+, 3+}$, $Al^{3+}$, and $PO_4^{3-}$ showed negative results.

The analytical data of the samples collected in 1973, 1974, and 1975 are summarized in Tables 12, 13, and 14.

Table 12. *Results of chemical analyses of water samples collected in Al Qatif area March 1973.* No. = sample number, W = well type, °C = water temperature in degrees centigrade, $O_2$ in mg/l (in 'ayns, 1 m below the water level), SK = sum of cations, SA = sum of anions; ion concentrations in μe/l, $H_3BO_3$ and $H_2SiO_3$ in mg/l. Well types: DA = drilled well, free outflow; HA = hand dug well, free outflow; HP = hand dug well, pumping; Dr = drainge channel, DP = drilled well, pumping; H = water level below land surface, not used for irrigation

| No. | W | °C | $O_2$ | $Na^+$ | $K^+$ | $Ca^{++}$ | $Mg^{++}$ | SK | $Cl^-$ | $Br^-$ | $SO_4^{--}$ | $HCO_3^-$ | $NO_3^-$ | $F^-$ | SA | $H_3BO_3$ | $H_2SiO_3$ |
|---|---|---|---|---|---|---|---|---|---|---|---|---|---|---|---|---|---|
| 1 | HA | 33.5 | 2.0 | 28,491 | 725 | 14,800 | 8,320 | 52,336 | 37,285 | 75.6 | 11,799 | 2,901 | 179 | 40 | 52,280 | 3.82 | 15.5 |
| 2 | HA | 33.5 | 3.5 | 23,271 | 605 | 12,920 | 7,360 | 44,156 | 30,102 | 58.4 | 10,924 | 2,876 | 232 | 44 | 44,236 | 3.41 | 16.2 |
| 3 | HA | 33.0 | 3.3 | 25,076 | 633 | 13,320 | 7,580 | 46,609 | 32,576 | 64.3 | 10,823 | 2,876 | 218 | 42 | 46,599 | 3.46 | 16.1 |
| 4 | HA | 32.5 | 3.0 | 25,672 | 661 | 13,400 | 7,600 | 47,333 | 33,374 | 65.7 | 10,771 | 2,876 | 208 | 42 | 47,337 | 3.58 | 16.1 |
| 5 | DA | 37.0 | 3.0 | 28,926 | 774 | 14,320 | 8,100 | 52,120 | 37,923 | 76.3 | 11,164 | 2,795 | 198 | 39 | 52,195 | 3.96 | 15.7 |
| 6 | HP | 28.0 | 4.2 | 18,399 | 506 | 11,640 | 7,120 | 37,665 | 21,801 | 39.0 | 11,237 | 3,431 | 163 | 42 | 36,713 | 3.03 | 16.2 |
| 7 | HA | 31.0 | 5.4 | 16,746 | 506 | 10,920 | 6,600 | 34,772 | 20,528 | 31.5 | 10,875 | 3,138 | 155 | 42 | 34,770 | 2.82 | 17.0 |
| 8 | HP | 27.5 | 1.2 | 15,694 | 492 | 10,960 | 6,700 | 33,846 | 19,648 | 31.5 | 10,839 | 3,093 | 165 | 42 | 33,818 | 2.96 | 16.7 |
| 9 | HP | 27.5 | 9.0 | 15,949 | 492 | 10,560 | 6,140 | 33,141 | 19,727 | 32.7 | 10,040 | 2,962 | 260 | 42 | 33,064 | 2.78 | 16.4 |
| 10 | DP | 33.2 | 2.7 | 13,274 | 458 | 9,800 | 5,680 | 29,212 | 16,215 | 25.3 | 9,554 | 3,022 | 278 | 42 | 29,136 | 2.46 | 16.8 |
| 11 | DP | 36.5 | 2.6 | 13,377 | 476 | 9,720 | 5,400 | 28,973 | 16,376 | 23.8 | 9,136 | 3,037 | 282 | 42 | 28,897 | 2.53 | 17.4 |
| 12 | HP | 28.0 | 4.5 | 14,499 | 560 | 9,480 | 5,160 | 29,699 | 17,257 | 23.0 | 8,602 | 3,370 | 254 | 42 | 29,548 | 2.31 | 19.0 |
| 13 | DA | 36.5 | 2.1 | 28,298 | 753 | 14,140 | 7,940 | 51,131 | 37,120 | 74.2 | 10,835 | 2,861 | 226 | 40 | 51,156 | 3.77 | 15.6 |
| 14 | HA | 33.8 | 2.8 | 19,332 | 568 | 11,720 | 6,600 | 38,220 | 25,037 | 44.5 | 9,980 | 2,896 | 250 | 42 | 38,250 | 2.90 | 16.5 |
| 15 | DA | 35.8 | 1.9 | 22,705 | 628 | 12,620 | 6,820 | 42,773 | 29,344 | 56.0 | 10,152 | 2,851 | 250 | 42 | 42,695 | 3.29 | 15.8 |
| 16 | HA | 32.5 | 4.3 | 17,872 | 526 | 11,240 | 6,380 | 36,018 | 22,760 | 39.8 | 10,000 | 2,891 | 272 | 42 | 36,005 | 2.85 | 16.0 |
| 17 | H | 23.0 | 6.8 | 19,910 | 568 | 12,080 | 6,880 | 39,438 | 25,157 | 43.2 | 10,843 | 3,017 | 244 | 44 | 39,348 | 3.09 | 16.4 |
| 18 | H | — | — | 19,090 | 540 | 11,640 | 6,280 | 37,550 | 23,683 | 37.2 | 10,357 | 2,977 | 312 | 44 | 37,410 | 2.56 | 16.0 |
| 19 | DA | (35) | — | 13,652 | 469 | 9,760 | 5,640 | 29,521 | 16,655 | 24.5 | 9,369 | 3,063 | 278 | 44 | 29,434 | 2.35 | 17.2 |
| 20 | HA | 33.4 | — | 14,836 | 483 | 10,060 | 5,540 | 30,919 | 17,972 | 28.0 | 9,349 | 3,078 | 292 | 42 | 30,761 | 2.43 | 17.6 |

Table 13. *Results of chemical analyses of water samples collected in Al Qatif area March 1974*

| No. | W | °C | $O_2$ | $Na^+$ | $K^+$ | $Ca^{++}$ | $Mg^{++}$ | SK | $Cl^-$ | $Br^-$ | $SO_4^{--}$ | $HCO_3^-$ | $NO_3^-$ | $F^-$ | SA | $H_3BO_3$ | $H_2SiO_3$ |
|---|---|---|---|---|---|---|---|---|---|---|---|---|---|---|---|---|---|
| 55 | DA | 32.5 | 2.5 | 21,749 | 605 | 12,240 | 6,640 | 41,234 | 27,588 | 52.2 | 10,486 | 2,871 | 263 | 39 | 41,299 | 2.86 | 15.9 |
| 56 | DA | 34.8 | 2.7 | 21,314 | 577 | 11,960 | 6,680 | 40,531 | 26,790 | 49.9 | 10,482 | 2,851 | 269 | 39 | 40,481 | 2.86 | 15.9 |
| 57 | HA | 33.8 | 2.5 | 23,869 | 661 | 12,920 | 7,400 | 44,850 | 30,581 | 59.0 | 11,012 | 2,871 | 232 | 39 | 44,794 | 3.32 | 16.1 |
| 58 | HA | 33.7 | 3.0 | 25,620 | 675 | 13,360 | 7,520 | 47,175 | 33,294 | 66.2 | 10,952 | 2,851 | 213 | 39 | 47,415 | 3.44 | 17.2 |
| 59 | HA | 32.5 | 2.9 | 26,025 | 675 | 13,540 | 7,540 | 47,780 | 33,532 | 67.7 | 10,875 | 2,902 | 221 | 39 | 47,637 | 3.33 | 16.4 |
| 60 | Dr | 28.3 | 9.2 | 25,084 | 675 | 17,800 | 7,960 | 51,519 | 30,187 | 52.6 | 17,297 | 3,946 | 121 | 78 | 51,682 | 3.89 | 29.2 |
| 61 | DA | 35.7 | 2.8 | 23,488 | 664 | 12,500 | 7,260 | 43,912 | 30,182 | 58.2 | 10,538 | 2,851 | 250 | 40 | 43,919 | 3.40 | 17.0 |
| 62 | Dr | 24.7 | 9.3 | 24,576 | 703 | 14,700 | 8,580 | 48,559 | 32,181 | 58.9 | 12,658 | 3,795 | 126 | 55 | 48,874 | 3.94 | 17.1 |
| 63 | HA | 33.1 | 3.8 | 20,008 | 563 | 11,680 | 6,600 | 38,851 | 24,995 | 44.9 | 10,462 | 2,831 | 241 | 44 | 38,618 | 2.97 | 16.1 |
| 64 | DA | 36.1 | 3.1 | 28,998 | 776 | 14,360 | 8,160 | 52,294 | 37,843 | 76.9 | 11,293 | 2,821 | 191 | 40 | 52,265 | 4.06 | 16.1 |
| 65 | DA | 36.4 | 3.2 | 28,273 | 774 | 14,200 | 7,820 | 51,067 | 37,044 | 76.2 | 10,875 | 2,711 | 232 | 40 | 50,978 | 4.00 | 16.1 |
| 66 | DA | 34.0 | 3.1 | 13,871 | 492 | 9,920 | 5,440 | 29,723 | 17,250 | 29.5 | 9,229 | 2,932 | 296 | 42 | 29,779 | 2.60 | 17.8 |
| 68 | DP | 33.4 | 3.1 | 14,490 | 464 | 10,360 | 5,520 | 30,834 | 17,331 | 28.9 | 9,839 | 3,062 | 263 | 42 | 30,566 | 2.60 | 16.7 |
| 69 | DP | 32.8 | 3.2 | 14,354 | 470 | 10,160 | 5,560 | 30,544 | 17,015 | 24.9 | 9,920 | 2,962 | 260 | 42 | 30,224 | 2.60 | 16.7 |
| 70 | DP | 36.2 | 2.1 | 13,339 | 498 | 9,840 | 5,320 | 28,997 | 16,534 | 25.5 | 9,060 | 3,002 | 324 | 42 | 28,988 | 2.49 | 17.8 |
| 71 | DP | 35.8 | 3.1 | 13,397 | 492 | 9,800 | 5,440 | 29,129 | 16,694 | 25.5 | 9,213 | 2,982 | 315 | 42 | 29,272 | 2.49 | 17.8 |
| 72 | DP | 35.7 | 3.1 | 13,375 | 492 | 9,960 | 5,240 | 29,067 | 16,616 | 24.2 | 9,197 | 3,032 | 315 | 38 | 29,222 | 2.49 | 17.5 |
| 73 | HP | 26.3 | 7.1 | 15,933 | 506 | 10,720 | 5,940 | 33,099 | 19,729 | 30.5 | 9,903 | 2,942 | 240 | 40 | 32,885 | 2.74 | 16.1 |
| 74 | DP | — | 0.3 | 13,780 | 563 | 8,800 | 4,600 | 27,743 | 16,814 | 25.9 | 8,289 | 2,641 | 241 | 36 | 27,806 | 2.37 | 17.8 |
| 75 | DA | 35.3 | 2.6 | 19,139 | 563 | 11,600 | 6,460 | 37,762 | 24,356 | 43.9 | 10,261 | 2,932 | 237 | 42 | 37,872 | 2.86 | 16.1 |
| 76 | DA | 35.6 | 2.7 | 19,139 | 554 | 11,520 | 6,520 | 37,733 | 24,356 | 44.2 | 10,152 | 2,922 | 244 | 42 | 37,760 | 2.86 | 16.4 |
| 77 | HA | 33.8 | 3.2 | 19,139 | 563 | 11,520 | 6,440 | 37,662 | 24,277 | 42.9 | 10,249 | 2,892 | 275 | 42 | 37,778 | 2.74 | 16.4 |
| 78 | H | 22.3 | 6.0 | 20,357 | 591 | 12,360 | 7,000 | 40,308 | 25,955 | 45.0 | 11,108 | 3,082 | 223 | 43 | 40,456 | 3.11 | 16.7 |

Table 14. *Results of chemical analyses of water samples collected in Al Qatif area February 1975*

| No. | W | °C | $O_2$ | $Na^+$ | $K^+$ | $Ca^{++}$ | $Mg^{++}$ | SK | $Cl^-$ | $Br^-$ | $SO_4^{--}$ | $HCO_3^-$ | $NO_3^-$ | $F^-$ | SA | $H_3BO_3$ | $H_2SiO_3$ |
|---|---|---|---|---|---|---|---|---|---|---|---|---|---|---|---|---|---|
| 123 | DP | 33.9 | 3.5 | 13,919 | 478 | 9,800 | 5,680 | 29,877 | 16,776 | 24.4 | 9,839 | 2,970 | 256 | 38 | 29,903 | 2.45 | 16.9 |
| 124 | DP | 37.3 | 3.3 | 13,375 | 487 | 9,720 | 5,440 | 29,022 | 16,695 | 25.2 | 9,205 | 3,000 | 298 | 40 | 29,263 | 2.33 | 17.7 |
| 125 | HP | 26.2 | 9.1 | 15,768 | 487 | 10,480 | 6,000 | 32,735 | 19,770 | 30.2 | 10,096 | 2,880 | 264 | 40 | 33,080 | 2.59 | 14.4 |
| 126 | HA | 33.8 | 2.8 | 23,380 | 661 | 13,000 | 7,000 | 44,041 | 29,941 | 59.2 | 10,924 | 2,860 | 215 | 40 | 44,039 | 3.31 | 16.1 |
| 127 | HA | 33.2 | 4.2 | 25,555 | 703 | 13,400 | 7,280 | 46,938 | 32,335 | 64.8 | 11,036 | 2,840 | 214 | 40 | 46,530 | 3.41 | 16.1 |
| 128 | HA | 32.9 | 3.7 | 25,772 | 731 | 13,480 | 7,240 | 47,223 | 33,375 | 65.2 | 10,883 | 3,020 | 225 | 40 | 47,608 | 3.52 | 16.1 |
| 129 | DA | 36.8 | 3.0 | 29,361 | 760 | 14,320 | 7,760 | 52,201 | 37,448 | 72.2 | 11,349 | 2,800 | 187 | 39 | 51,895 | 3.83 | 15.5 |
| 130 | DA | 36.7 | 3.5 | 28,817 | 774 | 14,040 | 7,840 | 51,471 | 36,965 | 74.6 | 11,100 | 2,820 | 205 | 40 | 51,205 | 3.98 | 15.5 |
| 131 | DA | 36.1 | 2.9 | 22,836 | 605 | 12,680 | 7,120 | 43,241 | 29,584 | 55.6 | 10,450 | 2,780 | 248 | 40 | 43,158 | 3.36 | 16.1 |
| 132 | Dr | 25.6 | — | 25,555 | 703 | 18,280 | 9,720 | 54,258 | 32,345 | 54.9 | 17,654 | 3,840 | 74 | 68 | 54,036 | 3.72 | 17.2 |
| 133 | HA | 33.9 | 4.3 | 19,030 | 563 | 11,600 | 6,560 | 37,753 | 24,595 | 45.3 | 10,265 | 2,780 | 237 | 39 | 37,961 | 3.28 | 16.6 |
| 134 | Dr | 25.4 | — | 25,011 | 703 | 18,480 | 9,120 | 53,314 | 31,547 | 53.1 | 17,630 | 3,860 | 113 | 68 | 53,271 | 3.72 | 21.6 |
| 135 | Dr | 25.4 | — | 26,098 | 830 | 15,160 | 8,440 | 50,528 | 33,145 | 54.6 | 13,349 | 4,000 | 177 | 54 | 50,780 | 3.72 | 14.4 |
| 136 | Dr | — | — | 24,467 | 717 | 17,840 | 8,880 | 51,904 | 31,786 | 53.6 | 15,840 | 3,920 | 107 | 66 | 51,773 | 3.98 | 21.4 |
| 137 | Dr | 24.7 | — | 28,273 | 760 | 15,240 | 8,360 | 52,633 | 36,252 | 68.2 | 12,650 | 3,520 | 131 | 40 | 52,661 | 3.98 | 14.4 |
| 138 | Dr | 25.1 | — | 26,098 | 746 | 17,280 | 9,000 | 53,124 | 33,300 | 59.7 | 15,678 | 3,860 | 95 | 63 | 53,056 | 4.04 | 9.4 |
| 139 | Dr | — | — | 31,535 | 844 | 17,760 | 9,840 | 59,979 | 40,324 | 75.6 | 15,518 | 3,720 | 12 | 52 | 59,702 | 4.50 | 13.0 |
| 140 | Dr | — | — | 32,888 | 866 | 17,920 | 10,560 | 62,234 | 42,321 | 79.0 | 15,743 | 3,800 | 24 | 52 | 62,019 | 4.76 | 13.0 |
| 141 | DP | 36.2 | 3.7 | 13,593 | 478 | 10,000 | 5,400 | 29,471 | 16,896 | 24.5 | 9,494 | 3,050 | 272 | 40 | 29,777 | 2.69 | 17.7 |
| 142 | HA | 30.4 | 4.3 | 17,943 | 506 | 12,400 | 6,640 | 37,489 | 22,998 | 42.2 | 11,438 | 2,910 | 200 | 40 | 38,628 | 3.31 | 16.1 |
| 143 | H | 21.2 | 8.5 | 25,228 | 717 | 14,200 | 8,320 | 48,465 | 32,101 | 58.7 | 13,004 | 3,220 | 224 | 44 | 48,652 | 3.72 | 17.2 |
| 144 | — | — | — | 115,702 | 2,877 | 18,474 | 28,674 | 165,727 | 139,369 | 321.3 | 23,052 | 3,230 | 36 | 57 | 165,975 | 8.63 | 16.7 |
| 145 | — | 16.8 | — | 839,495 | 18,107 | 33,734 | 187,146 | 1,078,482 | 962,547 | 1,453.1 | 97,187 | 2,660 | 0 | 79 | 1,063,926 | 41.71 | 0.8 |

In these tables, column 1 shows the sampling number, column 2 indicates the kind of well (HA = hand-dug well, mostly a quadrangular widened karst fissure with a free discharge; HP = the same as HA, but with a piezometric level below the land surface, thus having no free outflow and from which the water has to be pumped; DA = drilled well with the piezometric level above the land surface and free surface runoff; DP = drilled well with the piezometric level below the land surface and a pumping station). Column 4 shows the oxygen content of the water in mg/l measured in the well. The following columns give the ion-concentrations and the sum of cations (SK) and anions (SA) in micro-equivalents per liter (µe/l). In the last two columns the concentrations of meta-boric acid and meta-silicic acid are given in mg/l.

## *Methods of evaluation*

The total mineralization and the relative concentration of the ion components (me %) of waters from a certain area are in general not very different. Very often the differences lie within the limits of precision of the analysis, which are about ± 2% for the total concentration, and ± 5% up to ± 10% for the individual ions. Differences within these limits are not significant and the waters may be combined into one uniform group or into a water family of the first order. If the values of the total concentrations show major differences, while the relative percentages of the individual ions (me %) are practically the same, we deal with a water family of the second order. This group also includes waters with an unchanged or decreasing content of some ions despite an increasing total salt concentration provided that this situation may be explained by means of the ion-product of a sparingly soluble salt (e.g., $CaCO_3$, CaF, $CaSO_4$). Waters with different total concentrations as well as different relative amounts of freely soluble ions belong to a water family of the third order, if these variations follow a certain rule. This is the case, whenever two different waters become mixed. Mixing changes the total salt concentration as well as the relative ion-content depending on the composition and the volume of the mixed components. In such a case there is no change in concentration of those ions which are present in the same concentration in both components. In some of these instances, both the composition and the volume of the mixed components may be calculated by means of the mixing rule.

With a lot of analytical data the detection of water groups is difficult as well as time consuming. The application of computer techniques, especially cluster analysis, may prove to be very useful, as R. J. Beckmann and J. B. Ramsay show in the following chapter. Likewise, diagrams showing the total salt concentration on the abscissa and the concentrations of the individual ions on the ordinate may be of great help. Whenever such a diagram shows the concentrations of the individual ions as linear groups, then there exists a water family of the second or third order. Through statistical methods the course and degree of correlation may be estimated. However, these methods sometimes produce only very obvious connections in spite of complicated calculations, e.g., different concentrations of nearly pure NaCl solutions show that the total salt content and the concentrations of the $Na^+$ and $Cl^-$ ions are significantly correlated. Therefore, the detected correlations should be carefully checked as to their meaning, keeping in mind that the total concentration may be implicitly dependent on the variable considered. Very often the use of the correlation coefficient ($r$) may be too simple and the problem requires a more sophisticated statistical analysis, e.g., a significant correlation coefficient may be obtained from a sample containing two bivariate distributions, each with different mean values for the two variables. Furthermore, the usual correlation coefficient presupposes that the data are normally distributed, a condition that is seldom fulfilled. The rank (date) correlation coefficient of SPEARMAN ($r_s$) is less sensitive to the assumptions of normality and separations of mean values and is used in this paper.

## *Results*

The total mineralizations of the investigated waters vary between about 29 me/l and 52 me/l (i.e., 1.8 g/l up to 3.2 g/l). Water samples from the same well collected in different years revealed no relevant changes in their individual ion composition (Table 15). The percentage difference ($\Delta$%) between the highest and lowest values amount to 3 percent for the total concentration and lies for most of the ion-analyses below 5 percent. More differentiated was the nitrate content; the reason for this may originate partly from analytical difficulties. The water samples No. 17, 78, and 143 were collected from an 'ayn which has no discharge in recent times; its water temperature is irregularly low and people believe in its healing power (Hammam Balousa). The total concentration of this water increased from about 39.4 me/l up to 48.5 me/l in the period of 1973 to 1975. The tendency to increase was shown by nearly all ions except fluoride, which remained constant and nitrate which decreased about 9 percent. The decrease in nitrate with an increase in total salt concentration was usually observed for the waters of Al Qatif region.

The waters are characterized by the fact that $Cl^-$ and $SO_4^{2-}$ are the dominant anions and $Na^+$ and $Ca^{2+}$ the most important cations (Na-Ca-Cl-$SO_4$-water). With an increasing total salt content, $Cl^-$ increases more than $SO_4^{2-}$, $Na^+$ and $K^+$ more than $Ca^{2+}$ and $Mg^{2+}$ (Fig. 33). The content of $HCO_3^-$ and $F^-$ is more or less constant, whereas the content of $NO_3^-$ decreases with an increasing salt content. Considering the non-ionized materials, meta-boric acid increases with an increasing total concentration, while the meta-silicic acid apparently decreases in this case. The concentration of $K^+$ increases in parallel with the $Na^+$ content; the same is true for the $Br^-$ with $Cl^-$.

In the investigation area, four different kinds of wells exist from which the water flows or is pumped from the upper limestone layer (HA- and HP-wells respectively) or else additional water is taken from lower strata (DA- and DP-wells respectively).

The average water temperature is almost equal for drilled wells showing a free discharge or for wells being pumped (DA or DP respectively; see Table 16). The waters of the HA- and HP-wells ('ayns) show a lower temperature, the average difference amounts in the HA-wells to about 2.3 °C.

Table 15. *Comparison of analyses in 1973, 1974, and 1975* (see Table 11). △% = percentage difference between the highest and the lowest value. Except for Nos. 17, 78, 143 larger fluctuations (> 10%) were found only for the $O_2$ content and in some cases for $K^+$, $NO_3^-$ and $F^-$ ions

| | No. | W | °C | $O_2$ | $Na^+$ | $K^+$ | $Ca^{++}$ | $Mg^{++}$ | SK | $Cl^-$ | $Br^-$ | $SO_4^{--}$ | $HCO_3^-$ | $NO_3^-$ | $F^-$ | SA | $H_3BO_3$ | $H_2SiO_3$ |
|---|---|---|---|---|---|---|---|---|---|---|---|---|---|---|---|---|---|---|
| 1973 | 2 | HA | 33.5 | 3.5 | 23,271 | 605 | 12,920 | 7,360 | 44,156 | 30,102 | 58.4 | 10,924 | 2,876 | 232 | 44 | 44,236 | 3.41 | 16.2 |
| 1974 | 57 | | 33.8 | 2.5 | 23,869 | 661 | 12,920 | 7,400 | 44,850 | 30,581 | 59.0 | 11,012 | 2,871 | 232 | 39 | 44,794 | 3.23 | 16.1 |
| 1975 | 126 | | 33.8 | 2.8 | 23,380 | 661 | 13,000 | 7,000 | 44,041 | 29,941 | 59.2 | 10,924 | 2,860 | 215 | 40 | 44,039 | 3.31 | 16.1 |
| | △% | | 1 | 40 | 3 | 9 | 1 | 6 | 2 | 2 | 1 | 1 | 1 | 8 | 13 | 2 | 3 | 1 |
| 1973 | 3 | HA | 33.0 | 3.3 | 25,076 | 633 | 13,320 | 7,580 | 46,609 | 32,576 | 64.3 | 10,823 | 2,876 | 218 | 42 | 46,599 | 3.46 | 16.1 |
| 1974 | 58 | | 33.7 | 3.0 | 25,620 | 675 | 13,360 | 7,520 | 47,175 | 33,294 | 66.2 | 10,952 | 2,851 | 213 | 39 | 47,415 | 3.44 | 17.2 |
| 1975 | 127 | | 33.2 | 4.2 | 25,555 | 703 | 13,400 | 7,280 | 46,938 | 42,335 | 64.8 | 11,036 | 2,840 | 214 | 40 | 46,530 | 3.41 | 16.1 |
| | △% | | 2 | 40 | 3 | 11 | 1 | 4 | 1 | 3 | 3 | 2 | 1 | 2 | 8 | 2 | 1 | 7 |
| 1973 | 4 | HA | 32.5 | 3.0 | 25,672 | 661 | 13,400 | 7,600 | 47,333 | 33,374 | 65.7 | 10,771 | 2,876 | 208 | 42 | 47,337 | 3.58 | 16.1 |
| 1974 | 59 | | 32.5 | 2.9 | 26,025 | 675 | 13,540 | 7,540 | 47,780 | 33,532 | 67.7 | 10,875 | 2,902 | 221 | 39 | 47,637 | 3.33 | 16.4 |
| 1975 | 128 | | 32.9 | 3.7 | 25,772 | 731 | 13,480 | 7,240 | 47,223 | 33,375 | 65.2 | 10,883 | 3,020 | 225 | 40 | 47,608 | 3.52 | 16.1 |
| | △% | | 1 | 28 | 1 | 11 | 1 | 5 | 1 | 0 | 4 | 1 | 5 | 8 | 8 | 1 | 8 | 2 |
| 1973 | 5 | DA | 37.0 | 3.0 | 28,926 | 774 | 14,320 | 8,100 | 52,120 | 37,923 | 76.3 | 11,164 | 2,795 | 198 | 39 | 52,195 | 3.96 | 15.7 |
| 1974 | 64 | | 36.1 | 3.1 | 28,998 | 776 | 14,360 | 8,160 | 52,294 | 37,843 | 76.9 | 11,293 | 2,821 | 191 | 40 | 52,265 | 4.06 | 16.1 |
| 1975 | 129 | | 36.8 | 3.0 | 29,361 | 760 | 14,320 | 7,760 | 52,201 | 37,448 | 72.2 | 11,349 | 2,800 | 187 | 39 | 51,895 | 3.83 | 15.5 |
| | △% | | 2 | 3 | 2 | 2 | 0 | 5 | 0 | 1 | 7 | 2 | 1 | 6 | 3 | 1 | 6 | 4 |
| 1973 | 15 | DA | 35.8 | 1.9 | 22,705 | 628 | 12,620 | 6,820 | 42,773 | 29,344 | 56.0 | 10,152 | 2,851 | 250 | 42 | 42,695 | 3.29 | 15.8 |
| 1974 | 61 | | 35.7 | 2.8 | 23,488 | 664 | 12,500 | 7,260 | 43,912 | 30,182 | 58.2 | 10,538 | 2,851 | 250 | 40 | 43,919 | 3.40 | 17.0 |
| 1975 | 131 | | 36.1 | 2.9 | 22,836 | 605 | 12,680 | 7,120 | 43,241 | 29,584 | 55.6 | 10,450 | 2,780 | 248 | 40 | 43,158 | 3.36 | 16.1 |
| | △% | | 1 | 53 | 3 | 9 | 1 | 6 | 3 | 3 | 5 | 4 | 3 | 1 | 5 | 3 | 3 | 8 |
| 1973 | 17 | H | 23.0 | 6.8 | 19,910 | 568 | 12,080 | 6,880 | 39,438 | 25,157 | 43.2 | 10,843 | 3,017 | 244 | 44 | 39,348 | 3.09 | 16.4 |
| 1974 | 78 | | 22.3 | 6.0 | 20,357 | 591 | 12,360 | 7,000 | 40,308 | 25,955 | 45.0 | 11,108 | 3,082 | 223 | 43 | 40,456 | 3.11 | 16.7 |
| 1975 | 143 | | 21.2 | 8.5 | 25,228 | 717 | 14,200 | 8,320 | 48,465 | 32,101 | 58.7 | 13,004 | 3,220 | 224 | 44 | 48,652 | 3.72 | 17.2 |
| | △% | | 8 | 42 | 27 | 26 | 18 | 21 | 23 | 28 | 36 | 20 | 7 | 9 | 2 | 24 | 20 | 5 |

| | | | | | | | | | | | | | | | | | |
|---|---|---|---|---|---|---|---|---|---|---|---|---|---|---|---|---|---|
| 1974 | 69 | DP | 32.8 | 3.2 | 14,354 | 470 | 10,160 | 5,560 | 30,554 | 17,015 | 24.9 | 9,920 | 2,962 | 260 | 42 | 30,224 | 2.60 | 16.7 |
| 1975 | 123 | | 33.9 | 3.5 | 13,919 | 478 | 9,800 | 5,680 | 29,877 | 16,776 | 24.4 | 9,839 | 2,970 | 256 | 38 | 29,903 | 2.45 | 16.9 |
| | △% | | 3 | 9 | 3 | 2 | 4 | 2 | 2 | 1 | 2 | 1 | 0 | 2 | 11 | 1 | 6 | 2 |
| 1974 | 71 | DP | 35.8 | 3.1 | 13,397 | 492 | 9,800 | 5,440 | 29,129 | 16,694 | 25.5 | 9,213 | 2,982 | 315 | 42 | 29,272 | 2.49 | 17.8 |
| 1975 | 124 | | 37.3 | 3.3 | 13,375 | 487 | 9,720 | 5,440 | 29,022 | 16,695 | 25.2 | 9,205 | 3,000 | 298 | 40 | 29,263 | 2.33 | 17.7 |
| | △% | | 4 | 6 | 0 | 1 | 1 | 0 | 0 | 0 | 1 | 0 | 1 | 6 | 5 | 0 | 7 | 1 |
| 1973 | 9 | HP | 27.5 | 9.0 | 15,949 | 492 | 10,560 | 6,140 | 33,141 | 19,727 | 32.7 | 10,040 | 2,962 | 260 | 42 | 33,064 | 2.78 | 16.4 |
| 1974 | 73 | | 26.3 | 7.1 | 15,933 | 506 | 10,720 | 5,940 | 33,099 | 19,729 | 30.5 | 9,903 | 2,942 | 240 | 40 | 32,885 | 2.74 | 16.1 |
| 1975 | 125 | | 26.2 | 9.1 | 15,768 | 487 | 10,480 | 6,000 | 32,735 | 19,770 | 30.2 | 10,096 | 2,880 | 264 | 40 | 33,080 | 2.59 | 14.4 |
| | △% | | 5 | 28 | 1 | 4 | 3 | 3 | 1 | 0 | 8 | 1 | 3 | 10 | 5 | 1 | 7 | 14 |
| 1973 | 11 | DP | 36.5 | 2.6 | 13,377 | 476 | 9,720 | 5,400 | 28,973 | 16,376 | 23.8 | 9,136 | 3,037 | 282 | 42 | 28,897 | 2.53 | 17.4 |
| 1974 | 70 | | 36.2 | 2.1 | 13,339 | 498 | 9,840 | 5,320 | 28,997 | 16,534 | 25.5 | 9,060 | 3,002 | 324 | 42 | 28,988 | 2.49 | 17.8 |
| 1975 | 141 | | 36.2 | 3.7 | 13,593 | 478 | 10,000 | 5,400 | 29,471 | 16,896 | 24.5 | 9,494 | 3,050 | 272 | 40 | 29,777 | 2.69 | 17.7 |
| | △% | | 0 | 76 | 2 | 20 | 3 | 2 | 2 | 3 | 7 | 5 | 2 | 19 | 5 | 3 | 8 | 3 |
| 1973 | 13 | DA | 36.5 | 2.1 | 28,298 | 753 | 14,140 | 7,940 | 51,131 | 37,120 | 74.2 | 10,835 | 2,861 | 226 | 40 | 51,156 | 3.77 | 15.6 |
| 1974 | 65 | | 36.4 | 3.2 | 28,273 | 774 | 14,200 | 7,820 | 51,067 | 37,044 | 76.2 | 10,875 | 2,711 | 232 | 40 | 50,978 | 4.00 | 16.1 |
| 1975 | 130 | | 36.7 | 3.5 | 28,817 | 774 | 14,040 | 7,840 | 51,471 | 36,965 | 74.6 | 11,100 | 2,820 | 205 | 40 | 51,205 | 3.98 | 15.5 |
| | △% | | 1 | 67 | 2 | 3 | 1 | 2 | 1 | 0 | 3 | 2 | 5 | 13 | 0 | 0 | 6 | 4 |
| 1973 | 14 | HA | 33.8 | 2.8 | 19,332 | 568 | 11,720 | 6,600 | 38,220 | 25,037 | 44.5 | 9,980 | 2,896 | 250 | 42 | 38,250 | 2.90 | 16.5 |
| 1974 | 63 | | 33.1 | 3.8 | 20,008 | 563 | 11,680 | 6,600 | 38,851 | 24,995 | 44.9 | 10,462 | 2,831 | 241 | 44 | 38,618 | 2.97 | 16.1 |
| 1975 | 133 | | 33.9 | 4.3 | 19,030 | 563 | 11,600 | 6,560 | 37,753 | 24,595 | 45.3 | 10,265 | 2,780 | 237 | 39 | 37,961 | 3.28 | 16.6 |
| | △% | | 2 | 54 | 5 | 1 | 1 | 1 | 3 | 2 | 2 | 5 | 4 | 5 | 13 | 2 | 13 | 3 |

In the HP-wells the water temperature is as much as 7.6 °C lower than that from the boreholes. Measurements carried out with a thermocouple during the field campaign revealed that the lower water temperature of the discharging HA-wells is almost independent of the diameter and the surface contact of the water with air. The water temperature of 'Ayn Al Labaniyah, with a diameter of 6 meters, increased only 0.5 °C from the surface down to a depth of 10 meters (32.8 to 33.3 °C). This difference does not seem to be affected by air, because the same difference was measured in a drilled well (35.9 to 36.4 °C), too. It is assumed that the water in boreholes remains warmer because of its rapid vertical ascent. In contrast to this, the discharge of the 'ayns comes from a pattern of joints allowing adjustment to the respective geothermal depth. 'Ayns with a piezometric water level below land surface (HP), of course, show a lower water temperature from atmospheric cooling (about 5.4 °C difference to the 'ayns having a free outflow).

Concerning the average of the total concentration, there is practically no difference between the water from 'ayns with free outflow (HA) and boreholes having a positive piezometric level (Table 16). On the other hand, there are remarkable differences as compared to the wells from which the water has to be pumped. On the average the HP-wells show a lower mineralization of about 8.3 me/l, the DP-wells of 12.2 me/l respectively. These two types of pumped wells (HP, DP) are mostly situated in the western part of the region near Al Ajam which has a higher altitude, whereas the artesian wells with a positive piezometric level lie in the east (between Al Awwamiyah and Al Qudaih). A small desert area about 3 kilometers wide divides these two groups of wells. The difference of total concentration is in agreement with the local trend described above. The oxygen content of the water of all four well types shows no significant differences, saturation reaches about 40 percent.

Table 16. *Comparison of well types (total number of wells), temperature in degrees centigrade, total dissolved solids (= TDS) me/l, and oxygen content in mg/l.* HA = hand-dug artesian well, HP = hand-dug pumping well, DA = drilled artesian well, DP = drilled pumping well

| | HA (9) | HP (4) | DA (8) | DP (6) |
|---|---|---|---|---|
| Temp | 32.8 ± 1.3 | 27.5 ± 0.8 | 35.2 ± 1.4 | 35.0 ± 1.7 |
| TDS | 41.75 ± 6.15 | 33.41 ± 3.11 | 41.61 ± 7.27 | 29.50 ± 0.67 |
| $O_2$ | 3.4 ± 1.0 | 4.2 ± 2.4 | 2.7 ± 0.4 | 3.0 ± 0.3 |

Concerning the individual total concentration the waters show three groups, i.e., between 28 and 31 me/l, 37 and 45 me/l and around 52 me/l (Fig. 33). Only three out of twenty-eight water samples show other values, one of these samples was collected from a pond (No. 9); in this case the higher salt content can be caused by leaching out the soil, evaporation and recycling. The three ranges of the total concentration agree with the groups of electrical conductivity described before (Fig. 26). The low-mineralized waters lie in the west around Al Ajam (samples No. 10, 11, 12, 68, 69, 71, 72, 74), in the south near Al Khuwailidiyah (19), south of Al Qatif (66) and near Hillat

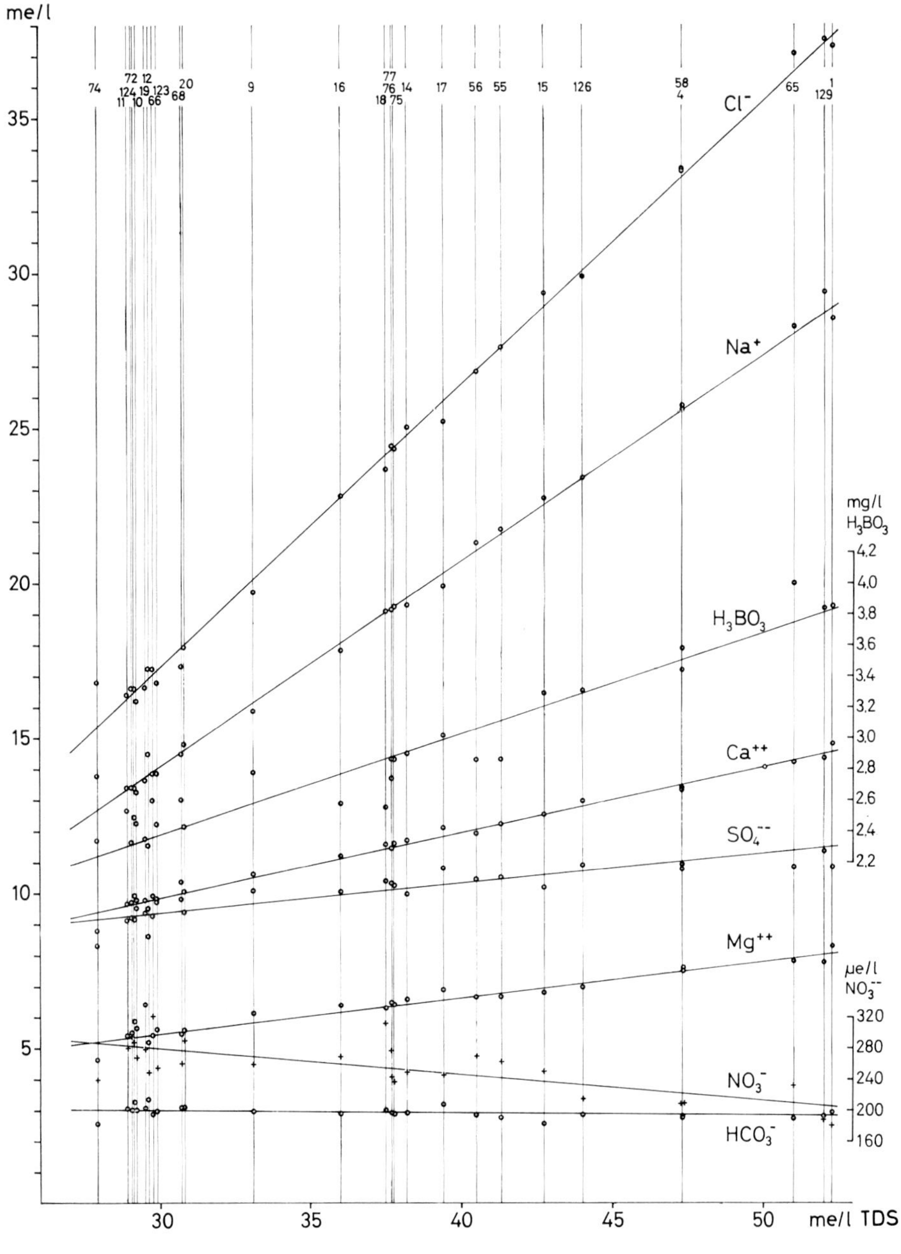


Fig. 33. Chemical characteristics of Al Qatif waters (see text)

Muhaish (20). These waters are perhaps at the margin of the large tongue of low-mineralized waters shown by A. I. NAIMI (1965) which infiltrate the coastal belt area from the south. The waters with the higher salt content discharge in the NE, where a groundwater stream of highly mineralized waters seems to come from the north towards the study area. The mixing of these two components results in a zone of waters with a concentration

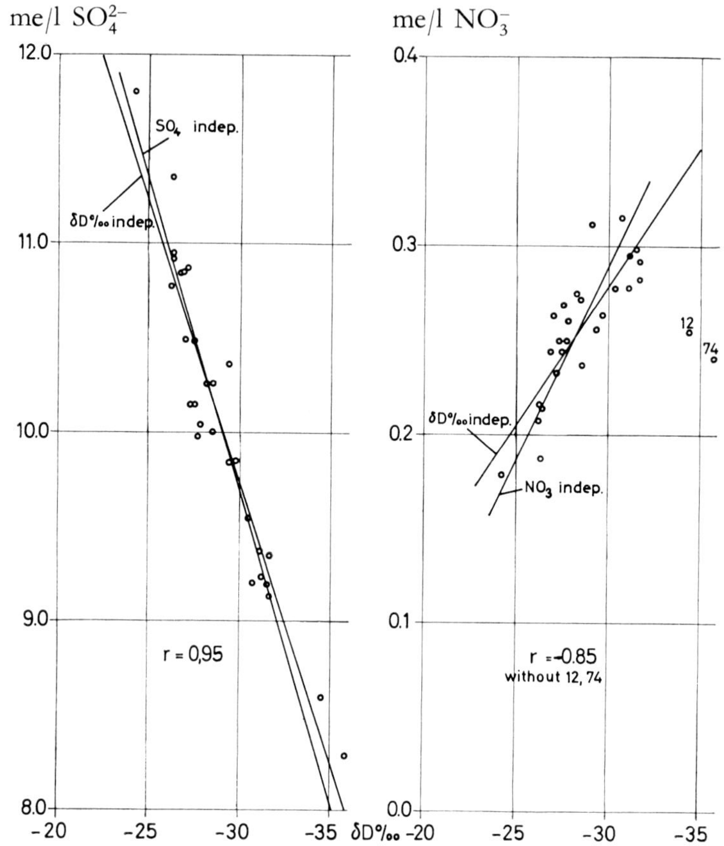


Fig. 34. The positive correlation of the deuterium and sulfate content and the negative correlation between nitrate and deuterium of Al Qatif waters

increasing towards the northeast (samples No. 1, 4, 14–18, 55, 56, 58, 65, 75–77, 126, 129). The existence of mixed waters is shown by the decreasing nitrate content parallel to the increase of total concentration (correlation coefficient $r = r_s = -0.84$)[1] from which it may be concluded that a component of salty water low in nitrate is mixed with a less salty one enriched with nitrate. Furthermore, the increasing content of stable isotopes combined with an increasing total salt concentration, reported also by T. DINÇER et al. (1974), supports this conclusion. The correlation of the deuterium and the

[1] Limit of significance for $n = 28$, $\alpha = 0.01$: $r = 0.48$, $r_s = 0.44$.

salt content of the water samples collected from Al Qatif region was even stronger than the well-known correlation for deuterium and $^{18}O$ ($r = 0.90$, $r_s = 0.93$ in comparison with $r = 0.81$, $r_s = 0.76$ respectively) [1]. There also exists a clear positive connection of the deuterium and sulfate content ($r = 0.95$, $r_s = 0.92$, Fig. 34). In contrast, there is a negative correlation between nitrate and deuterium ($r = -0.85$, $r_s = -0.87$, Fig. 34). The inversion of the variables, nitrate and the total salt concentration makes it possible to estimate the composition of the mixing components (Fig. 35).

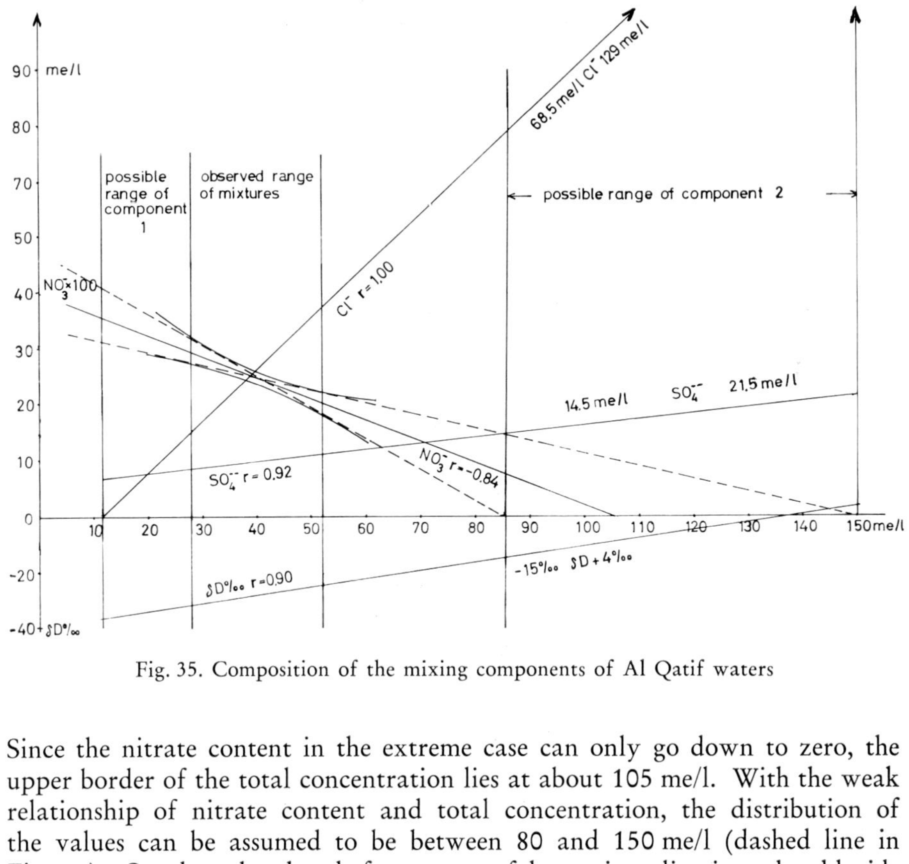


Fig. 35. Composition of the mixing components of Al Qatif waters

Since the nitrate content in the extreme case can only go down to zero, the upper border of the total concentration lies at about 105 me/l. With the weak relationship of nitrate content and total concentration, the distribution of the values can be assumed to be between 80 and 150 me/l (dashed line in Fig. 35). On the other hand, for waters of low mineralization, the chloride content in the extreme case will be zero; in this case the total concentration amounts to approximately 11 me/l with not much variation because the correlation of the total concentration and the chloride content is very strong ($r = 0.999$, $r_s = 0.995$).

Values reported from ITALCONSULT reveal that the concentration of waters originating directly from the Umm er Radhuma aquifer in some places are indeed higher than 80 me/l. The test well Q 15 U/C near Umm As Sahik produced waters of very different salt content between a depth of 263 down to 600 meters (Fig. 36). Water samples collected from depths of 454 and

[1] Limit of significance for $n = 28$, $\alpha = 0.01$: $r = 0.48$, $r_s = 0.44$.

600 meters respectively had a total concentration and ion distribution similar to water samples from Al Qatif, Al Ajam and the samples taken from 'ayns near Umm As Sahik (No. 6, 7, 8, 142). In contrast, water samples collected at depths of 363 and 404 meters reached mineralizations of 63 me/l and 92 me/l respectively. Also samples collected at depths of 500 and 540 meters had a salt concentration of three times more than the sample collected from a depth of 600 meters. As the waters from this rather deep borehole show the same relations between the individual ion content and the total concentration as the water samples collected in Al Qatif, Al Ajam, Tarut and Umm As Sahik, it is assumed, that they all belong to only one large water family of the third order (mixed waters). Since the borehole Q 15 U/C was closed with cement no samples could be collected for the measurements of nitrate and isotope content.

All the measurements mentioned above show that the various salt contents of the coastal belt waters have geochemical causes. Nevertheless, the possibility of infiltration of water from the drainage system as well as from the sea had to be investigated. Water from irrigation crosses in many small channels, infiltrating into the subsoil. This infiltration water is recollected by a drainage system carrying the waste water to the Gulf. If the drainage water infiltrates back into the wells (recycling), the wells discharging high-mineralized waters should have an ion distribution like the water in the drainage channels because in these also the overflow of wells mixes with the irrigation drainage. The salt concentration of the drainage waters (samples No. 60, 62, 132–140) is generally higher than in the well water and increases downstream. Very serviceable samples for the comparison are No. 60, 136, and 137 because their mineralization is approximately equal to the salt content of the most mineralized well waters (No. 1, 5, 13).

As can be seen from Table 17 there are various differentiating factors between the water of the drainage channels and the water collected from the wells. As a result of the uptake of biogenetic $CO_2$ from the soil, the content of hydrogen carbonate of the drainage water is about 25 percent higher than that of the well water. Of biogenetic relevance is the fact that the nitrate content of the drainage water is about 40 percent less than that of the well waters since nitrate is absorbed by plants and microorganisms. Through evaporation of irrigation waters, gypsum, calcium fluoride and siliceous acid become enriched in the soil and therefore also in the drainage water. Thus the chemical composition of the drainage water deviates in so many details from the highly mineralized well waters, that their high salt content cannot be explained with recycling. This is also proved by the fact that the deuterium content of the drainage and the well water of high mineralization is similar (— 25.1 and — 25.6 $\delta$D per mill respectively), although the sulfate content is different (15.3 me/l and 11.4 me/l respectively) [1]. Thus, the drainage water does not show the correlation of sulfate and deuterium content which characterizes the increase of mineralization in the well waters.

---

[1] The isotope data reported here were measured by W. Stichler, Institute for Radiohydrometry, Munich.

Any infiltration of the water from the Arabian Gulf and the Tarut Bay should primarily influence the wells near the beach. In the southern parts of Al Qatif oasis, however, wells situated near the beach discharge water with a low mineralization. Furthermore, Fig. 26 shows the mineralization increasing parallel to the coastal line and not towards the coast. The best proof, however, that sea water intrusion cannot be the cause of salination of Al Qatif wells is gained from the behaviour of sulphur isotopes ($\delta\,^{34}S$ permil) and chloride content respectively [1]. The Bay of Dawhat Zaloom south of

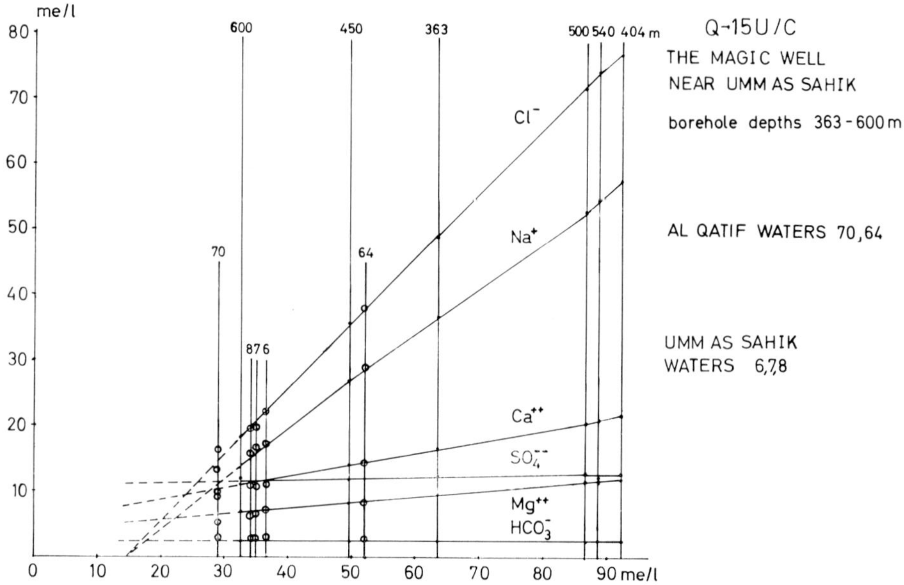


Fig. 36. Ion-concentration of waters originating from the Umm er Radhuma aquifer (reported from ITALCONSULT, 1969)

Dhahran is surrounded only by sand dunes and has no coastal springs. There, the sea water shows a chloride content of 967 me/l and a $\delta\,^{34}S$ content of 20.2 per mill. In Al Qatif wells the chloride concentration ranges between 16.2 and 37.2 me/l and the $\delta\,^{34}S$ between $9.0 \pm 0.2$ per mill and $9.8 \pm 0.2$ per mill respectively. A salt water infiltration from the Gulf into the wells should imply a parallel enrichment of chloride as well as $^{34}S$. This is not the case: 22 water samples showed no significant relation of chloride and the sulphur isotope ($r = 0.16$, $r_s = 0.09$) [2]. Thus, saltwater intrusion can be excluded.

The water of the Tarut Bay is a mixture of approximately 90 percent drainage water from the oasis channels and 10 percent sea water from the Gulf. This mixture relationship can be calculated with the help of the mixing

[1] The $^{34}S$ values used in this chapter were measured by E. Pak, Department of Nuclear Physics, Vienna. For details see chapter 2.1.8.

[2] Limit of significance for $n = 22$, $\alpha = 0.05$: $r = 0.42$, $r_s = 0.36$.

Table 17. *Differences of ion composition ($\mu e/l$) between three well waters of highest mineralization and drainage waters of similar total salt concentration.* $H_3BO_3$ and $H_2SiO_3$ in mg/l. Averages = relative composition in % of the total mineralization (see text)

| | $Na^+$ | $K^+$ | $Ca^{++}$ | $Mg^{++}$ | SK | $Cl^-$ | $Br^-$ | $SO_4^{--}$ | $HCO_3^-$ | $NO_3^-$ | $F^-$ | SA | $H_3BO_3$ | $H_2SiO_3$ |
|---|---|---|---|---|---|---|---|---|---|---|---|---|---|---|
| Drainage waters | | | | | | | | | | | | | | |
| 60 | 25,084 | 675 | 17,800 | 7,960 | 51,519 | 30,187 | 52.6 | 17,297 | 3,946 | 121 | 78 | 51,682 | 3.89 | 29.2 |
| 136 | 24,467 | 717 | 17,840 | 8,880 | 51,904 | 31,786 | 53.6 | 15,840 | 3,920 | 107 | 66 | 51,773 | 3.98 | 21.4 |
| 137 | 28,273 | 760 | 15,240 | 8,360 | 52,633 | 36,252 | 68.2 | 12,650 | 3,520 | 131 | 40 | 52,661 | 3.98 | 14.4 |
| Average | | | | | | | | | | | | | | |
| μe/l | 25,941 | 717 | 16,960 | 8,400 | 52,019 | 32,742 | 58.1 | 15,262 | 3,795 | 120 | 61 | 52,039 | 3.95 | 21.7 |
| μe % | 49.9 | 1.4 | 32.6 | 16.1 | 100 | 62.9 | 0.11 | 29.3 | 7.3 | 0.23 | 0.12 | 99.96 | | |
| Wells | | | | | | | | | | | | | | |
| 1 | 28,491 | 725 | 14,800 | 8,320 | 52,336 | 37,285 | 75.6 | 11,799 | 2,901 | 179 | 40 | 52,280 | 3.82 | 15.5 |
| 5 | 28,926 | 774 | 14,320 | 8,100 | 52,120 | 37,923 | 76,3 | 11,164 | 2,795 | 198 | 39 | 52,195 | 3.96 | 15.7 |
| 13 | 28,298 | 753 | 14,140 | 7,940 | 51,131 | 37,120 | 74.2 | 10,835 | 2,861 | 226 | 40 | 51,156 | 3.77 | 15.6 |
| Average | | | | | | | | | | | | | | |
| μe/l | 28,572 | 751 | 14,420 | 8,120 | 51,862 | 37,443 | 75.4 | 11,266 | 2,852 | 201 | 40 | 51,877 | 3.85 | 15.6 |
| μe % | 55.1 | 1.4 | 27.8 | 15.7 | 100 | 72.2 | 0.15 | 21.7 | 5.5 | 0.39 | 0.08 | 100.02 | | |

rule using the data of Table 18 [1]. The values agree for Na, K, Ca, Mg, Cl, $SO_4$ and metaboric acid content. Concerning $\delta\,^{34}S$ there is a much stronger enrichment in the Tarut Bay (14.4 ± 0.2) than could be expected from a mixture relation of 90 percent drainage water (9.4 per mill $\delta\,^{34}S$) and 10 percent sea water from the Gulf (20.2 per mill $\delta\,^{34}S$). The reason for this is probably a process of biologic assimilation prefering the (light)

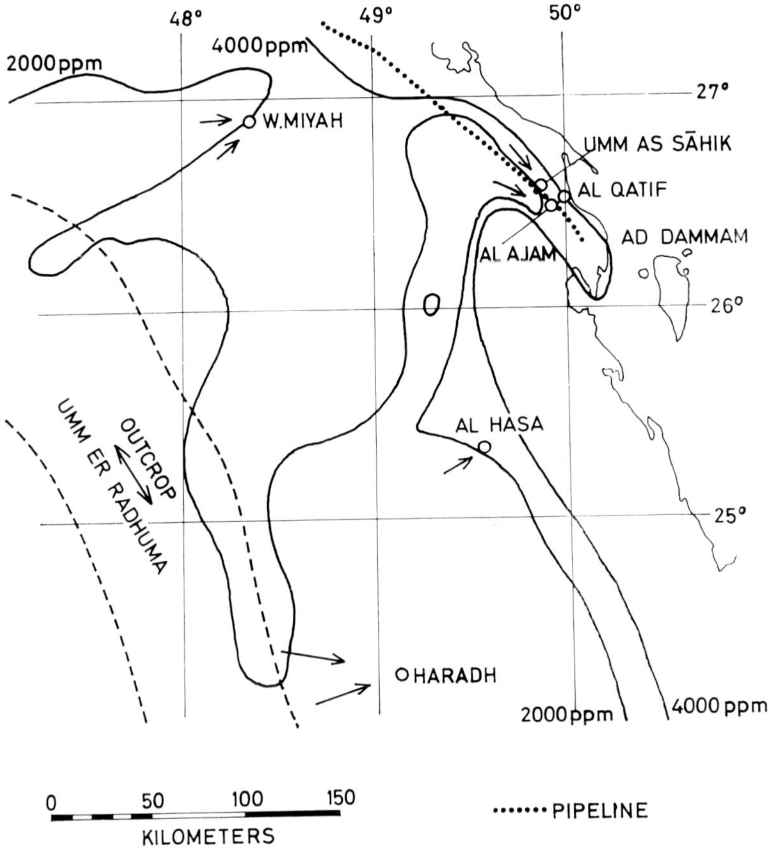


Fig. 37. Mineralization and flow directions of underground waters between the outcrop region of the Umm Er Radhuma Formation and the Gulf coastal area (according to A. I. NAIMI, 1965)

isotope $^{32}S$. It is assumed that the nitrate content is also influenced by biologic processes, as the concentration in Tarut Bay water is lower than predicted by the mixture relation discussed previously. Any saltwater intrusion from the Tarut Bay into the wells of the oases need not be considered due to the same arguments which were mentioned against a mixing with drainage or sea water from the Gulf.

[1] Volume percent drainage $= \frac{100\,(\text{me/l Gulf} - \text{me/l Tarut})}{\text{me/l Gulf} - \text{me/l drainage}}$.

Table 18. *Chemical composition of waters from Tarut Bay and sea water of the Bay of Dawhat Zaloom* (Arabian Gulf). Ion concentration in me/l. $H_2SiO_3$ and $H_3BO_3$ in mg/l. For comparison also the composition of the Standard Mean Ocean Water is given. The total salt concentration of the Gulf water is 1.8 times higher than for normal ocean water

| | $Na^+$ | $K^+$ | $Ca^{++}$ | $Mg^{++}$ | SK | $Cl^-$ | $Br^-$ | $SO_4^{--}$ | $HCO_3^-$ | $NO_3^-$ | $F^-$ | SA | $H_3BO_3$ | $H_2SiO_3$ |
|---|---|---|---|---|---|---|---|---|---|---|---|---|---|---|
| Tarut Bay | | | | | | | | | | | | | | |
| 144 | 115.70 | 2.88 | 18.47 | 28.67 | 165.72 | 139.37 | 0.23 | 23.05 | 3.23 | 0.04 | 0.06 | 165.98 | 8.63 | 16.7 |
| Mean Ocean | 459.33 | 9.72 | 19.96 | 104.44 | 593.45 | 535.25 | 0.81 | 55.16 | 2.29 | — | 0.07 | 593.58 | 27.4 | — |
| Gulf, Bay of Dawhat Zaloom | | | | | | | | | | | | | | |
| 1974 | | | | | | | | | | | | | | |
| 67 | 832.97 | 18.93 | 33.57 | 188.83 | 1,074.30 | 970.57 | 1.43 | 94.40 | 2.69 | 0.00 | 0.08 | 1,069.17 | 41.71 | 0.0 |
| 1975 | | | | | | | | | | | | | | |
| 145 | 839.50 | 18.11 | 33.73 | 187.15 | 1,078.49 | 962.55 | 1.45 | 97.19 | 2.66 | 0.00 | 0.08 | 1,063.93 | 41.71 | 0.8 |

With regard to the hydrochemical situation of Al Qatif region the following facts can be summarized: all aquifers of the limestone layers of the Tertiary strata are hydraulically connected and discharge water from the Umm er Radhuma Formation. The strata series contains, regionally as well as vertically, waters of different salt content due to local differences in geochemical composition of the rock material. Interspersed evaporite layers may account for the exceptionally high water mineralization at some depths. Seawater intrusion and seepage of drainage waters are of no significant influence on the salt content of the karst springs and wells. Salty underground waters showing an enrichment of heavy isotopes mix with waters of a lower salt and isotope content. The general flow direction of the first component is from the north, whereas the second comes from the south (Fig. 37). The water of the Umm er Radhuma aquifer is of older storage, which can be concluded from the low $^{14}C$ content (two samples with $3.5 \pm 2.4$ and $< 1.2$ percent modern respectively) [1]. This corresponds to an age greater than 22,000 years if the initial carbon content is assumed to be 85 percent of the modern standard. The fact that the deuterium content (average $\delta D$ per mill $= -28.9 \pm 2.6$) is considerably lower than in recent groundwaters ($< -11$ per mill) points in the same direction (see also Chapter 2.1.8.).

### *2.1.6.2. Al Hasa oasis*

*Introduction*

The Al Hasa oasis is situated some 60 kilometers inland of the Gulf coast, west of the vast sand desert of Al Jafurah. Plantations, although sometimes interrupted by stretches of sandy dunes, extend from the provincial town of Al Hofuf (150 meters above sea level) to approximately 30 kilometers north (northern oasis) and to approximately 20 kilometers east (eastern oasis). Some 200 square kilometers may be considered to be arable land. The western border is formed by a desert strip and the foremost escarpment of the Shedgum Plateau. The area is otherwise characterized by erosional remnants of the plateau, inselbergs, bizarre-looking rocky towers as well as caves in the east and the south of the oasis at some 240 meters above sea level (Jabal Burayqa, Jabal Duhnah).

*Results gained in earlier investigations*

The springs of the northern oasis originate in a NW-striking joint system, parallel to the oil-bearing Ghawar anticline; the springs of the eastern oasis have their origin along a fault system crossing the anticline rectangularly. The spring area forms the surface of a heavily karstified marly limestone of presumable Pliocene origin (Hofuf formation, Neogene), and the springs themselves discharge from vast solution caves and karst tubes (see 2.1.3.). The water partly rises from joints, and partly penetrates laterally through extended sheet joints into the caves. According to the divers' report these two sources sometimes have different temperatures.

---

[1] Measured by W. Rauert, Institute of Radiohydrometry, Munich.

Test boreholes made by ITALCONSULT reached the Alat and Khobar strata of the older Tertiary Dammam Formation after drilling through approximately 180 meters of Neogene. At a depth of some 280 meters the boreholes entered the Umm er Radhuma Formation which at this level has a thickness of about 320 meters. Drillings reaching down to an even greater depth (600 meters) hit the Aruma Formation, and at a depth of 900 meters the Wasia aquifer was finally met. Static water level, mineralization and temperatures of the test drillings are summarized in Table 19. The water of the Umm er Radhuma Formation has a relatively low mineralization and its temperature is approximately 39 °C. In the overlying Alat and Khobar

Table 19. *Geologic strata, water temperature, total ion-concentration (TDS in g/l) and static water level of test drillings in the Al Hasa oasis (data from ITALCONSULT reports)*

| Aquifer | Depth in meters | Temp. °C | T.D.S. g/l | Static water level, meters above sea level |
|---|---|---|---|---|
| Neogen | 0–180 | 29 | 1.415 | 145.6 [1] |
| Alat | 180–190 | 34 | 1.140 | 133.7 [1] |
| Khobar | 200–230 | 35 | 1.040 | 139.8 [1] |
| Umm er Radhuma | 280–600 | 39 | 0.900 | 154.0 [1] |
| Wasia | 900–1,330 | — | 17.651 | 196.0 |

[1] Average from two boreholes.

strata mineralization is slightly higher, whereas temperatures are lower. The temperature of 29 °C given for the Neogene is valid for a few boreholes only. In contrast, several springs discharging from these strata reach temperatures up to 40 °C. In the Neogene, the static water level is markedly higher than in the Alat and Khobar strata, but lower than in the Umm er Radhuma Formation. Chances are that both Alat and Khobar strata are partly interfused with major joints through which the Umm er Radhuma water rises up to the Neogene without any significant loss of pressure. T. Dinçer and his team indicate that the Alat and Khobar aquifer cannot significantly contribute to the discharge of the Al Hasa springs because the water pressure tends to be lower in these strata than in the Neogene from where the springs emerge. It is therefore assumed that the water originates from the Umm er Radhuma aquifer and that both Neogene and the Alat and Khobar strata act just as conducting ways. Water temperatures also say much in favour that the origin of the water must be in the Umm er Radhuma Formation. The warmest springs (37 ° to 40 °C) emerging mainly in the west of the oasis exceed the yearly average of air temperature by approximately 10 ° to 12 °C. Storage at a depth of 300 to 400 meters will account for this temperature difference if a geothermal gradient of 33 m/°C is assumed. According to the drilling profile this is the depth of the Umm er Radhuma. A few boreholes, however, produce considerably cooler waters (29 ° to 32 °C). The regional grouping of cooler waters in the east of the oasis leads to the assumption that these waters emerge from the vertical main joints and cool down while running laterally through the horizontal waterways in Neogene sediments.

Most of the boreholes reach 100 meters down into the Neogene. Data on static water level, temperature and mineralization are available on 18 of these wells. As exemplified by the left side of Fig. 38, the cool wells show a lower static water level. This further substantiates the assumption that the cooler springs are supplied from the large vertical joints some distance away from the point of emergence. Concerning mineralization as well as electrolytic conductivity, no differences were detected (Fig. 38, right side).

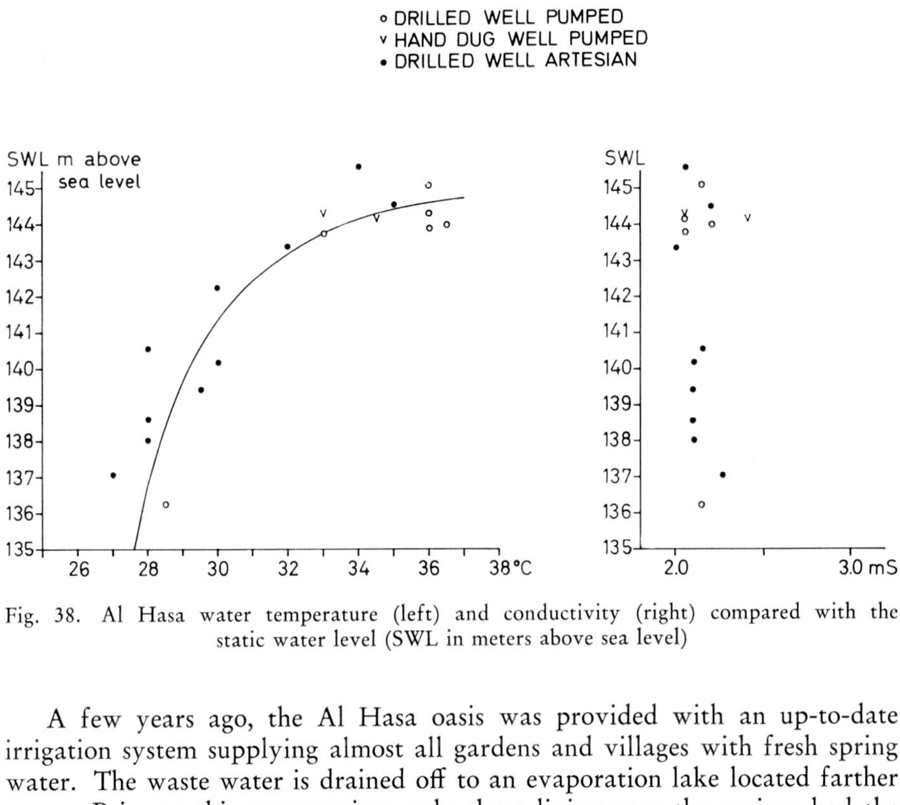


Fig. 38. Al Hasa water temperature (left) and conductivity (right) compared with the static water level (SWL in meters above sea level)

A few years ago, the Al Hasa oasis was provided with an up-to-date irrigation system supplying almost all gardens and villages with fresh spring water. The waste water is drained off to an evaporation lake located farther away. Prior to this construction, only those living near the springs had the benefit of being supplied with fresh water, whereas others had to put up with drainage water which finally oozed away in a strongly mineralized state at the outskirts of the oasis. In those areas and in the flat pans of the oasis the groundwater level rose close to the soil surface in cooler seasons since surface runoff channels were lacking. Sometimes, even salt lakes and salty swamps formed, depriving the inhabitants of large areas of arable land.

*Hydrochemical investigations*

Most waters investigated were taken from the middle section of the northern oasis in the vicinity of the villages of Al Mutairifi and Ash Sha'bah (Fig. 39). From the 18 water samples collected in this area, 4 were taken from hand-dug wells (HA 33, 38–40), 6 from drilled artesian wells

(DA 24—28, 32), 3 from hand-dug pumped wells (HP 35 and 35 a from the same well, respectively, 36) and 5 in wells pumped through boreholes (DP 29–31, 34, 37). Two samples were collected in the eastern oasis: one from 'Ayn Khudud (HA 41), the most productive spring of the oasis with an output of 1.7 cubic meters per second; and the second from 'Ayn Najm (HP 42), the warmest spring of the whole area known so far (40.2 °C). 'Ayn Najm is used as a remedy against muscle and joint diseases by the population.

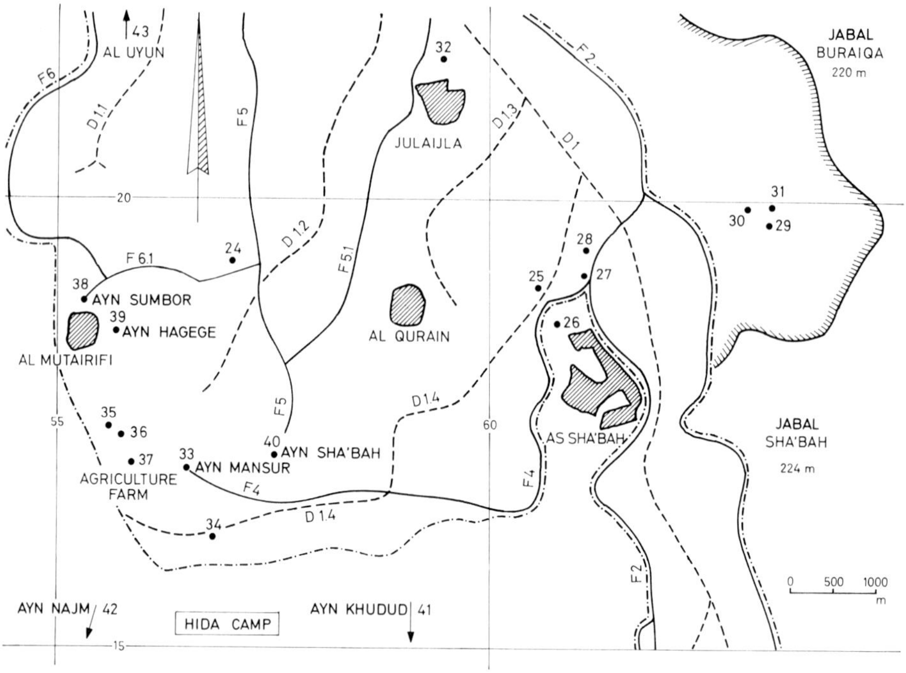


Fig. 39. Villages, wells and springs ('ayns), irrigation (full lines) and drainage channels (dashed lines) of Al Hasa area

Another sample was collected in the area of Al 'Uyun (DA 43) at the northernmost border of the oasis, where the cultivated area adjoins the desert where it is interspersed with salt swamps.

The analytical results are summarized in Table 20 and shown in Fig. 40. Apart from a few special cases discussed below, the chemistry of these waters is more or less the same so that these waters may be regarded as a group of the first order. Their total concentration fluctuates between 22.5 and 24.6 me/l. On the anion side chlorides and sulfates prevail, whereas on the cation side sodium and calcium ions dominate. Ion distributions are similar to the Al Ajam waters (Table 21), which is an argument in favor of the fact that both waters have their origin in the same formation. However, regarding the $\delta\,^{34}S$ values, the Al Hasa waters (two samples with 12.4 per mill) prove to be of a strikingly different character from Al Qatif group (9.0 to 9.8 per mill).

The water samples taken from wells 35 and 36 with their differing chemical properties furnish a good example of the reflux of irrigation water into the wells, a process that has become known as "recycling". Such a reflux may occur if the water level drops sharply during pumping and rises only

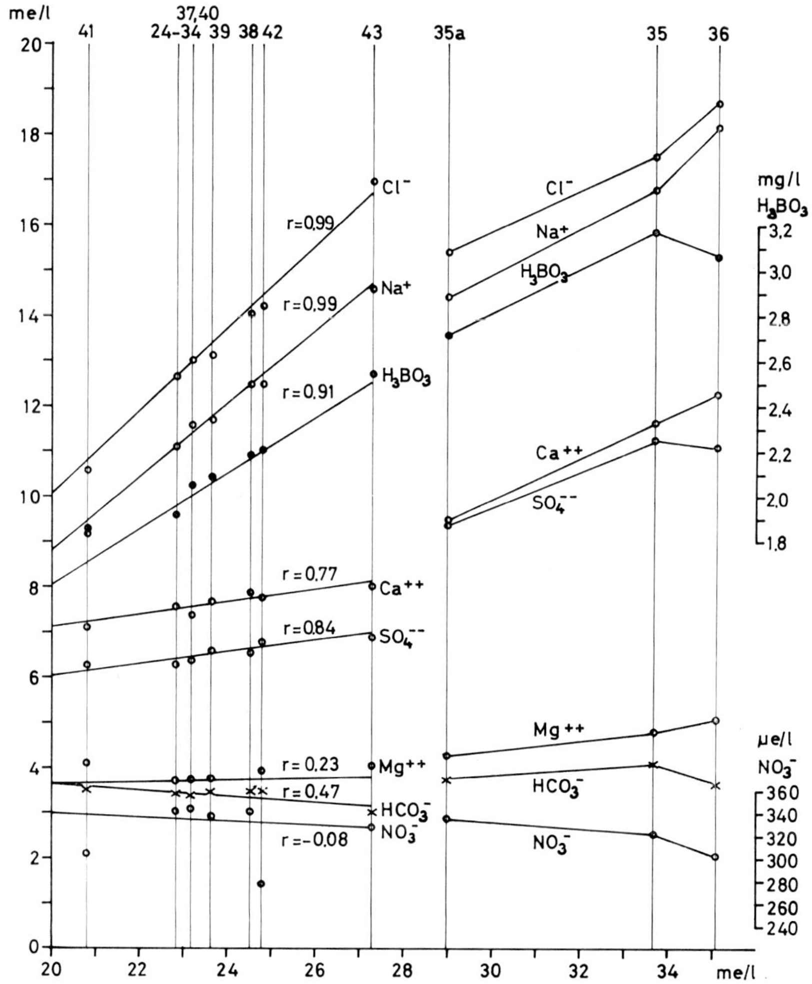


Fig. 40. Comparison of analytical results of Al Hasa water samples

very slowly after it, owing to a small recharge and overpumping. Insufficient drainage of irrigation water seeping into the soil favors the reflux. In contrast to the other 'ayns the water from wells 35 and 36 does not flow out freely and the water level is some 3 meters below the surface. Wells are pumped twice a day to irrigate small palm groves which are not connected to the general irrigation and drainage network. During pumping the water level dropped to 7 meters below the ground level. The water temperature of well

Table 20. *Chemical composition of waters of the Al Hasa area.* No. of sample, W = well type (see Table 12), °C = water temperature in degrees centigrade, $O_2$-content in mg/l, ion concentrations and sum of cations (SK) and anions (SA) in μe/l, $H_3BO_3$ and $H_2SiO_3$ in mg/l

| No. | W | °C | $O_2$ | $Na^+$ | $K^+$ | $Ca^{++}$ | $Mg^{++}$ | SK | $Cl^-$ | $Br^-$ | $SO_4^{--}$ | $HCO_3^-$ | $NO_3^-$ | $F^-$ | SA | $H_3BO_3$ | $H_2SiO_3$ |
|---|---|---|---|---|---|---|---|---|---|---|---|---|---|---|---|---|---|
| 24 | DA | 32.0 | 4.6 | 11,092 | 506 | 7,520 | 3,720 | 22,838 | 12,783 | 16.7 | 6,120 | 3,592 | 252 | 31 | 22,795 | 1.87 | 20.4 |
| 25 | DA | 31.6 | 4.5 | 10,874 | 478 | 7,440 | 3,680 | 22,472 | 12,563 | 17.3 | 6,257 | 3,285 | 316 | 33 | 22,471 | 1.87 | 20.4 |
| 26 | DA | 29.3 | 5.8 | 11,092 | 506 | 7,200 | 3,720 | 22,518 | 12,622 | 17.7 | 6,313 | 3,169 | 359 | 33 | 22,514 | 1.86 | 20.5 |
| 27 | DA | 29.0 | 6.0 | 11,092 | 506 | 7,500 | 3,720 | 22,818 | 12,622 | 18.4 | 6,317 | 3,451 | 363 | 33 | 22,804 | 1.79 | 20.4 |
| 28 | DA | 28.2 | 4.6 | 11,227 | 492 | 7,760 | 3,560 | 23,039 | 12,702 | 17.7 | 6,337 | 3,557 | 353 | 33 | 23,000 | 1.95 | 20.5 |
| 29 | DP | 29.7 | 4.4 | 10,874 | 492 | 7,600 | 3,760 | 22,726 | 12,542 | 17.7 | 6,305 | 3,552 | 359 | 33 | 22,809 | 1.88 | 20.4 |
| 30 | DP | 28.7 | 5.2 | 10,992 | 492 | 7,800 | 3,600 | 22,884 | 12,541 | 18.7 | 6,297 | 3,572 | 361 | 33 | 22,823 | 1.86 | 20.4 |
| 31 | DP | 29.0 | 5.2 | 11,109 | 492 | 7,680 | 3,720 | 23,001 | 12,701 | 18.7 | 6,281 | 3,587 | 355 | 33 | 22,976 | 1.96 | 20.7 |
| 32 | DA | 28.0 | 4.7 | 11,209 | 506 | 7,600 | 3,720 | 23,035 | 12,701 | 19.0 | 6,273 | 3,547 | 359 | 33 | 22,932 | 1.99 | 20.4 |
| 33 | HA | 38.6 | 5.0 | 11,309 | 520 | 7,600 | 3,580 | 23,009 | 12,781 | 19.0 | 6,385 | 3,411 | 333 | 33 | 22,962 | 2.04 | 20.4 |
| 34 | DP | 37.0 | 4.7 | 11,309 | 520 | 7,400 | 3,800 | 23,029 | 12,782 | 18.3 | 6,353 | 3,411 | 347 | 33 | 22,944 | 1.99 | 20.4 |
| 35 | HP | 29.0 | 5.4 | 16,790 | 774 | 11,320 | 4,800 | 33,684 | 17,497 | 22.7 | 11,654 | 4,134 | 323 | 33 | 33,664 | 3.17 | 21.6 |
| 35 a | HP | 33.7 | 5.2 | 14,441 | 661 | 9,520 | 4,340 | 28,962 | 15,419 | 21.0 | 9,434 | 3,744 | 335 | 33 | 28,986 | 2.72 | 20.8 |
| 36 | HP | 32.6 | 3.1 | 18,168 | 788 | 11,100 | 5,100 | 35,156 | 18,696 | 23.7 | 12,297 | 3,784 | 304 | 33 | 35,138 | 3.06 | 21.0 |
| 37 | DP | 39.0 | 5.3 | 11,644 | 520 | 7,280 | 3,760 | 23,204 | 13,102 | 17.7 | 6,385 | 3,280 | 341 | 32 | 23,158 | 2.05 | 20.4 |
| 38 | HA | 36.0 | 4.3 | 12,514 | 549 | 7,920 | 3,600 | 24,583 | 14,063 | 17.0 | 6,602 | 3,491 | 341 | 32 | 24,546 | 2.19 | 20.9 |
| 39 | HA | 37.0 | 4.3 | 11,744 | 512 | 7,680 | 3,760 | 23,696 | 13,103 | 16.8 | 6,634 | 3,471 | 339 | 32 | 23,596 | 2.08 | 20.4 |
| 40 | HA | 37.0 | 3.7 | 11,483 | 512 | 7,500 | 3,740 | 23,235 | 12,856 | 16.7 | 6,506 | 3,401 | 347 | 32 | 23,159 | 2.04 | 20.4 |
| 42 | HP | 40.2 | 4.0 | 12,514 | 563 | 7,760 | 3,960 | 24,797 | 10,631 | 8.7 | 6,305 | 3,542 | 304 | 32 | 20,823 | 1.86 | 18.9 |
| 41 | HA | 31.8 | 1.2 | 9,134 | 394 | 7,120 | 4,120 | 20,768 | 14,173 | 21.0 | 6,823 | 3,491 | 278 | 33 | 24,819 | 2.20 | 21.0 |
| 43 | DA | 28.5 | 7.2 | 14,615 | 593 | 8,040 | 4,080 | 27,328 | 16,933 | 26.5 | 6,932 | 3,058 | 327 | 37 | 27,314 | 2.54 | 20.4 |

Table 21. *Comparison of the average ion (me/l), $H_3BO_3$ and $H_2SiO_3$ (mg/l) concentrations of Al Ajam waters (Nos. 10, 11, 12, 68, 72, 74, 123, 124) and Al Hasa samples (Nos. 24–34, 37, 40)*

| | | $Na^+$ | $K^+$ | $Ca^{++}$ | $Mg^{++}$ | SK | $Cl^-$ | $Br^-$ | $SO_4^{--}$ | $HCO_3^-$ | $NO_3^-$ | $F^-$ | SA | $H_3BO_3$ | $H_2SiO_3$ |
|---|---|---|---|---|---|---|---|---|---|---|---|---|---|---|---|
| | $\bar{x}$ | 13.76 | 0.5 | 9.72 | 5.34 | 29.32 | 16.76 | 0.025 | 9.21 | 3.02 | 0.273 | 0.040 | 29.33 | 2.44 | 17.5 |
| Al Ajam | ± | 0.5 | 0.04 | 0.45 | 0.35 | 0.90 | 0.39 | 0.0018 | 0.55 | 0.20 | 0.025 | 0.0024 | 0.74 | 0.1 | 0.74 |
| | me % | 46.9 | 1.7 | 33.2 | 18.2 | 100 | 57.13 | 0.085 | 31.4 | 10.3 | 0.93 | 0.14 | 100 | | |
| | $\bar{x}$ | 11.18 | 0.5 | 7.53 | 3.70 | 12.72 | 22.91 | 0.018 | 6.32 | 3.45 | 0.340 | 0.033 | 22.87 | 1.93 | 20.4 |
| Al Hasa | ± | 0.22 | 0.04 | 0.17 | 0.07 | 0.24 | 0.15 | 0.0006 | 0.08 | 0.14 | 0.03 | 0.0006 | 0.21 | 0.08 | 0.09 |
| | me % | 48.8 | 2.2 | 32.9 | 16.1 | 100 | 55.6 | 0.079 | 27.6 | 15.1 | 1.49 | 0.14 | 100 | | |

No. 35 was 29 °C right after the beginning of pumping and rose to 33.7 °C during the next half hour. The sample collected immediately after having started the pump (35) showed a total mineralization of some 33.7 me/l, whereas the sample taken after a half hour (35 a) had just 29 me/l. The relative ion composition was almost the same for both samples. Therefore the high initial concentration is probably due to evaporation. Both dug-well samples (35, 36) also showed a slight enrichment of stable isotopes compared with the neighboring drilled well (37). However, the evaporation cannot have taken place in the well shaft because of the well shaft narrowness. Thus it has to be assumed that besides fresh spring water, evaporated drainage water gains access to the well. Similar conditions may be responsible for the strong mineralization of well No. 36. Also in this case the water sample was collected immediately after the beginning of pumping. The slightly increasing contents of both hydrogen carbonate and sulfate in samples 35 and 36 respectively are also symptoms of contamination with seeping irrigation water. This criterion was also taken into account during discussions on drainage waters of Al Qatif oasis.

The hand-dug well No. 41 ('Ayn Khudud) is situated in the south of the oasis, just about 1 kilometer east of Al Hofuf. Several rather productive springs rise in the vicinity; 'Ayn Haql (1.3 cubic meters per second), 'Ayn Barabi (0.57 cubic meters per second), 'Ayn Ta'adhid (0.29 cubic meters per second). No doubt, these wells would justify extra investigation, as 'Ayn Khudud is characterized by a particularly low oxygen content (1.2 mg/l) and the lowest mineralization level (20.77 me/l) of all Al Hasa wells.

The drilled artesian well No. 43 is located at the northwest border of the oasis, north of the village of Al 'Uyun. The total ion content (27.35 me/l) of this water is relatively high (if one leaves the wells 35 and 36 and their special conditions aside) which may presumably be attributed to the soil salinity that seems to be more intense at the outskirts of the oasis.

### *Regional differences of wells investigated*

A glance at the map (Fig. 39) reveals that the various wells are not regularly distributed across the area investigated, but accumulated in the west around Al Mutairifi and the east around Ash Sha'bah. If a comparison is made between the quality of these two groups, then the fact will clearly emerge that the eastern wells are drilled wells without casing, whereas the wells in the west are, with one exception, springs discharging from joints. Both groups produce water with the same chemical properties; their only difference lies in their temperatures (Table 22). The eastern water group is cooler by some 7.5 °C, which represents a striking difference in view of the closeness of the two groups. Considering their equal chemical properties it is hardly imaginable that they originate from different aquifers. Thus it may be assumed that the higher temperature of the western wells is caused by the water rising rapidly and directly from a deep aquifer, while the water present in the eastern wells is cooler because it has been transported laterally through more superficial sheet joints.

Table 22. *Statistical comparison of the water temperature (°C), $O_2$-content ($O_2$%) and total mineralization (me/l) of Al Hasa waters from the wells east and west of Al Qurain (see Fig. 39).* Samples No. 35 a, 35, 36 are not included because of their contamination with irrigation water. (7) = total number of wells. $O_2$ content in percentage of saturation at the water temperature and the mean air pressure (745 mm Hg) of the investigation area. No significant difference means $\alpha < 0.05$

| | east | west |
|---|---|---|
| °C | 29.2 ± 3.8% (8) | 36.7 ± 6.2% (7) |
| | $\alpha \ll 0.002$ | |
| $O_2$ % | 66.8 ± 12% (8) | 64.3 ± 17% (9) |
| | no significant difference | |
| me/l | 22.82 ± 1% (8) | 23.38 ± 2.5% (7) |
| | no significant difference | |

*Summary*

The waters of the Al Hasa oasis discharging from the Neogene (Hofuf Formation) belong to the Umm er Radhuma Formation. This stratum can be found at a depth of some 280 meters and displays a thickness of approximately 320 meters. The chemical composition of the Al Hasa waters is similar to those in the area of Al Ajam emerging from the same formation. This relationship agrees with A. I. NAIMI's map of water quality, according to which the area of Al Hasa lies at the edge of a tongue of water of low concentration which reaches the Al Ajam area further north (Fig. 37). According to the average deuterium value (— 39.6 ± 0.7), the Al Hasa waters have their origin in an old reservoir. This hypothesis is substantiated by the lack of tritium and the low content of $^{14}C$ (1.4 percent modern) which is too low to be measured accurately. The low temperatures of the wells drilled in the east of the investigated area suggest that these waters originate from the main joints in the west and flow across waterways of the Neogene which lie relatively near the surface. A seepage of irrigation water (recycling) could be observed in two nonfreely flowing 'ayns.

### *2.1.6.3. Wadi Al Miyah*

*Introduction*

Coming from the east one reaches the upper part of the Wadi Al Miyah near Al Wannan (lat 26° 37′ N, long 48° 28′ E) when one branches away from the Trans-Arabian pipeline at Jabal Al Hydaruk and follows up the Darb Al Kunhuri for some 200 kilometers. The broad and flat wadi flows from southwest towards northeast. Its rather inconspicuous flanks consist of sandy marls and clays of the Miocene Dam and Hadrukh Formations. The wadi fill consists of aeolian sand, silt, and fluviatile sand, and in places of very shallow ground water, sabhkahs cover the wadi floor. Sabkhahs are found often around spring and well fields where the landform prevents the surface runoff (see chapter 2.1.5.). The high soil salinity due to permanent evaporation implies that agricultural areas can mainly be found in somewhat higher places at the wadi borders. In such areas systematic drainage of the soil is at least as important for agriculture as the establishment of new wells.

*Hydrologic situation*

Most wells are hand-dug wells without free runoffs at depths ranging between 2 and 7 meters and a high mineralization (approximately 3 to 6 g/l). Only in larger villages, such as Al Wannan, As Sarar, Ghanwa and Al Khubrah can one find drilled wells supplying the area with better water (total mineralization of about 1.6 to 2 g/l). Near As Sarar ITALCONSULT made 12 test drillings. The geological profiles obtained show a similar stratum series as shown in Al Qatif and Al Hasa. Traces of the Alat Formation become visible at depths of 83 and 109 meters, and the Khobar Formation between 129 and 171 meters. The Umm er Radhuma Formation lies 241 meters below the surface, with a thickness of at least 500 meters. The lowest layers were not even hit by drillings reaching down to 750 meters. Table 23 shows the

Table 23. *Geological strata, water temperature in degrees centigrade, total mineralization (TDS in g/l), and static water level in test wells of Wadi Al Miyah (data from ITALCONSULT reports)*

| Aquifer | Depth in meters | Temp. °C | T.D.S g/l | Static water level, meters above sea |
|---|---|---|---|---|
| Neogene | 0–83 | — | 3.52 | 98.6 |
| Alat | 83–109 | 27 | 2.22 | 109.8 |
| Khobar | 129–171 | 29.5 | 2.19 | 111.5 |
| Umm er Radhuma | 241– | 34.5 | 1.82 | 117.3 |

static water level, temperature and mineralization. Apart from the Neogene, all other drillings met aquifers with the pressure increasing the deeper the drillings went. The temperature of the Umm er Radhuma waters exceed the annual average of air temperature (27 °C) by some 7 °C, which is in good agreement with the geothermal gradient for the formation (approximate depth of 240 meters). The Umm er Radhuma waters are the least mineralized; however, they show minor differences when compared to the Alat and Khobar waters. From the Alat Formation towards the Neogene, mineralization increases by some 40 percent. On the whole, these observations provide no clear answer to the question of whether or not one is concerned with separate aquifers in this case or whether the wells are primarily fed from the Umm er Radhuma Formation and their upper parts are just conducting pathways in which water temperature and pressure decreases and an additional solution of salts takes place. According to T. Dinçer and his team, the presence of a higher sulfate content of Umm er Radhuma waters appears to be a strong argument against a common origin. In our samples, too, most waters from the Umm er Radhuma Formation had a high relative sulfate content (Table 24, Nos. 119, 120). However, exceptions prove the rule (No. 121). A comparison of the sulfate concentration and temperature for wells 70 and 122 indicate that a more complex relationship may exist.

*Hydrochemical investigations*

Only five samples could be collected from the Wadi Al Miyah. Analysis results are given in Table 24 and Fig. 41. Sample No. 79 was taken from

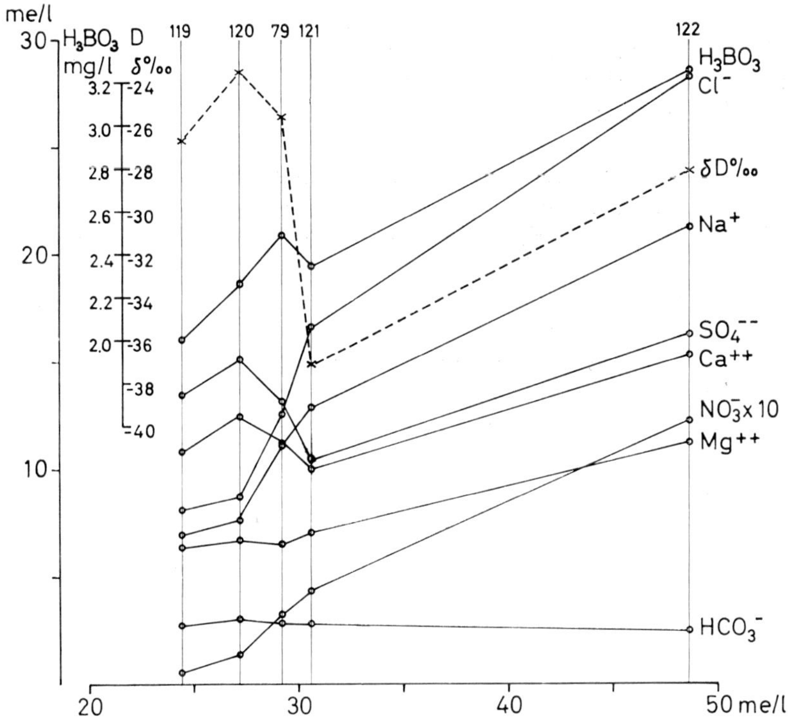


Fig. 41. Analyses of water samples collected in the area of Wadi Al Miyah

an artesian well of unknown depth by the road south of Al Wannan. Numbers 119–121 issue from drillings carried out in the north of As Sarar in the Umm er Radhuma Formation. Sample No. 122 was collected in a hand-dug well located 3.5 kilometers south of As Sarar. The waters do not represent a homogeneous population (Fig. 41). The sodium chloride as well as gypsum content of Nos. 79, 120, and 121 are divergent, and No. 79 might very well be a mixture of 120 and 121. The slope of the deuterium curve would also speak in favor of such a hypothesis. Generally, the impression is gained from the concentration diagram that two different water groups are involved, with gypsum dominating in samples Nos. 119 and 120 and the sodium chloride prevailing in Nos. 121 and 122. The chemical compositions of samples 119 and 120 from the Umm er Radhuma aquifers are similar. Minor differences in the total ion concentration (24.4 and 27.2 me/l) are primarily attributable to the prevailing $Ca^{+}$ and $SO_4^{--}$ ions. In contrast to this, sample No. 121 (another water sample from the Umm er Radhuma Formation) is characterized by its lower sulfate but higher chloride content. This water shows a very low deuterium content (— 37.1 $\delta$D per mill), whereas the remaining two water samples contain far more deuterium (119 : — 26.7 $\delta$D per mill, 120 : — 23.5 per mill). Just as in the Umm er Radhuma waters of the Qatif region a certain parallelism of both deuterium and sulfate values can be observed. The low temperature of the artesian well No. 79 and 'ayn No. 122 (with a depth of some 5 meters) indicates that their waters remain in the Neogene for some time. The 'ayn's strikingly high total concentration (48.6 me/l) is primarily due to the rising sodium chloride and nitrate content, which may be explained through the seepage of biogenically polluted irrigation water.

Our hydrochemical findings do not justify the assumption that the Neogene can be separated as an independent aquifer from the Umm er Radhuma Formation. According to the deuterium content the Neogene as well as the Umm er Radhuma waters have their origin in an old reservoir. Since it is most unlikely that old waters occur independently in the Neogene owing to the superficial position and the small thickness of this stratum, the real supplier of the Neogene waters is considered to be the powerful Umm er Radhuma reservoir. This hypothesis is substantiated by the fact that the waters of the Wadi Al Miyah show a similar deuterium—sulfate relationship to waters in other areas known to be supplied by the Umm er Radhuma Formation (e.g., Al Qatif, Haradh).

### *2.1.6.4. Wadi As Sah'ba near Haradh*

*Introduction*

Haradh is situated by the railway line from Ar Riyadh to Ad Dammam at approximately lat 24° 10′ N and long 49° 05′ E. Some 10 kilometers southeast of Haradh two series of a total of 52 wells were drilled into the Umm er Radhuma Formation a few years ago; they cross the broad Wadi As Sah'ba from northeast towards southwest. These wells supply artesian water to irrigate extended sheep pastures. The upper limits of the Umm er Radhuma Formation in this area lie between 87 and 223 meters below the surface. Its average thickness is given as some 170 meters. The formation reaches the surface about 70 kilometers westward, there only covered with permeable Quaternary sediments. The recharge area of the well field is estimated to be 2,600 square kilometers. With a yearly seepage of some 30 millimeters of precipitation a recharge of approximately 2.5 cubic meters per second might very well ensue. However according to T. Dinçer and his team (1974) the $^{18}O$ content of the Haradh wells is much lower than in recent precipitations, indicating that the water used today presumably comes from an old storage to a large extent.

Table 24. *Chemical composition of Wadi Al Miyah waters (μe/l; $O_2$, $H_3BO_3$, $H_2SiO_3$ in mg/l)*

| No. | W | °C | $O_2$ | $Na^+$ | $K^+$ | $Ca^{++}$ | $Mg^{++}$ | SK | $Cl^-$ | $Br^-$ | $SO_4^{--}$ | $HCO_3^-$ | $NO_3^-$ | $F^-$ | SA | $H_3BO_3$ | $H_2SiO_3$ |
|---|---|---|---|---|---|---|---|---|---|---|---|---|---|---|---|---|---|
| 79 | DA | 28.3 | 3.1 | 11,092 | 309 | 11,280 | 6,520 | 29,201 | 12,667 | 12.6 | 13,237 | 2,912 | 318 | 39 | 29,186 | 2.49 | 15.0 |
| 119 | DA | 34.0 | — | 6,916 | 239 | 10,812 | 6,442 | 24,409 | 8,028 | 10.0 | 13,466 | 2,691 | 45 | 46 | 24,286 | 2.00 | 13.9 |
| 120 | DA | 34.0 | — | 7,721 | 245 | 12,480 | 6,760 | 27,206 | 8,789 | 10.9 | 15,197 | 3,012 | 135 | 44 | 27,188 | 2.26 | 14.7 |
| 121 | DP | 34.0 | — | 12,941 | 345 | 10,160 | 7,060 | 30,506 | 16,626 | 14.1 | 10,466 | 2,912 | 439 | 46 | 30,503 | 2.34 | 15.0 |
| 122 | H | 25.0 | — | 21,314 | 484 | 15,440 | 11,360 | 48,598 | 28,300 | 22.2 | 16,337 | 2,510 | 1,230 | 46 | 48,445 | 3.23 | 15.0 |

Table 25. *Chemical composition of water from wells of Haradh (μe/l; $H_3BO_3$, $H_2SiO_3$ in mg/l)*

| No. | °C | $O_2$ | $Na^+$ | $K^+$ | $Ca^{++}$ | $Mg^{++}$ | SK | $Cl^-$ | $Br^-$ | $SO_4^{--}$ | $HCO_3^-$ | $NO_3^-$ | $F^-$ | SA | $H_3BO_3$ | $H_2SiO_3$ |
|---|---|---|---|---|---|---|---|---|---|---|---|---|---|---|---|---|
| 117 | 35.2 | 5.8 | 5,524 | 169 | 5,460 | 3,560 | 14,713 | 6,658 | 41.3 | 4,524 | 3,027 | 387 | 37 | 14,674 | 1.2 | 17.80 |
| 118 | 34.1 | 6.8 | 6,308 | 245 | 5,960 | 3,780 | 16,293 | 8,025 | 55.1 | 4,819 | 2,860 | 500 | 32 | 16,291 | 1.1 | 20.58 |
| 116 | 32.2 | 7.6 | 8,047 | 242 | 7,400 | 4,800 | 20,489 | 11,431 | 88.9 | 5,454 | 2,641 | 661 | 19 | 20,295 | 1.0 | 16.41 |

*Hydrochemical investigations*

During a short trip three water samples were collected. Sample No. 116 was taken from borehole No. 4 in the northwest, sample No. 118 from borehole No. 35 in the southeastern part and sample No. 117 from borehole No. 50 in the southeast border of the 45 kilometers well series. The analytical

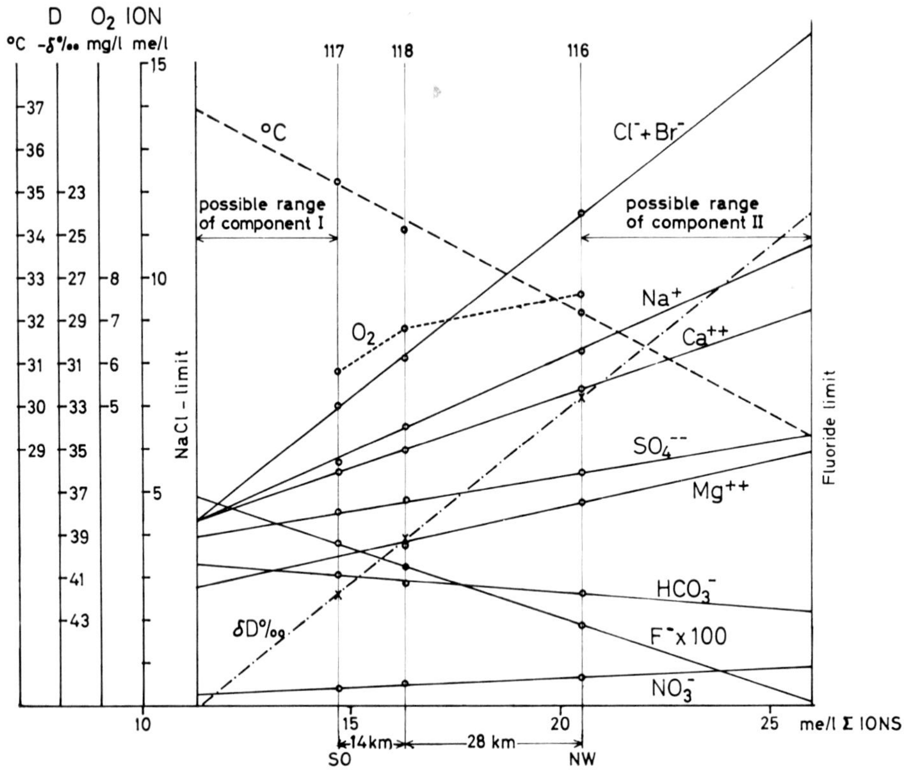


Fig. 42. Analyses of water samples collected in the area of Wadi As Sah'ba near Haradh

results show (Table 25, Fig. 42) that the water from the southeast border is far less mineralized than that at the northwest border. The water quality from the well in between (sample No. 118) can be seen as a mixture of both marginal waters. The linear decrease of the water temperature as well as the fluoride content together with an increasing total ion concentration speaks in favor of this assumption. Any precipitation of fluoride can be excluded, as the solubility product of calcium fluoride is far from being reached. Besides, the linear dependence of the deuterium content on the total concentration is a further argument for such a mixture. The real composition of the two mixing components can only be vaguely estimated because one has to take into account that both marginal waters are probably mixtures themselves. Supposing that alkali ions originate exclusively as dissolved sodium and potassium chloride, the rationale would be that water would not contain more alkali ions than chloride ions, because equivalent amounts of cation and anion are dissolved. Under these assumptions the lower limits of the total concentration of component I may be assumed at the point where both the alkali and the chloride concentrations intersect in Fig. 42. Accordingly, the potential scope of mineralization of component I ranges between 11.3 me/l and the value 14.7 me/l found in sample No. 117. An estimation of the upper limits of component II can be made from the graph of fluoride content. Since no water can contain less than zero fluoride the maximum total concentration

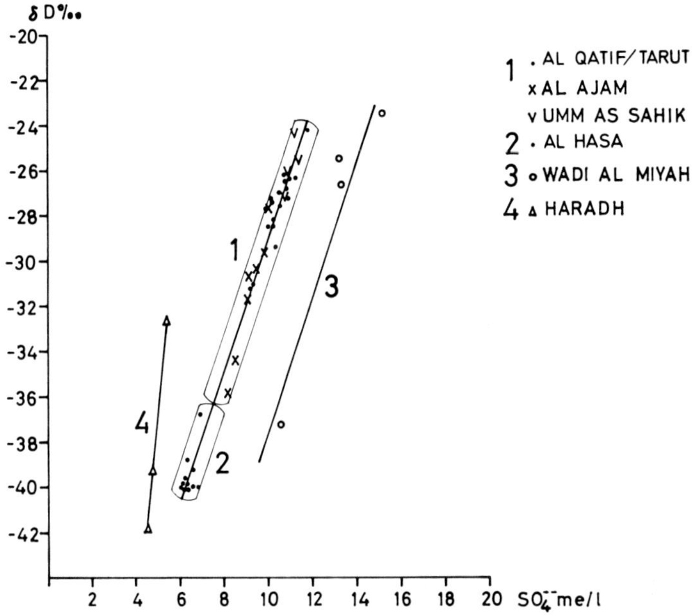


Fig. 43. Relationship between isotope enrichment and sulfate content in water samples from Al Qatif area, Al Hasa, Wadi Al Miyah and Wadi As Sah'ba (Haradh)

of component II is 26 me/l. The same limiting value is furnished by the water temperature if one starts out from the assumption that component II cannot be cooler than the yearly average of air temperature (29 °C) in this area. These considerations are of practical value in defining the concentration limits within which the water quality may change because of any disturbances in the natural mixture ratio by exploitation. Component I has to be regarded as an old primary Umm er Radhuma water due to its lower isotope content and its similarity in the ion distribution to waters of Al Hasa. A tongue of saltier water is considered to be the origin of component II, which according to A. I. Naimi (1965) pushes forward into the recharge area from the north (Fig. 37). The administrative directors of the King Faisal Project in Haradh report that a fault system exists in the underlying strata which may act as the pathway for the component II.

### *2.1.6.5. Final remarks*

The geological conditions of the northeast of Saudi Arabia are determined by a series of sedimentary formations dipping flatly towards the Arabian Gulf. The uppermost stratum of consolidated sediments has its origin in the Plio-Miocene (Neogene) and overlies the Dammam Formation (middle Eocene). The latter is subdivided into five strata provided that the series is fully developed: the Alat member, with porous, mostly dolomitic limestones containing fine marls in their lower zones; the Khobar member consisting of crystalline dolomites predominantly, limestone with yellow and grey marls as basal layers; Alveolina limestone member which is a narrow bed of porous limestones with an internal mould of *Alveolina elliptica;* two other thin-bedded, fossil-bearing layers consisting of creviced marl and clay limestones (Saila and Midra shales, respectively). The latter formation displays gypsum crystals [1] and ferruginous concretions. Below these strata one comes across the Rus Formation of early Eocene with porous limestones in its upper parts, marly limestones together with gypsum [2], anhydrite in the middle of the complex and crystalline limestone containing quartz geodes. Below it lies the Umm er Radhuma Formation (Paleocene) with a thickness of some 300 to 600 meters consisting mainly of crystalline dolomite in its upper parts and of various kinds of limestones in its lower zones. Below are limestones and sandstones (Wasia Formation) of Cretaceous age.

All more extended strata mentioned above include aquifers. The rate of salinity increase, according to A. I. NAIMI (1965), from southwest towards northeast, depends on the dip of the strata, the depth of the aquifers and the main flow direction. In the zones of Wasia sandstone, isolines of salinity form regular belts, whereas in Umm er Radhuma limestones, which are subject to karstification, as well as in the above-lying limestones, several tongues of weakly mineralized waters penetrate far into the northern and eastern regions (Fig. 37). In this case, more or less privileged waterways flowing towards the pressure-releasing springs near the Gulf coast appear to be involved. These are cleft rocks and washed-out, widened, karstified joints where less salt is dissolved than when the water flows through porous rocks. Huge karst tubes and solution caves have been observed in the Umm er Radhuma Formation of the As Summan Plateau (see Chapter 2.2.2.) and the Neogene of Al Hofuf.

All well fields investigated are supplied by the Umm er Radhuma Formation. The overlying strata are nothing more than conducting ways. The changing permeability of the Umm er Radhuma rocks caused by karstification and the regional difference in geochemical composition (evaporite horizons, etc.) lead to marked differences in the mineralization of water and favor the occurrence of mixed waters. Successful attempts were made to roughly estimate the concentration of mixing components in Al Qatif as well as

[1] Two gypsum samples taken from the Midra layer which lies exposed near Dhahran by the Dammam Dome had strikingly low $\delta^{34}S$ values (— 18.7 and — 16.4 per mill).

[2] Gypsum from the Rus Formation on the campus of The University of Petroleum and Minerals at Dhahran contained + 13.9 per mill $\delta^{34}S$.

Haradh. The arrows in Fig. 37 stand for the directions from which the mixing components presumably flow to the well fields.

Judging from isotope data, the waters of the Umm er Radhuma Formation belong to an old reservoir filled up in a more rainy time than at present (Chapter 2.1.8.). The remarkable relationship between isotope enrichment and salt content (especially sulfate content) in the mixed waters is of course a consequence of mixing (Fig. 43). However this process does not explain why just the more salty component of the mixture shows isotopic enrichment. This was also reported by T. Dinçer and his co-workers (1974) for $^{18}O$. Theoretically three principal possibilities for this correlation are imaginable: slow evaporation from the subterranean reservoir or infiltration of precipitation enriched with heavy isotopes across the salty soil often encountered in arid zones, or else the uptake of fossil water isotopically enriched by evaporation in earlier times. The apparent equal degree of enrichment of both deuterium and $^{18}O$, respectively, together with the increase of salt content speaks in favor of a slow subterranean evaporation under near equilibrium conditions (for theoretical explanations see chapter 2.3.2.2.). However, no clear cut evidence for such a process could be gained until now. Nevertheless, the long storage in a more or less fissured reservoir, together with the temperature gradient towards the surface, makes a subterranean evaporation quite possible.

## 2.1.7. Cluster Analyses of Water Wells of the Al Qatif and Al Hasa Areas

(R. J. Beckman, J. B. Ramsay) [1]

### *2.1.7.1. Intruduction*

After the 1973 field campaign Professor C. Job furnished his data to us on the composition of water from wells in the Eastern Province in Saudi Arabia to assist in the study of the Sabkhat Ar Riyas brines (D. H. Johnson, M. R. Kamal, G. D. Pierson, and J. B. Ramsay, see 2.1.5.). Utilizing these data we performed a cluster analysis which indicated that the Eastern Province water could be considered as three clusters, one in the Al Hasa area and two in the Al Qatif region. These results were helpful in the formulation of the interpretation presented by C. Job (2.1.6.1. and 2.1.6.2.). He later supplied his full set of data for use in this paper to describe the application of cluster analysis to hydrogeologic data. Cluster analysis may be described as a systematic method of grouping individual members of a sample population into clusters based on some definition of the measure of similarity between individuals. We have also examined the data using a statistical technique known as discriminate analysis which attempts to identify the variables that are most significant in describing the clusters.

---

[1] Work done under the auspices of the U.S. Atomic Energy Commission.

In the full data set reported by C. JOB there are approximately 100 samples with 16 different components measured for most samples. The task of separating this many samples based on the similarity of the chemical analyses alone is formidable. Experience, knowledge of the hydrology, skill, and luck all assist in any manual grouping of data containing this many variables. Some standard techniques reduce the number of variables by forming combinations. The "total dissolved solids", "(Na + K)/Cl", and "(Ca + Mg)/$SO_4$" are commonly used variables for classifying water. The social and biological scientists are plagued with problems involving many measurements on large sample sizes and various statistical procedures have been developed for examining large data arrays for underlying correlations and groupings. In this paper we describe the application of two of these techniques to the examination of water samples.

### *2.1.7.2. Cluster analysis*

A number of reference books discuss the subject of cluster analysis (M. R. ANDERBERG, 1973; B. S. DURAN et al., 1974; B. EVERITT, 1974; J. A. HARIGAN, 1975) but on a much broader base than is required for the present problem. The book by J. G. DAVIS (1973) is specifically directed toward the earth sciences and includes discussions of both cluster analysis and discriminate analysis among other multivariate techniques. A nonmathematical discussion of cluster analysis is given by J. E. DORAN and F. R. HODSON (1975). As with any statistical procedure the interpretation of the results of a cluster analysis requires both an understanding of the statistical procedure and the underlying science.

Cluster analysis provides an investigator with information on how the data can be grouped based on some measure of similarity among his samples. It makes no statistical tests as to the significance of the grouping and in fact may impart some constraints on the application of statistical tests of significance between groups because the data have been preselected to be similar. The order in which the data are formed into clusters also provides some guide as to the structure of the data. If this type of analysis provides insight into the underlying similarities and differences in a large data set then it has done all that can be expected.

A number of different algorithms are available for performing a cluster analysis. The selection of the method depends somewhat on the type of data to be examined and the prejudices of the investigator. In this study a "hierarchical amalgamative technique based on the Euclidean distance between samples in $N$-dimensional space" was used. Amalgamative means that the samples were initially treated as individuals and then collected together as the clustering operation took place. Hierarchical means that once two samples had been amalgamated they could not later be separated. The measure of similarity was simply the distance as defined in Euclidean space.

The data arrays examined in this study consisted of 13 variables. Each variable was defined in its own units; i.e., $\mu e\ Cl^-/l$, °C, or mg $H_3BO_3/l$. There is nothing in the mathematics which requires that each direction have the same unit of measure, however, the wide range in the values among the variables would cause differences in one variable to have a larger effect than other variables of lower concentration if the absolute values were used. For example the minimum and maximum values for $Cl^-$ in the samples discussed here were 10,630 and 37,450 µe $Cl^-$/l while the range in $Br^-$ was from 8.7 to 76.2 µe $Br^-$/l. Differences in the $Cl^-$ concentration overwhelm the $Br^-$ differences if absolute units are used to compute the distance. To overcome this problem data are usually standardized before a cluster analysis is performed. A usual procedure for standardizing data is to subtract the mean and divide by the standard deviation derived from the entire data set (B. EVERITT, 1974, p. 48), i.e.,

$$C_{ij} = \frac{X_{ij} - \bar{X}_i}{s_i} \tag{1}$$

where $C_{ij}$ is the standardized $i$th variable for the $j$th sample, $X_{ij}$ is the data value, $\bar{X}_i$ is the mean value of the $i$th variable in the set being examined, and $s_i$ is the standard deviation. This procedure explicitly applies equal weights to each variable, which in many cases is desirable. However, since the experimenter believes *a priori* that there may be more than one population in the data, this procedure is somewhat questionable from a statistical viewpoint. For this data set the standard deviations of the analytical procedures for each component were available and it was decided to use it rather than the one based on the sample population. The use of the chemical-analysis standard deviation also incorporates an unequal distribution of weights among the variables. Those whose range is within the expected chemical errors will range within approximately $\pm 3$ while those variables which vary for geologic reasons will have a wider range and contribute more to the distance between samples.

The measure of similarity used was the Euclidean distance between samples in $N$-dimensional space, where $N$ is the number of different variables (components) considered. The distance between two samples ($j$, $k$) is given by

$$d_{jk} = \left( \sum_{i=1}^{N} (C_{ij} - C_{ik})^2 \right)^{1/2} \tag{2}$$

The technique used to cluster the samples was to compute the distance among all possible combinations of samples and then select the closest two to form the first cluster. These two were then combined by averaging each component and recalculating the inter-sample distances and again looking for the closest pair. This next pair does not necessarily include the previously clustered pair. As a new sample is added to a pre-existing cluster each component is computed as the arithmetic mean of all samples in the cluster so that each sample maintains equal weight in defining the new vectors. The procedure described above is continued until all samples (vectors) have been combined into a single cluster. The appendix presents a step-by-step procedure using the data for three variables and five samples. The computations presented in this report were performed using a clustering program provided by the UCLA Biomedical

Parameter statistical package (BMDP, 1973) as modified for use at the Los Alamos Scientific Laboratory.

Deciding whether there are clusters in the data and where the boundaries lie can be perplexing with some data sets. In this problem, where the standard deviation used for standardization has a statistical interpretation, a reasonable guideline can be developed. The variance of a function of two or more measured variables with known variances can be estimated from the theory of propagation of errors as follows:

$$g = f(x, y, \ldots)$$

$$\sigma_g^2 = \sigma_x^2 \left(\frac{\partial g}{\partial x}\right)^2 + \sigma_y^2 \left(\frac{\partial g}{\partial y}\right)^2 \ldots \tag{3}$$

After two or more vectors have been amalgamated one of the two variables of Eq. (2), say $C_{ik}$, corresponds to the mean value of the component $i$ for the amalgamated vectors. The variance of $C_{ik}$ will decrease relative to the variance of an individual vector ($C_{ij}$). This is not strictly true for all amalgamation steps since $C_{ij}$ may also be an average of several vectors particularly near the end of the procedure, but early in the procedure this appears to be a reasonable assumption. Also covariance terms have not been included in Eq. (3) but they are not readily evaluated and have been deleted in order to obtain a simple quide. With these assumptions only the partial derivitives of $C_{ij}$ and the associated variances, $\sigma^2_{Cij}$, are required and substituting into Eq. (3) gives

$$\sigma_d^2 = \sum^{N} \sigma_{C_{ij}}^2 \left(\frac{(C_{ij} - C_{ik})^2}{d^2}\right)$$

Since all $\sigma^2_{Cij} = 1$ this reduces to

$$\sigma_d^2 = N, \tag{4}$$

where $N$ is the number of components. From this it appears that a reasonable guide for deciding on the boundaries between clusters would be somewhere between 2 or 3 $\sigma_d$ (2 or 3 $N^{1/2}$).

Using the analogy between the mean value of a one-dimensional data set and the mean values of the components for an amalgamated vector, Eq. (2) can be described as the "sum of squares of the deviations" of sample $j$ from the mean values over all components. In other words this algorithm clusters by minimizing a measure closely related to the variance in a one-dimensional problem.

### *2.1.7.3. Combined Al Hasa and Al Qatif areas*

During the period 1973–1975 C. Job obtained samples from 52 wells in the Al Hasa and Al Qatif areas. Some wells were sampled each year while others were only sampled once. The set used by C. Job for his correlation studies was used for this study. This set consisted of one sample from each well except for Well 35 for which both samples 35 and 35 a were included. The sample numbers for the set used are shown in Table 26 together with

Table 26. *Sample numbers for eastern province wells*

| Selected sample number | Well type[1] | Year collected | Other sample numbers |
|---|---|---|---|
| Al Qatif[2] | | | |
| 1 | HA | 1973 | |
| 126 | HA | 1975 | 2 57 |
| 58 | HA | 1974 | 3 127 |
| 4 | HA | 1973 | 59 128 |
| 129 | DA | 1975 | 5 64 |
| 65 | DA | 1974 | 13 130 |
| 14 | HA | 1973 | 63 133 |
| 15 | DA | 1973 | 61 131 |
| 16 | HA | 1973 | |
| 17 | H | 1973 | 78 143 |
| 18 | H | 1973 | |
| 19 | DA | 1973 | |
| 20 | HA | 1973 | |
| 66 | DA | 1974 | |
| 75 | DA | 1974 | |
| 76 | DA | 1974 | |
| 77 | HA | 1974 | |
| Al Ajam[2] | | | |
| 9 | HP | 1973 | 73 125 |
| 10 | DP | 1973 | |
| 11 | DP | 1973 | 70 141 |
| 12 | HP | 1973 | |
| 68 | DP | 1974 | |
| 69 | DP | 1974 | |
| 123 | DP | 1974 | 71 |
| 124 | DP | 1974 | |
| 74 | DP | 1974 | |
| Umm As Sahik[2] | | | |
| 6 | HP | 1973 | |
| 7 | HA | 1973 | |
| 8 | HP | 1973 | |
| 142 | HA | 1975 | |
| Tarut[2] | | | |
| 55 | DA | 1974 | |
| 56 | DA | 1974 | |

| Selected sample number | Well type | Year collected | Other sample numbers |
|---|---|---|---|
| Al Hasa | | | |
| 24 | DA | 1973 | |
| 25 | DA | 1973 | |
| 26 | DA | 1973 | |
| 27 | DA | 1973 | |
| 28 | DA | 1973 | |
| 29 | DP | 1973 | |
| 30 | DP | 1973 | |
| 31 | DP | 1973 | |
| 32 | DA | 1973 | |
| 33 | HA | 1973 | |
| 34 | DP | 1973 | |
| 35 | HP | 1973 | |
| 35 a | HP | 1973 | |
| 36 | HP | 1973 | |
| 37 | DP | 1973 | |
| 38 | HA | 1973 | |
| 39 | HA | 1973 | |
| 40 | HA | 1973 | |
| 41 | HA | 1973 | |
| 42 | HP | 1973 | |
| 43 | DA | 1973 | |

[1] Explanation see Table 12.

[2] The term Al Qatif will be used in the text to describe both the general area and to differentiate it from the Al Hasa area. During discussions of the Al Qatif area the terms Al Ajam, Tarut, Umm as Sahik, and Al Qatif will be used to discuss particular regions of the Al Qatif oasis. The meaning will be clear in the context.

the corresponding sample numbers from other years, the area, and the year the sample was collected.

Twelve chemical species ($Na^+$, $K^+$, $Ca^{2+}$, $Mg^{2+}$, $Cl^-$, $Br^-$, $SO_4^{2-}$, $HCO_3^-$, $NO_3^-$, $F^-$ $H_3BO_3$, and $H_2SiO_3$) were measured on all samples. On most, but not all, dissolved $O_2$, deuterium, $^{18}O$, and temperature were also measured. Cluster analysis requires that all variables be measurd for all samples. For the study of the two oases combined and for the Al Qatif oasis the 12 chemical species were used. Temperature was included when the Al Hasa area was considered separately because C. Job had reported a significant variation of temperature in that area.

As a preliminary study the data from the 53 samples were examined. The complete analysis is not reported here because the structure is somewhat complex for discussion. Basically this examination showed four main clusters consisting of the Al Hasa area, Al Ajam, and two clusters in Al Qatif. Rather than attempt to discuss this analysis the two major areas were analyzed separately and will be discussed individually. One interesting note on the combined analysis is that the Al Ajam wells cluster with the Al Hasa wells at an appreciably shorter amalgamation distance than when clustering with the main Al Qatif clusters.

### 2.1.7.4. *Al Hasa area*

In the Al Hasa area 20 wells were sampled with two samples obtained from Well 35. The twelve chemical species and the temperature for these wells were standardized by subtracting the mean and dividing by the estimated standard deviation of the chemical analysis procedure. The standard deviations for the chemical species were estimated as $0.05\bar{X}$. A variation of $0.5^\circ$ in temperature with the depth of the thermocouple was reported by C. Job and this value was used as the estimated value for the standard deviation of temperature. Deuterium and $^{18}O$ were not included in the cluster analysis but were standardized for inclusion in the table. Their standard deviations were estimated as $\pm$ 0.05‰ and $\pm$ 0.01‰ respectively. Table 27 presents the mean, standard deviation of the group of samples, minimum and maximum values, and the estimated standard deviation of the chemical analysis for each of the components. The cluster tree obtained for this area is shown in Fig. 44.

There appear to be three main clusters, three wells (41, 42, and 43) which cluster poorly and two (24 and 25) which form a subsidiary cluster. The group consisting of Wells 35, 35 a, and 36 was discussed in detail by C. Job (see 2.1.6.2.). The amalgamation distance for this group to the remaining 18 wells was about 24 standard deviation units indicating the extent that these 3 wells differ from the remainder. Cluster I contains Wells 26, 27, 28, 29, 30, and 31. They are all drilled wells lying in the eastern region of the area sampled. Cluster II contains Wells 33, 34, 37, 38, 39, and 40 which lie in the western area. Some of these wells are hand-dug and others are drilled. The mean values for the components for each cluster are shown in standardized units in Table 27. Also shown are the minimum and maximum values for each group. To illustrate the differences in the components among the clusters the mean values for each component have been plotted in Fig. 45. This figure clearly shows that there is no significant difference in the chemical composition of the two main clusters with the temperature being the only large difference between the two clusters. The small cluster consisting of 35, 35 a, and 36

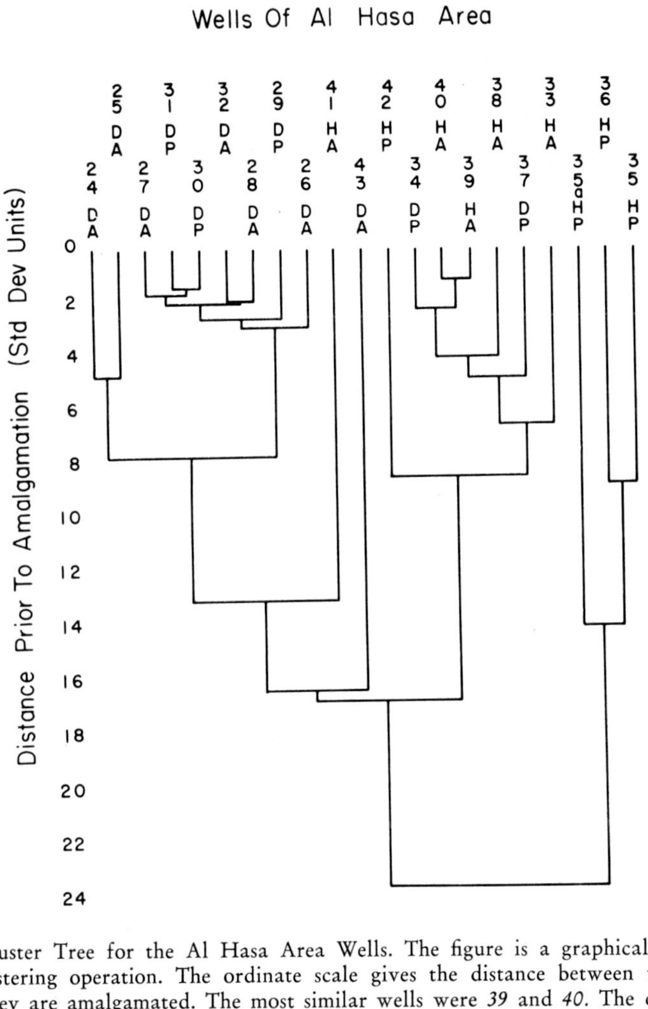


Fig. 44. Cluster Tree for the Al Hasa Area Wells. The figure is a graphical representation of the clustering operation. The ordinate scale gives the distance between two vectors at the step they are amalgamated. The most similar wells were *39* and *40*. The distance among the wells of the first cluster (*27*, *28*, *29*, *30*, *31*, and *32*) indicate that they are very similar in character. The three wells *35*, *35 a*, and *36* form a cluster but they do not have the same level of similarity as either of the other two clusters

has a much higher concentration of dissolved solids than the two larger clusters. Data for Wells 24 and 25 have not been plotted but they appear to have a chemical composition similar to the main clusters but the temperatures are intermediate between the two clusters. The standardized tempera-

Table 27. *Summary of Al Hasa wells*

*A. Overall summary for 21 samples in physical units*

| | Mean | Std. dev. [1] | Min. | Max. | Std. dev. of analysis (estimated) |
|---|---|---|---|---|---|
| Na (µe/l) | 12,153.6 | 2,138.7 | 9,134 | 18,168 | 607.7 |
| K (µe/l) | 541.7 | 93.8 | 394 | 788 | 27.09 |
| Ca (µe/l) | 8,016.2 | 1,164.5 | 7,120 | 11,320 | 400.8 |
| Mg (µe/l) | 3,897.1 | 401.9 | 3,560 | 5,100 | 194.9 |
| Cl (µe/l) | 13,610.1 | 1,944.7 | 10,631 | 18,696 | 680.5 |
| Br (µe/l) | 18.62 | 3.41 | 8.7 | 26.5 | 0.93 |
| $SO_4$ (µe/l) | 7,181.4 | 1,781.1 | 6,120 | 12,297 | 359.1 |
| $HCO_3$ (µe/l) | 3,501.4 | 223.8 | 3,058 | 4,134 | 175.1 |
| $NO_3$ (µe/l) | 333.1 | 29.0 | 252 | 363 | 16.6 |
| F (µe/l) | 32.8 | 1.1 | 31 | 37 | 1.64 |
| $H_3BO_3$ (mg/l) | 2.14 | 0.39 | 1.79 | 3.17 | 0.11 |
| $H_2SiO_3$ (mg/l) | 20.51 | 0.48 | 18.9 | 21.6 | 1.03 |
| Temperature (°C) | 32.66 | 4.11 | 28.0 | 40.2 | 0.50 |
| Dissolved $O_2$ (mg/l) | 4.68 | 1.17 | 1.2 | 7.2 | 0.23 |
| $\delta D$ ‰ | — 39.59 | 0.75 | — 40.1 | — 36.70 | 0.05 |
| $\delta^{18}O$ ‰ | — 5.17 | 0.18 | — 5.45 | — 4.85 | 0.01 |

[1] This is the standard deviation computed from the individual samples. It was not used for standardizing the variables, rather the standard deviation of the chemical analysis procedure was used.

*B. Standardized means, minima, and maxima for clusters*

| | Cluster I Wells 26, 27, 28, 29, 30, 31, 32 | | | Cluster II Wells 33, 34, 37, 38, 39, 40 | | | Cluster III Wells 35, 35 a, 36 | | |
|---|---|---|---|---|---|---|---|---|---|
| | Mean | Min. | Max. | Mean | Min. | Max. | Mean | Min. | Max. |
| Na | — 1.76 | — 2.10 | — 1.52 | — 0.80 | — 1.39 | 0.59 | 7.10 | 3.76 | 9.90 |
| K | — 1.61 | — 1.84 | — 1.32 | — 0.72 | — 1.10 | 0.27 | 7.36 | 4.40 | 9.09 |
| Ca | — 1.06 | — 2.04 | — 0.54 | — 1.13 | — 1.84 | — 0.24 | 6.56 | 3.75 | 8.24 |
| Mg | — 1.09 | — 1.73 | — 0.70 | — 0.98 | — 1.63 | — 0.50 | 4.36 | 2.27 | 6.17 |
| Cl | — 1.44 | — 1.57 | — 1.34 | — 0.73 | — 1.22 | 0.66 | 5.28 | 2.66 | 7.47 |
| Br | — 0.37 | — 0.99 | 0.41 | — 1.11 | — 2.06 | 0.41 | 4.13 | 2.56 | 5.46 |
| $SO_4$ | — 2.45 | — 2.53 | — 2.35 | — 1.03 | — 2.31 | 3.35 | 10.99 | 6.27 | 14.25 |
| $HCO_3$ | — 0.06 | — 1.90 | 0.49 | — 0.52 | — 1.27 | — 0.06 | 2.20 | 1.39 | 3.61 |
| $NO_3$ | 1.52 | 1.19 | 1.79 | 0.49 | — 0.01 | 0.83 | — 0.75 | — 1.75 | 0.11 |
| F | 0.09 | 0.09 | 0.09 | 0.32 | — 0.52 | 0.09 | 0.09 | 0.09 | 0.09 |
| $H_3BO_3$ | — 2.27 | — 3.28 | — 1.41 | — 0.71 | — 1.41 | 0.46 | 7.87 | 5.41 | 9.61 |
| $H_2SiO_3$ | — 0.04 | — 0.11 | 0.19 | — 0.03 | — 0.11 | 0.38 | 0.61 | 0.28 | 1.06 |
| Temperature | — 7.64 | — 9.32 | — 5.92 | 9.54 | 6.68 | 12.68 | — 1.79 | — 7.32 | 2.08 |
| Dissolved $O_2$ | 1.89 | — 1.22 | 5.61 | — 0.58 | — 4.21 | 2.62 | — 0.51 | — 6.77 | 3.05 |
| $\delta D$ | — 6.20 | — 10.20 | — 4.20 | — 3.20 | — 10.20 | 3.80 | 3.80 | — 6.20 | 9.80 |
| $\delta^{18}O$ | — 11.61 | — 27.9 | 10.10 | 1.77 | — 18.90 | 12.10 | 15.77 | 7.10 | 30.10 |

ture of Well 24 is — 1.32 and for Well 25 it is — 2.12 compared to the means of the two major clusters of — 7.6 and + 9.5. Well 24 lies in the western region near those wells of Cluster II and Well 25 lies in the east very near Wells 26, 27, and 28. Whether these deviations provide any additional insight into the hydrology of Al Hasa will require additional interpretation which

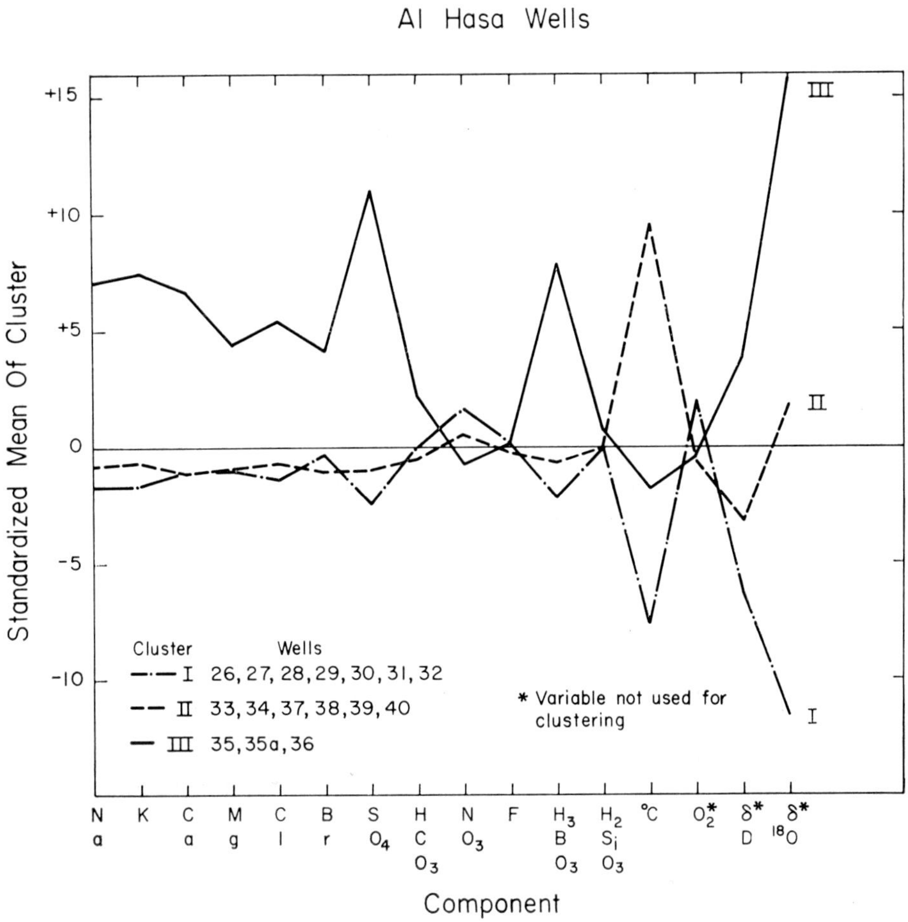


Fig. 45. Component Means for Al Hasa Wells. The graph shows the mean values for each of the measured components for the three clusters. The unit of measure is the standard deviation of the chemical analysis procedure: e.g. the $SO_4^{2-}$ values for Cluster *III* is + 11 standard deviation units from the mean value of all wells

is outside the scope of this paper. The three wells (41, 42, and 43) which fall outside the main clusters are located on the periphery of the oasis. Each appears somewhat different in the magnitude of the various components. These latter three wells are discussed by C. Job (2.1.6.2.).

Table 28. *Summary of Al Qatif wells*

*A. Overall summary for 32 wells in physical units*

| | Mean | Std. dev. [1] | Min. | Max. | Std. dev. of analysis (estimated) |
|---|---|---|---|---|---|
| Na (μe/l) | 18,667.7 | 4,874.9 | 13,274 | 29,361 | 933.4 |
| K (μe/l) | 558.5 | 87.1 | 458 | 774 | 27.93 |
| Ca (μe/l) | 11,408.8 | 1,553.1 | 8,800 | 14,800 | 570.4 |
| Mg (μe/l) | 6,383.1 | 881.4 | 4,600 | 8,320 | 319.2 |
| Cl (μe/l) | 23,512.3 | 6,706.6 | 16,215 | 37,448 | 1,175.6 |
| Br (μe/l) | 41.52 | 16.40 | 23 | 76.2 | 2.08 |
| $SO_4$ (μe/l) | 10,177.2 | 848.1 | 8,289 | 11,799 | 505.9 |
| $HCO_3$ (μe/l) | 2,963.7 | 157.2 | 2,641 | 3,431 | 148.2 |
| $NO_3$ (μe/l) | 245.2 | 43.4 | 155 | 315 | 12.3 |
| F (μe/l) | 41.1 | 1.9 | 36 | 44 | 2.06 |
| $H_3BO_3$ (mg/l) | 2.90 | 0.47 | 2.31 | 4.00 | 0.145 |
| $H_2SiO_3$ (mg/l) | 16.65 | 0.80 | 15.5 | 19.0 | 0.83 |
| Temperature (°C) | 32.95 | 3.31 | 23.0 | 37.3 | 0.50 |
| Dissolved $O_2$ (mg/l) | 3.34 | 1.63 | 0.3 | 9.0 | 0.17 |
| δD ‰ | — 28.64 | 2.58 | — 34.4 | — 24.2 | 0.05 |
| $\delta^{18}O$ ‰ | — 3.88 | 0.27 | — 4.4 | — 3.34 | 0.01 |

[1] This is the standard deviation computed for the individual samples. It was not used for standardizing the variables, rather the standard deviation of the chemical analysis procedure was used.

*B. Standardized means, minima, and maxima for clusters*

| | Cluster I Wells 1, 129, 65, 58, 4, 15, 126 | | | Cluster II Wells 9, 10, 11, 12, 19, 20, 66, 68, 72, 74, 123, 124 | | | Cluster III Wells 6, 7, 8, 14, 16, 17, 18, 55, 56, 75, 76, 77, 142 | | |
|---|---|---|---|---|---|---|---|---|---|
| | Mean | Min. | Max. | Mean | Min. | Max. | Mean | Min. | Max. |
| Na | 8.09 | 4.33 | 11.46 | — 4.97 | — 5.78 | — 2.91 | 0.23 | — 3.19 | 3.30 |
| K | 4.99 | 2.49 | 7.72 | — 2.35 | — 3.60 | 0.16 | — 0.51 | — 2.38 | 1.87 |
| Ca | 3.97 | 2.12 | 5.95 | — 2.77 | — 4.57 | — 1.49 | 0.42 | — 0.86 | 1.74 |
| Mg | 3.65 | 1.37 | 6.07 | — 2.90 | — 5.59 | — 0.76 | 0.71 | — 0.32 | 2.31 |
| Cl | 8.89 | 4.96 | 11.85 | — 5.42 | — 6.21 | — 3.22 | 0.22 | — 3.29 | 3.47 |
| Br | 12.42 | 6.98 | 16.71 | — 7.34 | — 8.92 | — 4.25 | 0.09 | — 4.83 | 5.15 |
| $SO_4$ | 1.57 | — 0.05 | 3.19 | — 1.71 | — 3.71 | — 0.27 | 0.74 | — 0.39 | 2.48 |
| $HCO_3$ | — 0.86 | — 1.71 | — 0.42 | 0.34 | — 2.18 | 2.74 | 0.15 | — 0.76 | 3.15 |
| $NO_3$ | — 2.71 | — 5.40 | 0.39 | 2.52 | — 0.34 | 5.70 | 0.87 | — 7.36 | 5.45 |
| F | — 0.38 | — 1.00 | 0.46 | — 0.11 | — 2.47 | 1.43 | 0.31 | — 1.00 | 1.43 |
| $H_3BO_3$ | 4.93 | 2.72 | 7.62 | — 2.91 | — 4.05 | — 0.80 | 0.03 | — 2.32 | 2.86 |
| $H_2SiO_3$ | — 0.73 | — 1.38 | 0.66 | 0.91 | — 0.30 | 2.83 | — 0.45 | — 0.90 | 0.42 |
| Temperature | 3.38 | — 0.91 | 7.69 | 0.99 | — 10.91 | 8.69 | — 2.87 | — 19.91 | 5.29 |
| Dissolved $O_2$ | — 3.82 | — 8.62 | — 0.83 | 1.09 | — 18.20 | 33.93 | 1.32 | — 12.81 | 20.74 |
| δD | 47.41 | 26.84 | 88.84 | — 46.91 | — 115.16 | 16.84 | 10.84 | — 15.16 | 36.84 |
| $\delta^{18}O$ | 12.03 | — 0.40 | 34.60 | — 21.40 | — 52.40 | 19.60 | 21.85 | 0.60 | 53.60 |

### *2.1.7.5. Al Qatif area*

As with the Al Hasa area the general area of Al Qatif is divided into a number of small villages which include the main town of Al Qatif. Samples of water from wells in a number of these villages were obtained and analyzed by

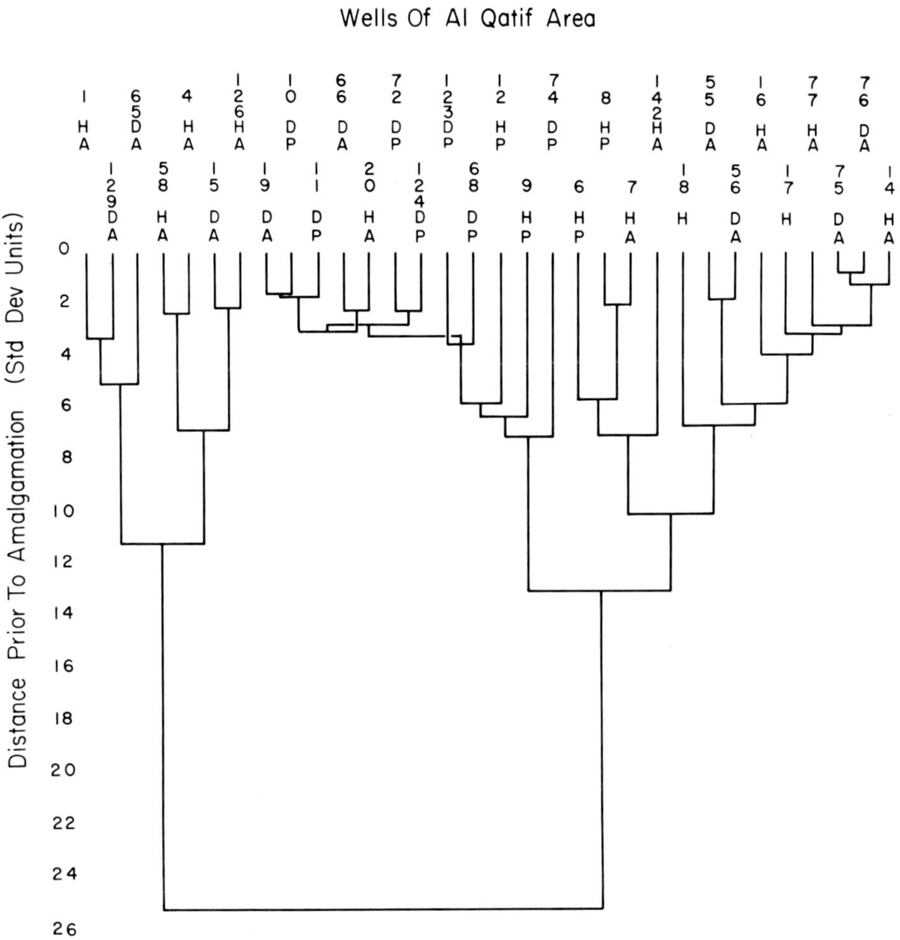


Fig. 46. Cluster Tree for the Al Qatif Area Wells. The branching after the amalgamation of wells *72* and *124* with the cluster consisting of wells *19, 10. 11, 66,* and *20* indicates that the amalgamated vector (*72–124*) was closer to the main cluster than either of the two individual vectors. Similar behavior is shown by wells *68* and *123*

C. Job. The same twelve chemical species used for the Al Hasa cluster analysis study were used for the Al Qatif area. Table 28 presents the means, standard deviations, minima, maxima, and estimated chemical standard deviations for all variables. The cluster tree for this area is shown in Fig. 46. It has a more complex structure than was found for the Al Hasa area. One cluster which consists of samples 1, 129, 65, 58, 4, 15, and 126 is clearly distinct. The mean of each component for three proposed clusters is shown in

Fig. 47. From this figure it is obvious that this first cluster has a much higher dissolved solid content than the other wells. This first cluster also illustrates the problem associated with interpreting cluster trees. This first cluster could be subdivided into two subclusters consisting of wells 1, 129, and 65 as one cluster and wells 58, 4, 15, and 126 as the second. For our later purposes the

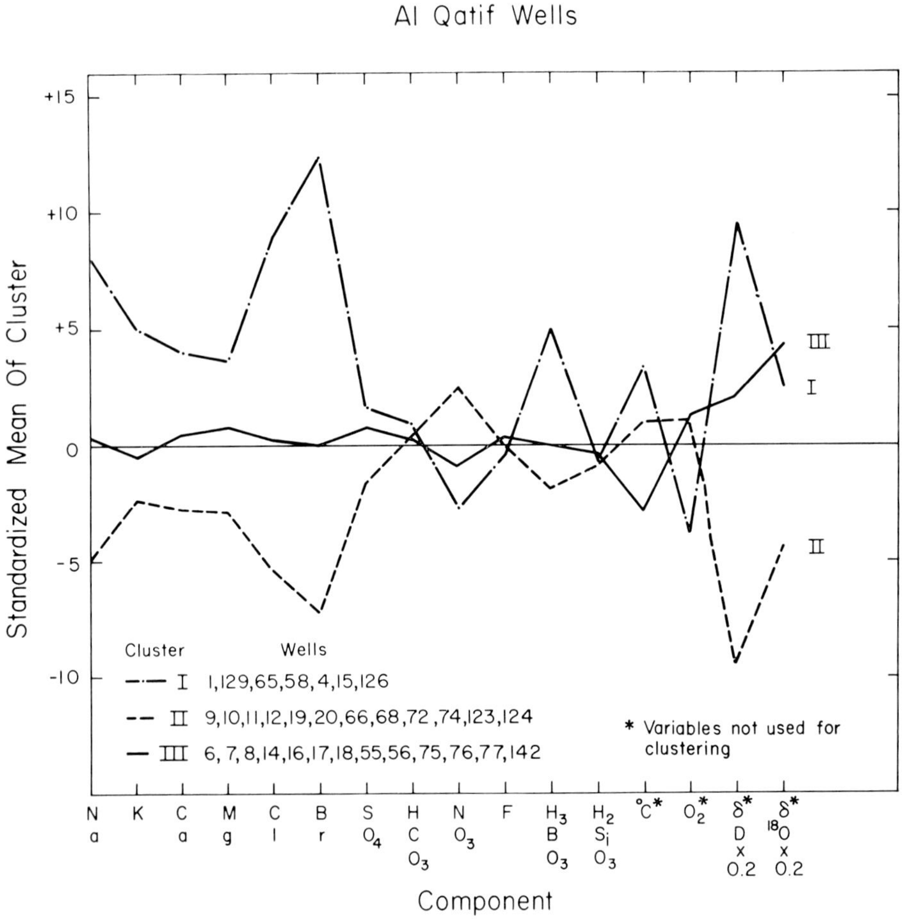


Fig. 47. Component Means for the Al Qatif Area Wells. This graph indicates that the major difference in the Al Qatif wells lies in the concentration of total dissolved salts

set of seven wells was selected. Two other clusters are distinguished. Cluster II is taken as 9, 10, 11, 12, 19, 20, 66, 68, 72, 74, 123, and 124. Wells 6, 7, 8, 14, 16, 17, 18, 55, 56, 75, 76, 77, and 142 are considered as Cluster III. The locations of these wells are shown on Fig. 48 which shows the geographic distribution of the clusters. Cluster II, which can be identified as a low-dissolved-solid water, lies southwest of a line running S 68° E just south of Well 18. This includes the wells of Al Ajam, and Wells 19 and 20 nearer the coast. The cluster tree shows Wells 9, 12, and 74 as the last to be clustered in this group and they fall on the outskirts of the area. The two wells on Tarut Island (Wells 55 and 56) are quite similar as might be expected. It is interesting to note that they join the mainland cluster after it formed but before the subcluster from Umm As Sahik located several kilometers to the northwest.

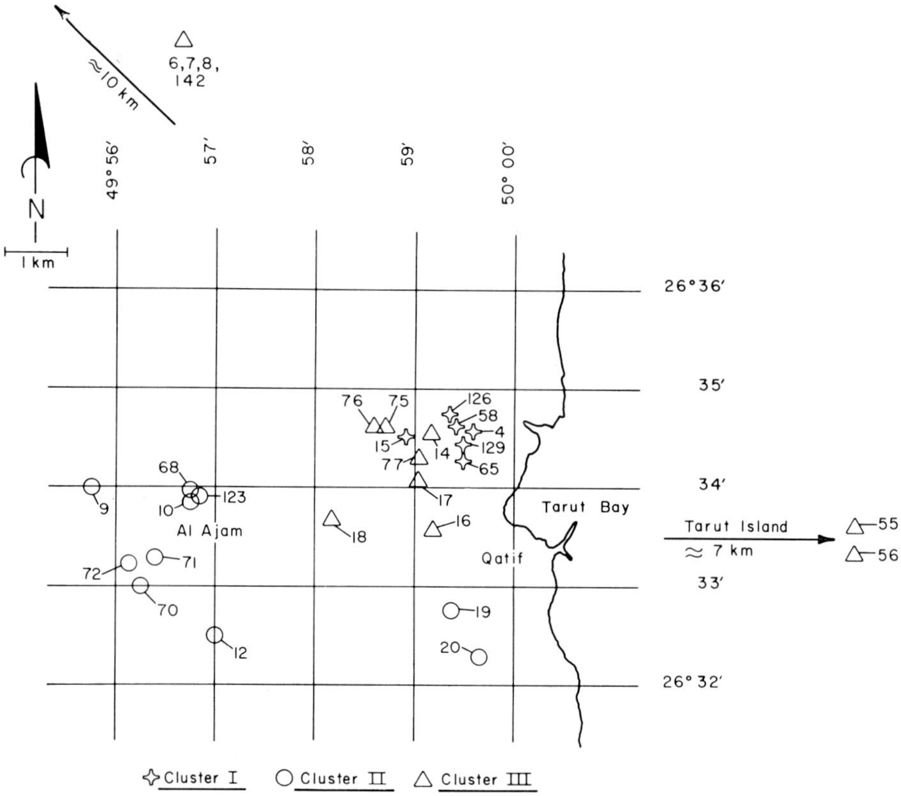


Fig. 48. Location of Wells in the Al Qatif Area. This sketch shows the approximate locations and identifying numbers of the wells studied in the Al Qatif area. Despite the close geographic proximity of the wells in clusters *I* and *III* they appear to be sufficiently dissimilar to warrant further study

The differentiation of Clusters I and III is somewhat more interesting with regards to the water quality. Cluster I, with the exception of Wells 1 and 15, lies just northeast (less than 500 meters) of the main group of wells of Cluster III. Well 15 appears to lie within the geographic area of Cluster III while Well 1 lies north of the main group. Cluster I has a high total-dissolved-solid content and has a higher $Br^-/Cl^-$ ratio than does Cluster III. The cluster tree also shows that the four wells in the vicinity of Umm As Sahik form a subsidiary cluster relative to Cluster III.

For the largest majority of amalgamation steps the amalgamation distance increases during the procedure. However this is not a requirement of the algorithm. Two instances where the amalgamation distance decreased are shown at the amalgamation of Wells 72 and 124 with Wells 19, 10, 11, 66, and 20. This is also illustrated when Wells 123 and 68 cluster with the previous seven wells.

### *2.1.7.6. Discriminant analysis*

Given two or more groups of data vectors each with $N$ components, stepwise discriminant analysis is a procedure in which components are selected one at a time which best separate the groups. In addition to picking the most significant components, discriminant analysis partitions the space of the selected components into classification regions. That is, if there are $g$ groups and $k$ selected components, then the $k$ dimensional space of the components is partitioned into $g$ regions, one region for each of the $g$ groups. Points which fall into a given region are then classified as belonging to the group corresponding to that region.

The significance of the components is in essence established by the distance of the group means from each other normalized by their standard deviations. Therefore the most significant components for discriminating between clusters of wells given in the preceding sections should be those components which add most to the normalized distances between the clusters.

A discriminant analysis between the 21 samples from the Al Hasa area and the 32 samples from the Al Qatif area shows that $H_2SiO_3$ is the most significant variable. The regions of classification based upon $H_2SiO_3$ are: 1. the Al Qatif region which consists of all wells whose $H_2SiO_3$ measurement is less than 18.6 mg/l, and 2. the Al Hasa region which includes those wells where the $H_2SiO_3$ measurement is greater than 18.6 mg/l. The scatter plot in Fig. 49 shows on the horizontal axis the $H_2SiO_3$ regions for Al Qatif and Al Hasa. It should be noted that only one observation from either of these sets of data is misclassified. Well 12 HP which comes from the Al Qatif region is classified as Al Hasa. However, this well was one of the three wells in the Al Qatif Cluster II (the other two being 9 HP and 74 DP) which were amalgamated one at a time into Cluster II at approximately 8 standard deviation units. This may indicate that these wells do not represent the "typical" wells from the Al Qatif area.

Considering only the 12 chemical species, the most significant variable in discriminating between the 3 clusters from the Al Qatif area was found to be $Cl^-$. The classification regions were given by: 1. Cluster II for values of $Cl^-$ less than 2,050.8 μe/l, 2. Cluster I for values of $Cl^-$ between 2,050.8 μe/l and 28,754.8 μe/l, and 3. Cluster III for values of $Cl^-$ greater than 28,754.8 μe/l. Under this classification scheme only Well 8 HP from Cluster I with a $Cl^-$ concentration of 19,648 μe/l was classified incorrectly, but even then its $Cl^-$ concentration was only 860 μe/l below the minimum necessary to correctly classify it as Cluster II.

In the case of discriminating between the three clusters of the Al Hasa area, the most significant variable among the chemical species and temperature was $H_3BO_3$. The regions of classification were: 1. Cluster I for values of $H_3BO_3$ less than 1.98 mg/l, 2. Cluster II for values between 1.98 mg/l, and 2.52 mg/l, and 3. Cluster III for all those values of $H_3BO_3$ greater than 2.52 mg/l. As was the case in the Al Qatif area, only one well is misclassified using the discriminant procedure. Well 32 DA with a $H_3BO_3$ value of 1.99 mg/l is classified as Cluster II, but is within 0.01 mg/l of correct classification.

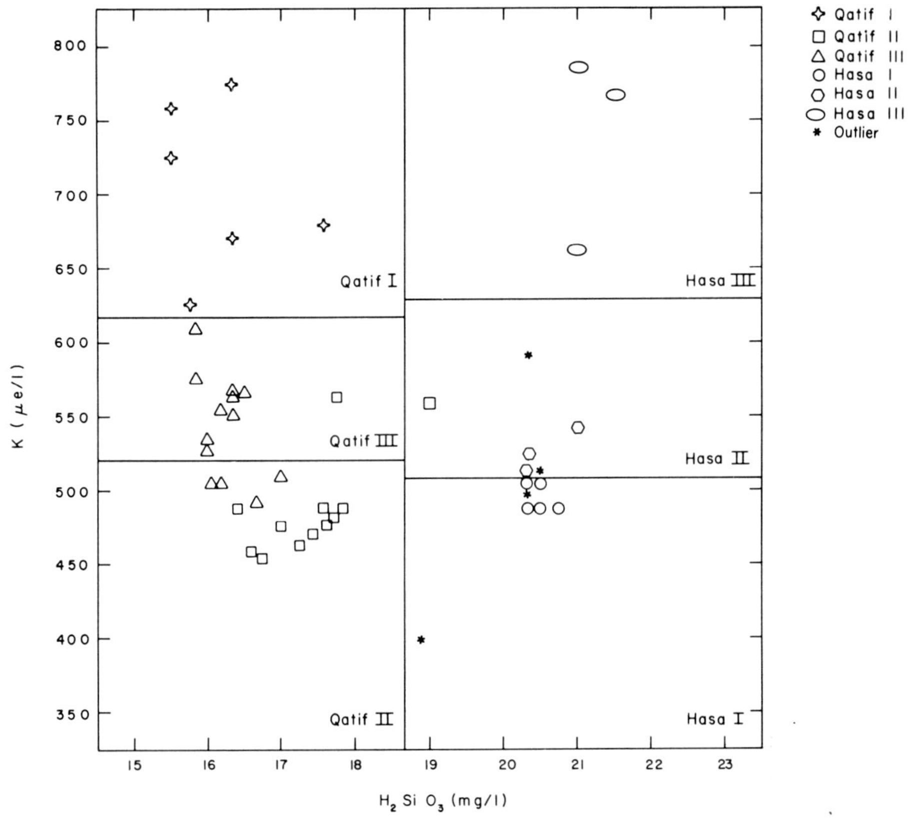


Fig. 49. Partition of the Wells by Discriminant Analysis on $K^+$ and $H_2SiO_3$. This scatter plot shows the partition between the Al Hasa and Al Qatif areas by the variable $H_2SiO_3$. The clusters within each area are partitioned by the variable $K^+$

It is of interest to note the most significant variables, in the order of their significance, for the two areas, as these variables probably add most to the distances between the clusters. For the Al Hasa area these variables (the most significant first) are $H_3BO_3$, $K^+$, temperature, $Mg^{2+}$, and $Na^+$[1]. The most significant variables for the Al Qatif area are $Cl^-$, $Br^-$, $Na^+$, $Ca^{2+}$, $Mg^{2+}$,

[1] It should be noted that temperature was cited by C. Job as the most significant discriminator between the eastern and western wells of the Al Hasa area. He excluded wells 35, 35 a, and 36 as contaminated and included wells 24 and 25. We excluded wells 24 and 25 since they formed a small cluster, and included wells 35, 35 a, and 36 as a third and separate cluster. For discriminating among the three clusters $H_3BO_3$ is the most significant variable, while temperature is the most important if only Clusters I and II are considered. The discriminators reported here do not conflict with those reported by C. Job.

and $K^+$. If a single variable from the above groups were to be chosen as the one which contributes the most distance between the three clusters of the Al Hasa area, while at the same time it contributes the most to the distances between the Al Qatif area, then it would be the most significant variable for simultaneously discriminating between the two groups of three clusters. In the data presented here the most significant variable for simultaneously discriminating between the three clusters of the Al Qatif area and the three clusters of the Al Hasa area is $K^+$. The vertical axis of Fig. 49 shows the classification regions based on $K^+$ for the clusters of the two areas. No wells from the Al Hasa area were misclassified. Wells 24 DA, 25 DA, 41 HA, 42 HP, and 43 DA were excluded from the analysis as outliers, and where possible these are shown on Fig. 49 as *. For the Al Qatif area four wells from Cluster III are classified as Cluster II, while two wells of Cluster II were classified as Cluster III. This discrimination would have been better if the three wells 9 HP, 12 HP, and 74 DP of Cluster II, each with a $K^+$ value in excess of 492 μe/l, would have been deleted from the analysis. In this case only one value from Cluster III would have been misclassified.

### *2.1.7.7. Discussion*

The application of computer-assisted cluster and discriminate analysis to the results of water samples has been used to elucidate the similarities and differences existing among wells in the Eastern Province of Saudi Arabia. The clusters were shown to be correlated to the geographical location similar to that presented by C. Job. In the Al Hasa area two wells (wells 24 and 25) were identified which did not cluster with their respective geographical neighbors. The peripheral wells of Al Hasa were the last to be clustered in that area indicating that they are the least similar to the remaining Al Hasa wells. Three clusters were identified in the Al Qatif area with two clusters being geographically close.

Discriminate analysis was used to establish which variables contributed most to the differences among the clusters. It was found that $H_2SiO_3$ concentrations higher than 18.6 mg/l indicated that the water was from an Al Hasa source. Additionally, $H_2SiO_3$ was the most significant for this discrimination. Within each oasis $K^+$ distinguished the different clusters. For discriminating the three clusters at Al Hasa the $H_3BO_3$ levels were most significant and temperature becomes the most important if only the two major clusters are considered.

It is not within the province of this paper to consider the geochemical implications of these discriminators but only to provide a guide for further examination and field investigation. The general structure of the data as analyzed in this paper fit within the general hydrogeologic model proposed by C. Job (2.1.6.).

Discriminant analysis after a cluster analysis can be useful in determining those variables which tend to separate the clusters. In the case presented here the normalized distance between the clusters of the Al Hasa area is probably most influenced by $H_3BO_3$, and $Cl^-$ is the variable which adds the most to the normalized distances between the clusters of the Al Qatif area.

### 2.1.7.8. *Appendix*

Clustering based on the distance between vectors in a space of more than three dimensions cannot be visualized but is a simple extension of vector algebra of three dimensions. The procedure used in the cluster analysis presented in this study can be illustrated by following the steps for a set of data consisting of three variables for five samples. Two samples (30 and 31) were selected from the Al Hasa area and three (14, 75, and 76) were selected from the Al Qatif area and the three variables used were $Cl^-$, $K^+$, and $H_2SiO_3$ as follows:

| Case No. | Sample No. | Potassium (μe/l) | Chloride (μe/l) | $H_2SiO_3$ (mg/l) |
|---|---|---|---|---|
| 1 | 14 HA | 568 | 9,980 | 16.5 |
| 2 | 75 DA | 563 | 10,261 | 16.1 |
| 3 | 76 DA | 554 | 10,152 | 16.4 |
| 4 | 30 DP | 492 | 6,297 | 20.4 |
| 5 | 31 DP | 492 | 6,281 | 20.7 |
| Mean | | 533.8 | 8,594.2 | 18.02 |

The first step of the procedure is to standardize the individuals as

$$V_{ij} = \frac{X_{ij} - \bar{X}_j}{s_j} \tag{1}$$

where $V_{ij}$ is the standardized variable for the $i$th case and the $j$th component, $X_{ij}$ is the reported value, $\bar{X}_j$ is the mean value of the $j$th component and $s_j$ is the standard deviation. For the three chemical species examined the estimated standard deviation of the chemical analysis procedure was approximately 5% and $s_j$ was taken as 0.05 $X$ and was not estimated from the individual samples. This calculation yields the following:

| Case | $V_{i1}$ (K) | $V_{i2}$(Cl) | $V_{i3}(H_2SiO_3)$ |
|---|---|---|---|
| $i = 1$ | 1.281 | 5.292 | — 1.687 |
| 2 | 1.094 | 4.604 | — 2.131 |
| 3 | 0.757 | 4.604 | — 1.798 |
| 4 | — 1.566 | — 7.331 | 2.642 |
| 5 | — 1.566 | — 7.169 | 2.975 |

The distance between two vectors (samples) $k$ and $l$, is computed as

$$d_{kl} = \left\{ \sum_{j=l}^{N} (V_{kl} - V_{lj})^2 \right\}^{1/2} \tag{2}$$

where the summation is made over the $N$ components which gives the following distances between the vectors.

| Case/Case | 1 | 2 | 3 | 4 | 5 |
|---|---|---|---|---|---|
| 1 | 0.0 | 0.84 | 0.87 | 13.65 | 13.61 |
| 2 | 0.84 | 0.0 | 0.47 | 13.13 | 13.11 |
| 3 | 0.87 | 0.47 | 0.0 | 12.94 | 12.91 |
| 4 | 13.65 | 13.13 | 12.94 | 0.0 | 0.37 |
| 5 | 13.61 | 13.11 | 12.91 | 0.37 | 0.0 |

Vectors 4 and 5 are the closest and are selected as the first pair to be clustered. They are amalgamated by averaging each component and reducing the number of vectors to four. The new table of standardized variables is as follows:

| Case | $V_{i1}$ | $V_{i2}$ | $V_{i3}$ |
|---|---|---|---|
| $i = 1$ | 1.281 | 5.292 | — 1.687 |
| 2 | 1.094 | 4.604 | — 2.131 |
| 3 | 0.757 | 4.604 | — 1.798 |
| (4–5) | — 1.566 | — 7.250 | 2.808 |

The distances among vectors 1, 2, and 3 remain unchanged but the distance between (4–5) and the others is computed giving the following distance matrix.

| Case/Case | 1 | 2 | 3 | (4–5) |
|---|---|---|---|---|
| 1 | 0.0 | 0.84 | 0.87 | 13.62 |
| 2 | 0.84 | 0.0 | 0.47 | 13.11 |
| 3 | 0.87 | 0.47 | 0.0 | 12.93 |
| (4–5) | 13.62 | 13.11 | 12.93 | 0.0 |

The closest pair is 2–3 which are then amalgamated and the total number of vectors reduced to three giving the following standardized variables.

| Case | $V_{il}$ | $V_{i2}$ | $V_{i3}$ |
|---|---|---|---|
| $i = 1$ | 1.281 | 5.292 | — 1.687 |
| (2–3) | 0.925 | 4.604 | — 1.965 |
| (4–5) | — 1.566 | — 7.250 | 2.808 |

The distances between the vectors are now:

| Case/Case | 1 | (2–3) | (4–5) |
|---|---|---|---|
| 1 | 0.0 | 0.82 | 13.62 |
| (2–3) | 0.82 | 0.0 | 13.02 |
| (4–5) | 13.62 | 13.02 | 0.0 |

The closest pair is 1–(2–3). To maintain an equal weighting of all the initial vectors the amalgamation is performed by averaging the three initial values of each component. This gives the following components for the two remaining vectors.

| Case | $V_1$ | $V_2$ | $V_3$ |
|---|---|---|---|
| (1–2–3) | 1.044 | 4.834 | — 1.872 |
| (4–5) | — 1.566 | — 7.250 | 2.808 |

The distance between these two vectors is 13.22 and this is the last amalgamation.

The sequence of cluster formation can be illustrated by a cluster tree which shows the order and the distance between the vectors amalgamated at that step. The cluster tree for this example is illustrated as follows:

| Amalgamation distance | Case 1 | | 2 | | 3 | 4 | | 5 |
|---|---|---|---|---|---|---|---|---|
| | I | | I | | I | I | | I |
| 0.37 | I | | I | | I | | + | |
| | I | | I | | I | | I | |
| 0.47 | I | | | + | | | | |
| | I | | | I | | | I | |
| 0.82 | | + | | | | | I | |
| | | I | | | | | I | |
| 13.22 | | | | + | | | | |

In the example used there was a monotonic increase in the amalgamation distance through the procedure. However, this is a function of the particular data set and not imposed by the method. The procedure is hierarchical because once two vectors have been amalgamated they cannot be separated at a later step. Except for an increase in the size of the distance matrix the operations for an N-dimensional space are identical to those illustrated here.

## 2.1.8. Isotopic Composition of Waters of Al Qatif and Al Hasa Areas

(H. Moser, E. Pak, W. Rauert, W. Stichler, J. G. Zötl)

### *2.1.8.1. Introduction*

Concerning the natural isotopes used for Quaternary studies in Saudi Arabia, the measurement of the so-called environmental isotopes was an important tool of research. For hydrological investigations, samples for the determination of the principal environmental isotopes, deuterium ($^2H$ or D), oxygen-18 ($^{18}O$), tritium ($^3H$ or T), and, in a few cases, carbon-14 ($^{14}C$) were collected. Carbon-14 measurements of solid materials will be described in chapter 2.2.2. The decisive results of sulfur-34 ($^{34}S$) determinations can also be seen from chapter 2.1.6.

### *2.1.8.2. Deuterium and oxygen-18* [1]

Out of every $10^6$ water molecules in natural waters of normal isotopic composition $H_2\,^{16}O$ there are about 2,000 molecules of $H_2\,^{18}O$, containing the heavy-oxygen isotope $^{18}O$, and about 320 molecules of $HD\,^{16}O$, in which one of the two ordinary hydrogen atoms is replaced by the heavy-hydrogen isotope deuterium ($^2H$, D). It has been found that sea water contains more molecules with heavy isotopes than fresh water. Systematic investigations over the last 20 years have revealed that the proportions of $H_2\,^{18}O$ and $HD\,^{16}O$ fluctuate within the ranges 1,880–2,010 and 180–340 ppm respectively (S. Epstein, T. Mayeda, 1953) [2]. This is due essentially to the lower

[1] All the measurements of D, $^{18}O$, T and $^{14}C$ in water samples were carried out in the Institut für Radiohydrometrie der Gesellschaft für Strahlen- und Umweltforschung in Munich, Germany.

[2] For summary reviews with literature references, see, for example, S. Epstein, T. Mayeda, 1953; I. Friedman et al., 1964; W. Dansgaard, 1964; H. Craig, 1961; H. Moser, W. Stichler, 1971; J. R. Gat, 1971; W. Drost et al., 1974.

vapour pressures of $H_2\,^{18}O$ and $HD\,^{16}O$, as a result of which isotopic fractionation takes place during every phase transformation, e.g., during condensation and evaporation. Isotopic exchange, diffusion and dispersion ensure that isotopic fractionations, which initially take place at the phase boundaries, become measurable volume effects in a water sample.

It is usual to establish the deuterium and oxygen-18 content as the relative deviation $\delta$ D or $\delta\,^{18}O$ from a standard, the so-called SMOW (Standard Mean Ocean Water):

$$\delta\,D, \delta\,^{18}O \text{ respectively} = \frac{R_{\text{sample}} - R_{\text{standard}}}{R_{\text{standard}}} \cdot 1000 \text{ per mill}$$

where R are the isotope ratios D/H and $^{18}O/^{16}O$, respectively, in the sample and the standard, respectively. The accuracy (2 $\sigma$-criterion) of mass spectrometric measurements of $\delta$ D and $\delta\,^{18}O$ is $\pm$ 1 per mill and $\pm$ 0.2 per mill, respectively. Concerning the analytical techniques see W. Drost et al. (1974). With regard to isotopic fractionation in the natural cycle, the isotopic composition of water is determined mainly by the following physical factors:

a) During evaporation there is heavy-isotope enrichment of the water, whereas during condensation there is heavy-isotope depletion of the water vapour. Similarly, during freezing there is heavy-isotope depletion of the liquid phase.

b) Other things being equal, isotopic fractionation increases with decreasing temperature.

c) The ratio of the deuterium content to the oxygen-18 content is influenced by the kinetics of the phase transformation. Oxygen-18 enrichment is more marked than deuterium enrichment, particularly with increasing evaporation rate. Under certain conditions, there can also be a change in this ratio as the result of a specific isotope exchange (e.g., with the aquifer).

Because of such isotopic effects, precipitations are labelled in a characteristic manner.

Water vapour from the sea and lakes has a lower content of heavy isotopes than the evaporating surface water. In precipitations, the heavy-isotope content decreases with increasing distance from the coast (continental effect); moreover, with increasing quantities of rain there is decreasing heavy-isotope content (amount effect). Finally, the heavy-isotopes content in precipitations also decreases where a moist air mass rises because of orographic elevations (altitude effect); here, in addition to the increased raining out of the heavy isotopes, the decrease in the condensation temperature with altitude also contributes to the heavy-isotope depletion of the precipitations. The temperature effect in isotopic fractionation has the further consequence of seasonal changes in the isotope content, as well as heavy-isotope depletion with increasing geographical latitude (latitude effect).

The relationship between the $\delta$ D and $\delta\,^{18}O$ values has been extensively discussed (H. Craig, 1961; W. Dansgaard, 1964). If the annual mean $\delta$ D and $\delta\,^{18}O$ values in precipitations from continental areas in the northern hemisphere are plotted against each other, the linear relationship satisfies the equation $\delta\,D = 8 \cdot \delta\,^{18}O + 10$ (see Fig. 50).

In the equation $\delta D = m\ \delta^{18}O + t$ for the $\delta$ values of precipitation, $m$ and $t$ are essentially determined by the meteorological conditions prevailing during the formation of water vapours in the course of the hydrologic cycle. $t$ depends essentially on the thermodynamic situation and on the transport phenomena during the primary evaporation from the sea-surface. $m$, however, depends on the kinetics of secondary evaporation of the precipitation water. Its normal value of 8 decreases with increasing evaporation rates.

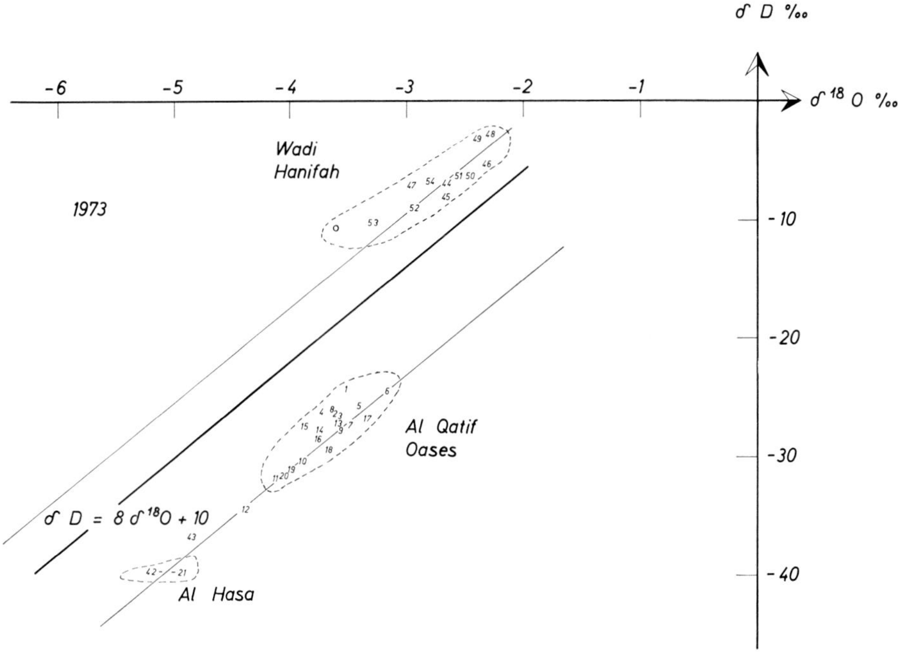


Fig. 50. Relationship between $\delta D$ and $\delta^{18}O$ content in water samples from Wadi Hanifah, Al Qatif oases and Al Hasa area

W. DANSGAARD (1964) established that in the Eastern Mediterranean the primary evaporation rate is high ($t \cong 22$), whereas in the northern hemisphere it amounts to generally $t \cong 10$.

In Al Hasa area of eastern Saudi Arabia water samples for D and $^{18}O$ measurements were collected from March 28 until April 4, 1973. Table 29 shows the results. With regard to the question whether or not one sample collection provides significant data, it is important to mention that in Al Qatif area samples were collected from the same springs between March 8 and March 19, 1973, February 28 and March 6, 1974, and February 9 and 14, 1975. The results (Table 30) show very clearly that the obtained data are equivalent. Consequently, the conclusion can be drawn that this is also true for the water samples taken from Al Hasa.

Table 29. *Stable isotope data from Al Hasa springs 1973*

| Location | No of sample | δ D ‰ | δ $^{18}$O ‰ |
|---|---|---|---|
| springs in Al Hasa area | 24 | — 39.8 | — 5.36 |
| | 25 | — 40.1 | — 5.24 |
| | 26 | — 39.9 | — 5.27 |
| | 27 | — 39.8 | — 5.26 |
| | 28 | — 39.9 | — 5.07 |
| | 29 | — 40.1 | — 5.30 |
| | 30 | — 39.8 | — 5.40 |
| | 31 | — 40.0 | — 5.45 |
| | 32 | — 39.8 | — 5.26 |
| 'Ayn Mansur | 33 | — 39.6 | — 5.07 |
| various | 34 | — 40.1 | — 5.12 |
| | 35 | — 39.1 | — 5.07 |
| | 35 a | — 39.9 | — 5.10 |
| | 36 | — 39.2 | — 4.87 |
| Agricultural farm | 37 | — 39.8 | — 5.36 |
| 'Ayn Sumbor | 38 | — 40.0 | — 5.05 |
| 'Ayn Haql | 39 | — 39.4 | — 5.12 |
| 'Ayn Shabah | 40 | — 39.6 | — 5.20 |
| 'Ayn Khudud | 41 | — 38.8 | — 4.85 |
| 'Ayn Najm | 42 | — 40.0 | — 5.32 |
| Al 'Uyun | 43 | — 36.7 | — 4.86 |

Looking at the δ D–δ $^{18}$O relation of Al Qatif and Al Hasa waters (Fig. 50) it can be seen that the values lie distinctly below the line $\delta D = 8 \cdot \delta\, ^{18}O + 10$, valid for temperate climates, e.g., Middle Europe. There can be no doubt that the equation line of the Al Hasa and Al Qatif water samples is based upon a climatic effect. The waters recently discharging from springs in Al Hasa and Al Qatif oases originate from precipitations evaporated from the sea under other climatic conditions than today's. The low *t*-values lead to the conclusion that the water vapour of this condensate was in a better equilibrium with the sea water than is the case at present. The evaporation rate was evidently lower than today, and this could be due to a more humid climate than at present. This conclusion is confirmed by two other circumstances: the carbon-14 measurements (see 2.1.8.4.) and the absence of tritium (see 2.1.8.3.).

### *2.1.8.3. Tritium*

It is barely 25 years since the discovery of radioactive carbon-14 (see chapter 2.1.8.4.) and tritium in nature (E. C. Anderson et al., 1947; V. Faltings, P. Harteck, 1950). Both radionuclides are predominantly produced in the upper atmosphere through the action on nitrogen atoms of neutrons generated by cosmic rays.

Tritium combines with oxygen to form water in the atmosphere and reaches the earth, and hence the groundwater, in precipitation and in the form of water vapour. The average annual concentration of naturally pro-

Table 30. *Stable isotope data of Al Qatif area 1973–1975*

| Location | No. of sample | δ D ‰ | δ $^{18}O$ ‰ |
|---|---|---|---|
| 'Ayn Al Labaniyah I | 1973: 2 | — 26.3 | — 3.63 |
| | 1974: 57 | — 26.4 | — 3.77 |
| | 1975: 126 | — 26.3 | — 3.82 |
| 'Ayn Al Labaniyah II | 1973: 3 | — 26.5 | — 3.59 |
| | 1974: 58 | — 26.4 | — 3.83 |
| | 1975: 127 | — 26.3 | — 3.84 |
| 'Ayn Hussain | 1973: 4 | — 26.2 | — 3.72 |
| | 1974: 59 | — 26.3 | — 3.70 |
| | 1975: 128 | — 26.5 | — 3.76 |
| 'Ayn Subia I | 1973: 5 | — 25.7 | — 3.41 |
| | 1974: 64 | — 26.2 | — 3.68 |
| | 1975: 129 | — 26.3 | — 3.76 |
| 'Ayn Subia II | 1973: 13 | — 26.9 | — 3.59 |
| | 1974: 65 | — 27.2 | — 3.75 |
| | 1975: 130 | — 26.6 | — 3.78 |
| 'Ayn Al Jalamiyah | 1973: 15 | — 27.3 | — 3.88 |
| | 1974: 61 | — 27.1 | — 3.75 |
| | 1975: 131 | — 27.2 | — 3.89 |
| Drainage near 'Ayn Al Jalamiyah | 1973: n.s. | n.s. | n.s. |
| | 1974: 62 | — 25.8 | — 3.60 |
| | 1975: 132 | — 25.6 | — 3.69 |
| 'Ayn Muhairig | 1973: 14 | — 27.7 | — 3.75 |
| | 1974: 63 | — 28.2 | — 3.83 |
| | 1975: 133 | — 27.2 | — 3.95 |
| 'Ayn Obede (Al Ajam) | 1973: 11 | — 31.7 | — 4.11 |
| | 1974: 70 | — 31.7 | — 4.40 |
| | 1975: 141 | — 31.4 | — 4.41 |
| Hammam Balousa | 1973: 17 | — 26.8 | — 3.34 |
| | 1974: 78 | — 27.3 | — 3.44 |
| | 1975: 143 | — 24.6 | — 3.44 |
| Sea water Dawhat Zaloom | 1973: n.s. | n.s. | n.s. |
| | 1974: 67 | + 24.2 | + 5.04 |
| | 1975: 145 | + 27.6 | + 4.84 |
| As Sabacha II (Al Ajam) | 1973: n.s. | n.s. | n.s. |
| | 1974: 69 | — 30.0 | — 4.16 |
| | 1975: 123 | — 29.4 | — 4.18 |
| 'Ayn Hussain (Al Ajam) | 1973: n.s. | n.s. | n.s. |
| | 1974: 71 | — 32.2 | — 4.29 |
| | 1975: 124 | — 31.5 | — 4.36 |
| 'Ayn Towairit (Al Ajam) | 1973: 9 | — 27.8 | — 3.68 |
| | 1974: 73 | — 28.1 | — 3.92 |
| | 1975: 125 | — 28.5 | — 3.89 |

n.s. = no sample.

duced tritium is very low, e.g., about 6 T.U.[1] in Central European precipitation. Since 1953, however, the tritium concentration in rain has increased all over the world as a result of nuclear weapons tests and, in 1963, reached a peak of up to 1,000 times the natural level in the northern hemisphere. In the southern hemisphere the peak tritium concentrations did not occur until 1964 and 1965 and were considerably lower than in the northern hemisphere (e.g., B. TH. VERHAGEN et al., 1970).

Table 31. *Tritium content in Al Qatif waters in 1973, 1974, 1975*

| Location | Sample number | | | Tritium content (T.U.) | | |
|---|---|---|---|---|---|---|
| | 1973 | 1974 | 1975 | 1973 | 1974 | 1975 |
| 'Ayn Al Labaniyah I | 2 | 57 | 126 | $< 0.8$ | $< 3.5$ | $0.1 \pm 1.3$ |
| 'Ayn Al Labaniyah II | 3 | 58 | 127 | $< 0.8$ | $< 4.5$ | $3.3 \pm 1.3$ |
| 'Ayn Hussain | 4 | 59 | 128 | $< 2.3$ | $< 2.6$ | $1.0 \pm 1.4$ |
| 'Ayn Subia I | 5 | 64 | 129 | $< 0.9$ | $< 2.2$ | $0.6 \pm 1.3$ |
| 'Ayn Subia II | 13 | 65 | 130 | $< 2.3$ | $< 4.4$ | $1.0 \pm 1.3$ |
| 'Ayn Al Jalamiyah | 15 | 61 | 131 | $< 2.7$ | $< 4.5$ | $1.3 \pm 1.2$ |
| Drainage near 'Ayn Al Jalamiyah | n.s. | 62 | 132 | n.s. | $3.4 \pm 2.2$ | $1.5 \pm 1.3$ |
| 'Ayn Muhairig | 14 | 63 | 133 | $< 2.7$ | $< 4.0$ | $0 \pm 1.2$ |
| 'Ayn Obede | 11 | 70 | 141 | $< 0.9$ | $< 2.2$ | $0.6 \pm 1.0$ |
| Hammam Balousa | 17 | 78 | 143 | $< 2.2$ | $2.5 \pm 1.8$ | $1.2 \pm 1.0$ |
| Sea water Dawhat Zaloom | n.s. | 67 | 145 | n.s. | $24.4 \pm 2.7$ | $13.2 \pm 1.3$ |
| As Sabacha II (Al Ajam) | n.s. | 69 | 123 | n.s. | $< 2.2$ | $0.4 \pm 1.4$ |
| 'Ayn Hussain (Al Ajam) | n.s. | 71 | 124 | n.s. | $< 3.4$ | $0.5 \pm 1.4$ |
| 'Ayn Towairit (Al Ajam) | 9 | 73 | 125 | $< 2.7$ | $< 4.6$ | $0.3 \pm 1.4$ |

Precipitation that infiltrated into the soil before 1953 has today a tritium concentration of less than 2 T.U. due to radioactive decay with a half-life of 12.3 years. Tritium concentrations in precipitations that have been recharging groundwater during the last few years are higher by 1 to 2 orders of magnitude. Thus the hydrological evaluation of tritium measurements in water is now based on the determination of bomb-produced tritium in order to distinguish more recent from older water within a time-scale of a few years to a few decades. The identification of more recent waters with the aid of groundwater recharge models presupposes a knowledge of the chronology of bomb-induced tritium input into the groundwater as a result of precipitation etc. The tritium concentration in precipitations shows seasonal fluctuations, as well as an increase from the Atlantic coast of Europe towards the interior of the continent (continental effect; see W. ROETHER, 1967).

Out of a total number of 20 sampling locations in 1973 the tritium measurements were repeated, in 1974 and in 1975, in 14 places. The results for Al Qatif waters are shown in Table 31. Table 32 shows the tritium values of the Al Hasa waters and of wells in Wadi Hanifah.

[1] A concentration of one tritium atom per $10^{18}$ hydrogen atoms is known as one "tritium unit" (abbreviated to T.U.).

From hydrogeological reasoning and some carbon-14 analyses (see chapter 2.1.8.4.) one can expect that the groundwater in the areas of Al Qatif and Al Hasa is more than 20 years old and therefore does not contain tritium. In fact most of the water samples have no or only very little tritium, below the detection limit. Also no tritium was found in additional samples not reported in Table 31, with detection limits varying between 1.1 and 3.8 T.U. As for the small tritium contents which were measured in a few groundwater samples, it should be taken into account that most of the springs discharge from wells captured in open basins of a diameter ranging from a few meters up to more than 20 meters. Therefore, rainfalls in short but intensive showers can contaminate the springs directly or through seepage from the surrounding farm land.

Table 32. *Tritium content in Al Hasa waters and of wells in Wadi Hanifah (1973)*

| | | Al Hasa | | Wadi Hanifah | |
|---|---|---|---|---|---|
| Sample No. | T.U. | Sample No. | T.U. | Sample No. | T.U. |
| 24 | < 0.7 | 35 | < 2.5 | 44 | < 12 |
| 25 | < 2.6 | 35 a | < 2.6 | 45 | < 4 |
| 26 | < 2.3 | 36 | < 2.3 | 46 | 7.0 ± 2 |
| 27 | < 2.5 | 37 | < 2.6 | 47 | < 3.8 |
| 28 | < 0.5 | 38 | < 0.3 | 48 | 74 ± 10 |
| 29 | < 2.5 | 39 | < 2.6 | 49 | 67 ± 8 |
| 30 | < 1.2 | 40 | < 1.2 | 50 | 16 ± 2 |
| 31 | < 2.8 | 41 | < 2.4 | 51 | < 4 |
| 32 | < 0.9 | 42 | < 1.9 | 52 | < 0.9 |
| 33 | < 0.9 | 43 | < 2.9 | 53 | < 3.8 |
| 34 | < 2.6 | | | 54 | < 3.4 |

In contrast to this is the situation in Wadi Hanifah. There are some wells or springs with higher tritium concentrations, thus showing a more or less pronounced contribution of present rainfalls to the feeding of shallow groundwater wells (Table 32; see also chapter 2.4.4.). These water samples (No. 46, 48, 49, and 50) have the highest D and $^{18}O$-contents. Fig. 50 exemplifies these special isotopic conditions very well, especially by the fact that $t$ of the equation $\delta D = m \delta ^{18}O + t$ is higher than 10 (16) thereby characterizing recent water of this region.

### *2.1.8.4. Carbon-14*

The radiocarbon method for determining the age of organic samples and sinter precipitated from water is briefly described in chapter 2.2.3. With regard to groundwater dating according to the model of K. O. Münnich (1957, 1968; J. C. Vogel, D. Ehhalt, 1963; I. Wendt et al., 1967) carbon-14 enters the groundwater owing to the dissolution of fossil ($^{14}C$-free) calcium carbonate by biogenic ($^{14}C$-containing) $CO_2$ in water as calcium bicarbonate:

$$CaCO_3 + CO_2 + H_2O \rightleftharpoons 2\,HCO_3^- + Ca^{++}$$

Experience to date shows that recent groundwater predominantly has $^{14}C$ concentrations of between 70 and 90 percent modern (K. O. Münnich, 1968). Pointers to the origin of the carbon in groundwater, and hence to deviations from the dating model outlined can be obtained from mass-spectrometric determination of the isotopic ratio of $^{13}C/^{12}C$. So far experience would seem to refute any appreciable isotopic exchange of $^{14}C$ between the dissolved bicarbonate and the carbonates of the matrix of natural aquifers. Considering the uncertainties, however, interpretation of carbon-14 data should generally be treated with caution.

Nevertheless the four $^{14}C$-measurements of Al Qatif and Al Hasa waters (see Table 33) show without any doubt old waters which infiltrated under climatic conditions different from those prevailing at present.

Table 33. *Results of Carbon-14, Carbon-13 and tritium measurements of water samples collected from Al Qatif and Al Hasa areas (1973)* [1]

| Place or No. of sample | $^{14}C$-content [2] (% modern) | $^{14}C$-age (years B.P., uncorrected) | $^{13}C$ [3] (‰) | Tritium content (T.U.) |
|---|---|---|---|---|
| 'Ayn Al Labaniyah I (Al Qatif) | < 5.9 | > 22,000 | — 9.0 | < 0.8 |
| 'Ayn Al Labaniyah II (Al Qatif) | < 1.2 | > 34,500 | — 8.8 | < 2.7 |
| No. 32 (Al Hasa) | < 1.4 | > 33,000 | — 10.3 | < 0.9 |
| 'Ayn Mansur (Al Hasa) | < 1.4 | > 33,000 | — 10.0 | < 0.9 |

[1] Sampling through carbonate precipitation by C. Job, 1973.

[2] An initial $^{14}C$-content of 85% modern was assumed.

[3] The $^{13}C$-content is given as the relative per mill deviation from the limestone standard PDB.

### *2.1.8.5. Sulfur-34*

The abundances of the stable isotopes of sulfur are known to vary slightly in nature, due to isotope fractionation effects occurring in numerous geochemical and biological processes. Accordingly, the determination of sulfur isotope abundances has been widely used for the investigation of several geological problems (ore genesis, evaporite dating, etc.). For a review see H. G. Thode, 1970, and for a bibliography, P. G. Coomer et al., 1970.

Isotope analyses are carried out, after conversion of the samples to $SO_2$, with a mass spectrometer, and the results are expressed as $\delta\,^{34}S$ values, defined as permil relative deviations of the ratio $^{34}S/^{32}S$ (the two most abundant isotopes) from a standard, for which troilite sulfur of the Canon Diablo meteorite is used conventionally.

Sulfates from precipitation, springs, rivers, and lakes have shown wide variations in $\delta\,^{34}S$, indicating rather different origins of their sulfur [1]. However, for springs with sulfate concentrations as high as those of the Saudi Arabian springs investigated, it seems likely that the dominant source of sulfate is dissolution of evaporites. It is known that neither during evaporation of sea water nor during dissolution of evaporites any considerable fractionation of the sulfur isotopes occurs, and that $\delta\,^{34}S$ of evaporites

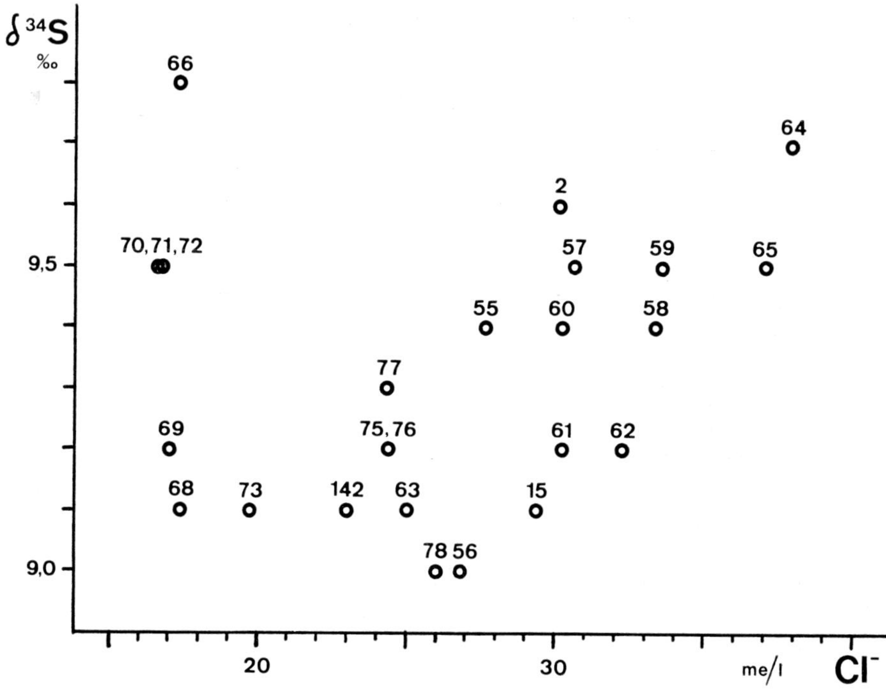


Fig. 51. $\delta\,^{34}S$ and chloride content of Al Qatif waters

of different ages shows characteristic variation corresponding to that of $\delta\,^{34}S$ of coeval ocean-water sulfate which changed due to changing bacterial reduction of oceanic sulfate and other changes of the turnover rates in the sulfur cycle. Whereas $\delta\,^{34}S$ of recent ocean water is $+\,20 \pm < 1$ per mill, it fluctuated around this value ($\delta\,^{34}S > +\,14$ per mill) since the Triassic, had an absolute minimum in the Permian ($\delta\,^{34}S$ between

[1] The sulfates of the present investigation were precipitated as $BaSO_4$ and submitted by C. Job; the sample numbers correspond to his article (chapter 2.1.6.). The sulfur isotope determinations were carried out in the Institut für Radiumforschung und Kernphysik of the Austrian Academy of Science in Vienna, Austria. Experimental details of the analytical procedure have been published by E. Pak, H. Felber (1974). The measured $\delta\,^{34}S$-values of the investigated waters are compiled in Table 34; the accuracy is 0.2 per mill. Considering the typical $\delta\,^{34}S$-values cited above, the results suggest that the sulfate of the springs originates from Permian evaporites.

+ 8 and + 12 per mill), and was higher in earlier epochs (H. G. THODE, J. MONSTER, 1965; H. NIELSEN, W. RICKE, 1964; W. T. HOLSER, I. R. KAPLAN, 1966).

In the Qatif area, excluding sample 74, $\delta\,^{34}S$-values of all 25 springs lie in the remarkably narrow range from + 9.0 to 9.8 per mill which is only slightly wider than the variation due to analytical error. Nevertheless, an attempt was made to get information about the possibility of infiltration of

Table 34. *$\delta\,^{34}S$ values of the investigated waters of Al Qatif and Al Hasa*

| | | | No. | Lab.-No. | $\delta\,^{34}S$ (‰) |
|---|---|---|---|---|---|
| Al Qatif | 'Ayn Al Labaniyah I | 1973 | 2 | VRI-S-19 | + 9.6 |
| | 'Ayn Al Labaniyah I | 1974 | 57 | -53 | + 9.5 |
| | 'Ayn Al Labaniyah II | 1974 | 58 | -168 | + 9.4 |
| | 'Ayn Hussain | 1974 | 59 | -54 | + 9.5 |
| | Drainage near 'Ayn Hussain | 1974 | 60 | -55 | + 9.4 |
| | 'Ayn Al Jalamiyah | 1973 | 15 | -20 | + 9.1 |
| | 'Ayn Al Jalamiyah | 1974 | 61 | -56 | + 9.2 |
| | Drainage near 'Ayn Al Jalamiyah | 1974 | 62 | -57 | + 9.2 |
| | 'Ayn Muhairig | 1974 | 63 | -68 | + 9.1 |
| | 'Ayn Subia I | 1974 | 64 | -58 | + 9.7 |
| | 'Ayn Subia II | 1974 | 65 | -69 | + 9.5 |
| | 'Ayn Al Qatif | 1974 | 66 | -169 | + 9.8 |
| | 'Ayn Quhaih | 1974 | 75 | -171 | + 9.2 |
| | 'Ayn Sidery | 1974 | 76 | -65 | + 9.2 |
| | 'Ayn Quarra | 1974 | 77 | -172 | + 9.3 |
| | Drainage channel | 1975 | 136 | -90 | + 9.3 |
| | Drainage channel | 1975 | 140 | -91 | + 9.6 |
| | Hammam Balousa | 1974 | 78 | -173 | + 9.0 |
| Al Ajam | As Sabacha I | 1974 | 68 | -60 | + 9.1 |
| | As Sabacha II | 1974 | 69 | -170 | + 9.2 |
| | 'Ayn Obede | 1974 | 70 | -61 | + 9.5 |
| | 'Ayn Hussain | 1974 | 71 | -62 | + 9.5 |
| | 'Ayn Sifa | 1974 | 72 | -70 | + 9.5 |
| | 'Ayn Towairit | 1974 | 73 | -63 | + 9.1 |
| | Al Gasma | 1974 | 74 | -64 | + 10.7 |
| Umm As Sahik | | 1975 | 142 | -92 | + 9.1 |
| Tarut "A" | | 1974 | 55 | -52 | + 9.4 |
| Tarut "B" | | 1974 | 56 | -167 | + 9.0 |
| Tarut Bay | | 1975 | 144 | -93 | + 14.4 |
| Sea water Dawhat Zaloom | | 1974 | 67 | -59 | + 20.3 |
| Sea water Dawhat Zaloom | | 1975 | 145 | -94 | + 20.1 |
| Al Hasa 'Ayn Mansur | | 1973 | | -21 | + 12.4 |
| Al Hasa No. 32 | | 1973 | | -22 | + 12.4 |

sea water (or Tarut Bay water) which would cause a correlated increase of $\delta\,^{34}S$ and chloride content. By statistical evaluation, C. JOB (cf. chapter 2.1.6.) stated that a significant correlation cannot be found. In spite of that, by view of Fig. 51 a positive correlation is suggested after omission of samples 66, 70, 71, and 72, and even the gradient would match with a hypothetical mixture with Tarut Bay water. Then however, the concentrations of all other ions ought to have gradients corresponding to the composition of the Bay water, which is not the case as can be proven from the data cited in chapter 2.1.6. Moreover, a reason for omission of the four samples cannot be adduced. Still, in our opinion, the misfit of the ion concentration gradients is a stronger argument against the possibility of sea water infiltration than the lack of demonstrability of a correlation between $\delta\,^{34}S$ increase and chloride content in this case.

## 2.2. As Sulb Plateau

### 2.2.1. General Geology

(E. Schyfsma)

Like the Shedgum Plateau, As Sulb Plateau forms part of As Summan Plateau, a vast, undulating area consisting of continental clastics of Miocene and Pliocene age (Fig. 2, No. 28).

As Sulb area extends roughly from lat 25° 19′ N to 26° 41′ N and from long 47° 15′ E to 48° 30′ E. Its southern extremity is situated about 100 kilometers due west of Al Hofuf, and from there it extends about 190 kilometers to the northwest.

In the northwestern part of As Sulb Plateau, the continental Mio-Pliocene clastics overlie unconformably the Umm er Radhuma Formation of Paleocene to basal Eocene age (Fig. 53). In the southeastern part, they overlie unconformably the Rus Formation of Lower Eocene (Ypresian) age and the Dammam Formation of Lower to Middle Eocene (Ypresian to Lutetian) age (Fig. 52). The thickness of the Umm er Radhuma Formation here is 105 to 110 meters, that of the Rus Formation 25 to 30 meters, and that of the Dammam Formation, of which only the basal part is exposed, about 15 meters. All these formations show a very gentle dip to the east, rarely exceeding one degree.

The continental Mio-Pliocene clastics form a wedge that in the northern part of As Summan Plateau thickens from 0 meters in the west to about 100 meters in the east, where they grade into marine facies. In the south, the wedge thickens to about 300 meters before passing under the sands of Ar Rub' Al Khali. Surface and subsurface data show that the Mio-Pliocene clastics are everywhere in discordant contact with underlying units. Paleocene and Eocene strata are involved in As Sulb area, but away from this area, Cretaceous and even Paleozoic beds are affected as well. The contact in each case, however, is relatively simple to identify, as the varicolored lenticular sandy carbonates and sandstones of the Mio-Pliocene usually contrast sharply with the pure carbonates and bedded clastics of older formations.

The Miocene and Pliocene continental clastics of As Sulb Plateau grade into marine equivalents to the east. Along the Arabian Gulf this part of the Tertiary can be subdivided into three formations which from the bottom up are Hadrukh, Dam, and Hofuf. The most western margin of marine sedimentation seems to be generally less than 100 kilometers from the present Gulf shoreline. From this shoreline to the west, marine marker beds grade to continental sediments, and the coastal subdivisions can no longer be recognized beyond the 100 kilometers borderline. Beyond the western limit of the continental clastics, patches of lake deposits, plastered on the Sulaiy outcrop belt, have been named the Kharj Formation. It is suspected that they are, at least in part, equivalent to the Miocene and Pliocene continental clastics and the marine Hadrukh, Dam and Hofuf Formations.

The Miocene and Pliocene continental clastics in As Sulb Plateau consist of a heterogeneous series of red to reddish-brown, gray and rarely off-white, calcareous sandstones, sandy marls, poorly-sorted sandy shale, and impure, sandy, lacustrine limestone. In these continental deposits, no subdivision into formations is possible. The nearest point of these deposits to the present Arabian Gulf shore is about 150 kilometers. There is no doubt about their Mio-Pliocene age since they grade into shallow marine formations with diagnostic fossils near the coast.

Fossils recovered from the Hadrukh Formation include *Chlamys* cf. *senatoria* (Gmelin), *Clementia papyracea* (Gray), *Diplodonta* cf. *incerta* D'Archiac et Haime, *Ostrea latimarginata* Vredenburg, *Turritella gradata, Ostrea* spp., *Turritella* spp., *Chara,* crab claws, sharks' teeth, echinoid remains, and fossil wood. The stratigraphic position below the Dam Formation postulates an early Miocene age for the Hadrukh.

Fossils from the Dam Formation include: *Ostrea latimarginata* Vredenburg, *Echinocyamus* sp., *Archaias* sp., similar to the Lower Fars fauna of Iran, Iraq and Kuwait, which would point to a middle Miocene age for the Dam Formation. Other fossils recovered from the Dam include *Balanus* sp., *Agassizia* cf. *scrobiculata* Valenc var. *persica, Brisus latitudinensis* Clegg, *Echinodiscus desori* Duncan et Sladen, *Lovenia elongata* Gray, *Moira adamthi* Clegg, *Schizaster* sp., *Temnopleurus persica* Clegg, *Temnopleurus* sp., *Archaias angulatus* (Fichtel et Moll), *Elphidium* sp., *Operculina* sp., *Peneroplis* spp., *Quinqueloculina* sp., *Triloculina* sp., *Anomia* sp., *Arca* sp., *Bullaria* sp., *Chlamys senatoria* (Gmelin), *Clavatula* sp., *Clementia papyracea* (Gray), *Conus* sp., *Coralliophila* sp., *Diplodonta* cf. *incerta* D'Archiac et Haime, *Epitonium* sp., *Ostrea angulata* Sowerby, *Ostrea virleti* Deshayes, *Ostrea* cf. *virleti* Deshayes, *Ostrea* spp., *Placenta* sp., *Turritella* cf. *gradata, Turritella* spp., corals, fossil wood, vertebrate fragments, crab claws, and ostracodes. This assemblage points to a Mio-Pliocene age.

Fossils from the Hofuf Formation include: *Chara* sp., *Lymnaea* sp., *Planorbis* sp., *Thiara* sp., ostracodes, and indeterminate gastropods. The presumed age is late Miocene or Pliocene.

The only fossils so far discovered in the Kharj deposits are *Corbicula* sp., and *Melania* sp. These fossils are not diagnostic of any specific age but are Neogene. Stratigraphically, however, there is no doubt about the age of the Kharj deposits. They are post-Eocene and pre-Quaternary, i.e., most certainly Tertiary. If the channel gravels in the area are about the same age, they may be roughly contemporaneous with the great gravel flood of the basal Hofuf Formation in the area to the east.

The Umm er Radhuma Formation consists of a reptitious series of light-colored foraminiferal aphanitic and calcarenitic limestone, dolomitic limestone, and brown dolomite. Local silicification is common, and chert occurs sporadically throughout the formation. Calcarenitic and siliceous beds are more common in the upper part of the formation, and partially dolomitized aphanitic

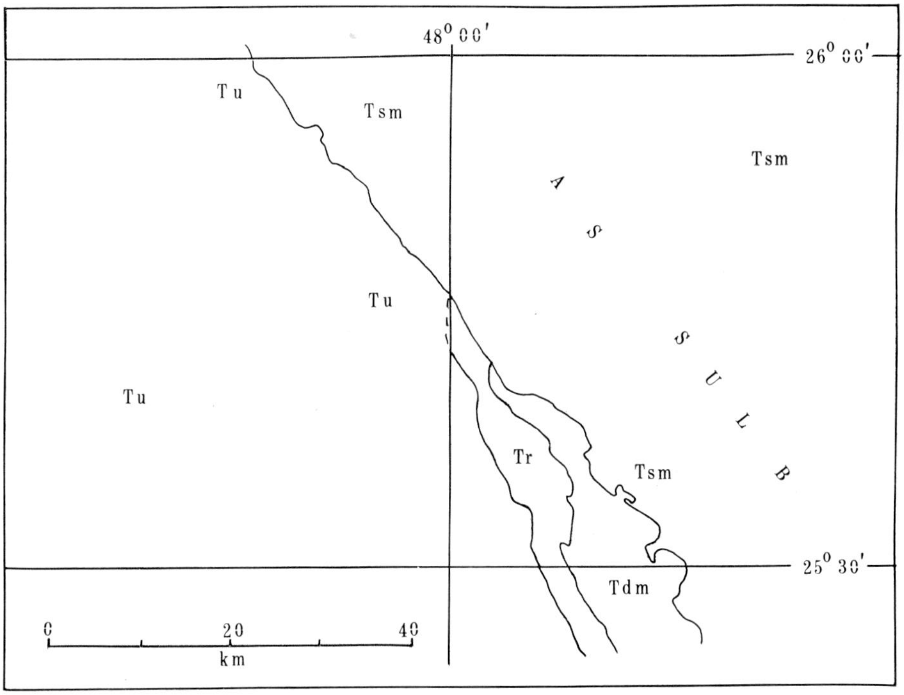


Fig. 52. Geological map of the southeastern part of the As Sulb Plateau. *Tsm* Mio-Pliocene continental clastics; *Tdm* Dammam Formation (Lower-to Middle Eocene); *Tr* Rus Formation (Lower Eocene); *Tu* Umm er Radhuma Formation (Paleocene to basal Eocene)

limestone in the lower. The dolomitization of the limestones seems to be irregular.

Fossiliferous beds occur at many levels, but the fossils have usually been leached out, leaving small pits on the weathered surface. In recrystallized and dolomitized parts, the fossils appear as "phantoms" at the best; they are in this case usually indeterminable. In well-preserved sections, however, the Umm er Radhuma Formation shows three main biozones.

The upper biozone contains *Sakesaria cotteri* Davies, *Lockhartia tipperi* (Davies), and *Lockhartia hunti* Ovey var. *pustulosa* Smout. This assemblage correlates with the Lower Eocene (Ypresian) of Qatar and the Laki beds of India. The upper part of the Umm er Radhuma Formation is, therefore, considered basal Eocene in age.

The two lower biozones are Paleocene in age as proved by the following microfossil assemblage: *Anomalina dorri* Cole var. *aragonensis* Nuttall, *Bulimina semicostata* Nuttall, *Globorotalia velascoensis* (Cushman), *Loxostoma applinae* (Plummer), *Siphogenerina eleganta* Plummer, *Nummulites curvispira* D'Archiac et Haime, *Nummulites* spp., *Operculina* cf. *discoidea* Schwager, *Valvulina* spp., *Valvulineria* sp., *Uvigerina* sp., miliolids.

The Rus Formation in As Sulb area represents an unbroken transition from the Umm er Radhuma below to the Dammam Formation above. The brown Umm er Radhuma dolomite is conformably overlain by pale colored, off-white, soft, chalky, porous limestones and marls, which in turn give way at the top to tan, calcareous and gypsiferous shale and interbedded marl of the basal Dammam.

Diagnostic fossils are not known from the Rus Formation. They were dissolved leaving abundant indeterminate molds and casts as evidence of their presence in the chalky, porous limestones. The age of the Rus must be Lower Eocene (Ypresian) since the formation is underlain and overlain by rocks of proven Lower Eocene age.

The basal part only of the Dammam Formation is exposed in As Sulb area. It consists of brown shale and light-gray marl, called the Midra member, overlying conformably the chalky limestone of the Rus Formation. The presence of *Nummulites globules* Leymerie definitely dates the lower part of the Dammam Formation as Lower Eocene (Ypresian).

No evidence has been found of the Oligocene. Since the deposition of the Paleocene and Lower Eocene rocks, the sea has never reached the area again. Prior to the deposition of the Mio-Pliocene continental clastics, the Umm er Radhuma, Rus, and basal Dammam Formations were subjected to intense weathering and erosion. This weathering lasted during the major portion of the Eocene and the Oligocene (approximately 30 million years) and resulted in pronounced karstification in such a way that the Umm er Radhuma Formation today is one of the most important aquifers in the Eastern Province.

### 2.2.2. Karstification and Geomorphogeny of As Sulb Plateau

(H. Felber, H. Hötzl, H. Moser, W. Rauert, J. G. Zötl)

As Sulb Plateau, the largest part of As Summan Plateau, occupies its central regions (Figs. 8, 53). It has one of the most peculiar landscapes of Saudi Arabia.

As Sulb Plateau is not a flat plain but is rather hilly with ridges of Umm er Radhuma limestone (Paleocene) partly overlain by Miocene clastic sediments. The thick complex of limestones is strongly karstified, but the karst phenomena usually lie below the surface which just shows the openings of shafts and caves (*dahl*), very frequently almost imperceptible to the eye (Fig. 54). However, numerous shafts are open to depths of more than fifty meters. In these shafts horizontal cave systems extend on various levels. A remarkable feature of some of the horizontal caves, e.g., the cave "Dahl Abu Marwah" southwest of Ma'aqala, is that the roof of their upper systems is just about 20 centimeters thick (Fig. 55). This condition is very well characterized by J. B. Philby (1952), who speaks of a sounding land surface.

It should be mentioned that the shaft openings (*jama, aven*) are predominantly found in flat, shallow troughs or remainders of shallow valleys. Horizontal cave systems at the level of the recent valley floor are rare. One

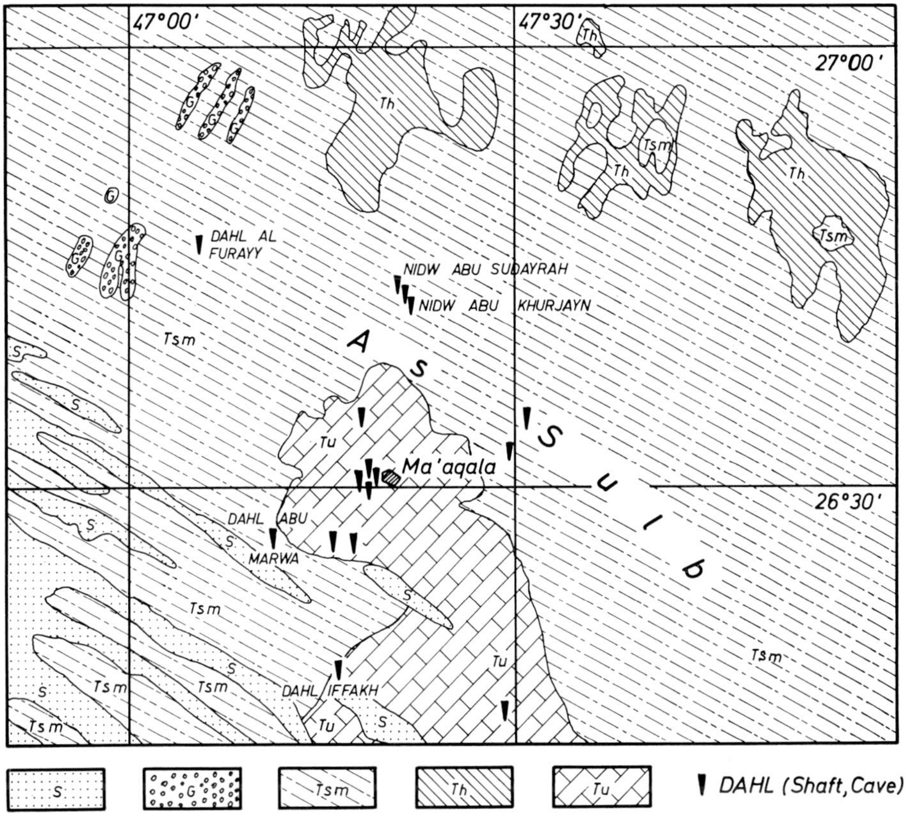


Fig. 53. Generalized geologic map of the northwestern part of As Sulb Plateau. *S* Eolian sand, *G* sheetlike lag gravel deposits of quartz and other pebbles from the basement complex, *Tsm* Sandstone, marl, and limestone discontinuously overlying Umm er Radhuma Formation, *Th* sandy limestone of Hadrukh Formation, *Tu* Umm er Radhuma Formation (from R. A. Bramkamp and L. F. Ramirez, 1958)

system of this kind, which was found only with the assistance of natives, had a shaft connecting the horizontal cave systems with a small plain at a higher altitude. Through steady solar irradiation of the plateau this system eventually became a typical "wind cave", with air being blown out with such intensity that the sand was whirled upward.

In the caves and aven of the largest region of the plateau neither stalactites nor stalagmites were found, only small warty deposits consisting of calcite. Carbon-14 measurements revealed a $^{14}C$ age of 5,060 ± 250 years for these calcite crusts (Table 35, No. 9). From these data it may be concluded that the precipitation of the Holocene "Neolithic pluvial" (and earlier Pleistocene phases?) infiltrated into the underground, thus proving the activity of karstification until the Holocene.

Table 35. *Results of Carbon-14 measurements of calcite samples (stalactites, duricrust) collected from caves (Dahls) of the As Sulb Plateau*

| | Locality | Lab. | Reg.-No. | Material | $^{14}C$-age years (years B.P.) [3] | $\delta_{PDB}$ $^{13}C$ (‰) |
|---|---|---|---|---|---|---|
| 1 | Dahl Abu Marwah | IRM [1] | Lab.-No. 3,657 | calcareous tufa | > 40,000 | — 4.0 |
| 2 | Dahl Abu Marwah | IRM | Lab.-No. 3,658 | stalagmite | > 37,500 | — 1.7 |
| 3 | Dahl Abu Marwah | IRM | Lab.-No. 3,652 | stalactite in duricrust | > 33,000 | — 6.1 |
| 4 | Dahl Abu Marwah | IRM | Lab.-No. 3,662 | duricrust | 30,500 + 2,400 — 1,900 | — 2.5 |
| 5 | Dahl Abu Marwah | IRKW [2] | VRI-450 | stalactite in duricrust | > 37,000 | |
| 6 | Dahl Abu Marwah | IRKW | VRI-451 | duricrust | 32,000 + 1,900 — 1,500 | |
| 7 | Dahl Abu Marwah | IRM | Lab.-No. 3,659 | stalactite | > 37,000 | |
| 8 | Dahl Abu Marwah | IRM | Lab.-No. 3,661 | stalactite | > 37,000 | |
| 9 | Nidw Abu Khurjayn | IRM | Lab.-No. 3,660 | calcareous tufa | 5,060 ± 250 | + 1.7 |

[1] Institut für Radiohydrometrie, Munich.
[2] Institut für Radiumforschung und Kernphysik, Vienna.
[3] Uncorrected.
>: older than the age measurable by the analysis technique used.

While any climbing into shafts and avens in the northern and eastern part of As Sulb Plateau in search for calcareous tufa was by no means successful, in Dahl Abu Marwah and the caves in the vicinity of this system southwest of Ma'aqala, very well developed stalactites and stalagmites were found. Stalactites cemented into duricrust in the roof of Dahl Abu Marwah (Fig. 55) were most helpful in the interpretation of the morphogeny (see below). Carbon-14 measurements and thin sections were used.

From the cave "Dahl Abu Marwah" two samples of duricrust were collected for thin sections; one sample directly from the top of the local hill which is about 20 meters higher than the land surface of the cave's roof, and the second from the land surface above the cave, cementing the old stalactites (Fig. 55, Table 35, No. 4 and 6, both the second sample).

The thin sections do not show much difference between these two duricrust samples. Well-rounded quartz grains with diameters of 0.1 to 1.5 millimeters, effectively marked by wind transportation, prevail. The percentage of these quartz grains changes in the various parts of the thin sections, the average being 40 percent and they are cemented by fine-grained and microcrystalline calcite. The calcite pattern is very different. Parts with large crystals are changing with microcrystalline calcite. The calcite matrix is enriched at the margins of the macroscopic distinguishable nodule. However, no difference concerning the maturity of the two duricrusts can be inferred from the thin sections. The only differences, then, are macroscopic ones or those which may be seen from bedding into different strata.

Concerning the stalactites cemented into the duricrust, the following genetic development may be deduced:

1. Development of the stalactites on the roof of an underground cave.
2. Complete filling of the open parts of the cave by eolian drifted sand.

Fig. 54. Dahls in the area of Ma'aqala. The uppermost step of the landscape is a long ridge with a uniform top niveau covered by dark and resistant duricrust (Photo: J. G. ZÖTL, 1974)

3. Cementation of this sand by carbonate solutions (from below?) and crystallization as well as recrystallization of calcite.
4. Erosion of the limestone overlying the caves and re-activation of solution forms (caves).

Stalactites and duricrust of the roof of the cave "Dahl Abu Marwah" were investigated with regard to their carbon-14 content (Table 35, No. 4, 5, 6). As was expected, no carbon-14 could be measured in the stalactites. Thus the stalactites are obviously older than 37,000 years (Table 35, No. 1, 2, 3, 5, 7, 8).

In comparison, the duricrust showed a small content of carbon-14. The $^{14}C$-measurements (samples Nos. 4, 6) show an uncorrected age younger than 35,000 years. Assuming a recent activity of 85 percent modern and supposing that the duricrust contains not a large amount of fossil carbonate, a $^{14}C$ age of 30,500 to 32,000 years may be calculated. In spite of some uncertainties—content of fossil carbonate, volume and kind of recrystallization—these carbon-14 measurements are nevertheless of high value, showing a partly progressive development of duricrust during Late Pleistocene and/or Holocene. The general conditions demonstrate a repeated reactivation of the cave system of Abu Marwa.

Fig. 55. Stalactites cemented into duricrust in the roof of Dahl Abu Marwah. (Photo: J. G. Zötl, 1974)

With the exception of karstification, the series of strata and their origin are the most important factors from which the morphogeny can be investigated.

The series of strata found their origin in marine transgression and regression in the region of the shallow eastern Arabian Shelf during the Tertiary Period. After accumulation of Umm er Radhuma limestone (Paleocene), which forms the recent surface of large parts of As Sulb Plateau, the marine regression from the whole Arabian Shelf took place. East of As Sulb Plateau the flat basin had another marine transgression (deposition of anhydrite, limestone, dolomite, marl of the Rus and Dammam Formations) during the Eocene, but this transgression did not reach As Sulb Plateau. In this terrestrial phase the Umm er Radhuma limestone was partly eroded. One may assume that during this long period (Eocene and Oligocene) the limestone areas underwent their first intensive karstification. This early karstification may be the reason that the Umm er Radhuma Formation is an excellent aquifer reaching far into the Arabian Gulf, although it is covered by Miocene strata of more than 100 meters thickness.

The uplift of the Arabian Shield during the Tertiary Period induced intensive erosion in the western part of the Arabian Peninsula and the accumulation of terrestrial strata to the east (Hadrukh Formation). At the margin of As Sulb Plateau, where these sediments can also be found now, they are conglomerates and sandstones with carbonate cement. The pebbles in the conglomerate are quartz, and Jurassic and Triassic limestones. Towards the east the grain size becomes smaller. In the Lower Miocene the sea coast was approximately in the area of the present beach. The following sedimentation of the Dam Formation (Miocene) was of marine origin; the region of the present As Sulb Plateau, however, was not reached by this transgression. During the Pliocene again terrestrial sedimentation prevailed.

It is assumed that in the area of As Sulb Plateau denudation has been active ever since the Pliocene (probably since the Upper Miocene). Clastic deposits of the early Miocene were preponderantly denuded, and by this erosion the Umm er Radhuma limestone became the land surface in large areas around Ma'aqala. This land surface dips slightly towards the east and is surrounded by sediments dating from the early Miocene. The wide arch of the old pre-Miocene landscape can be inferred. It is worth mentioning that the karstification of the Umm er Radhuma limestone in the area of the arch and its surroundings is particularly intensive.

The post-Miocene morphogeny produced a landscape of flat steps with altitudes differing just about 30 meters. From a profile the following elements of the landscape may be distinguished: the uppermost step is a long ridge, partly existing as inselbergs, with a uniform top niveau covered by a dark and resistant duricrust (Fig. 54).

Below the top niveau there is a wide surface of the next step divided into two levels (10 to 15 meters below the top niveau). In some areas only the level of the higher altitude can be seen clearly; on the other hand, in some valleys the lower level may be dominant.

The lowest elements of these steps are the flat floors of small valleys which, towards the north and northeast, become one uniform plain with some outliers towering above it.

All niveaus are covered by duricrust. There is a difference in colour and resistance of the duricrust on the top plain, where the crust is dark and hard, and the lower niveau, where the duricrust is not so dark and compact. However, no differences can be seen in thin sections.

R. P. Miller (1935) describes ridges comparable to those of As Sulb Plateau in the northwest of As Summan Plateau. He holds the opinion that these ridges are old valley floors, where the groundwater was near the surface, and that this was the reason for the origin of a hard duricrust. This duricrust, in turn, protected the underlying rock against denudation finally creating an inversion of relief. A similar morphogeny of As Sulb Plateau cannot be excluded, but there is no evidence concerning the period of this development. With regard to the carbon-14 measurements mentioned before it may be concluded that the lowest niveau, i.e., the recent floor and land surface in this area, is of pre-Würm origin.

The strong karstification makes As Summan Plateau a most important recharge area for underground aquifers. In recent arid climatic conditions, episodic rainfall of very great intensity has led to infiltration and surface runoff into the numerous vertical shafts. There is no doubt that during the pluvials large quantities of water enriched the underground aquifers. Intensive rainfalls also at present feed the karstwater at depth.

Carbon-14 dating of organic matter and inorganic materials such as calcareous tufa yields much more precise data than does radiocarbon dating of water. The carbon-14 measurements carried out on shells (2.1.2.2., 2.1.2.3.) have produced absolute classifications for the fluctuations in sea level. Radiocarbon dating is a very important part of the investigations of the geomorphogeny of As Sulb plateau, Wadi Ar Rimah, and the cuesta region of the Tuwayq Mountains. As Sulb Plateau, having neither organic matter (fossils) nor morphologic features (terraces, etc.) is an especially good example of an area where radiocarbon measurements are an essential investigative procedure.

Natural radiocarbon is continuously produced in the upper atmosphere by a nuclear reaction on nitrogen initiated by cosmic-ray-produced neutrons. Oxidized, the radiocarbon finally appears in the lower atmosphere and is homogeneously distributed world-wide by meteorologic mixing processes. In the form of carbon dioxide radiocarbon enters into the biosphere by assimilation of the plants in the same way as the stable carbon isotopes ($^{12}C$ and $^{13}C$) do. As soon as the metabolism of the organism ceases by death, the radiocarbon concentration $C_t$ in the sample decreases with time $t$ according to the law of radioactive decay

$$C_t = C_o \cdot 2^{-t/T}.$$

$C_o$ is the initial radiocarbon concentration represented by a standard with the specific recent activity of biogenetic carbon A.D. 1950 (= 100 percent modern). The half-life $T$ is taken by agreement to be 5,568 years. By measurement of $C_t$ and $C_o$, the time elapsed after death, that means the age $t$, can be calculated.

The manner in which radiocarbon gets into calcareous sinter material is quite different from the one for biogenetic matter. Calcareous sinter is an incrustation precipitated from water containing calcium bicarbonate. Calcium bicarbonate is formed by dissolution of limestone in presence of carbon dioxide: Limestone is too old to have any measurable radiocarbon activity. The carbon dioxide required for its dissolution, however, is modern with regard to radiocarbon concentration because it is taken up by the water mainly from the air in the porous soil enriched in carbon dioxide by root respiration and decomposition of organic matter. Measurements on hard waters precipitating sinter revealed an average initial radiocarbon concentration of about 85 percent modern (K. O. Münnich, J. C. Vogel, 1959; see Section 2.1.8.4.).

Conventionally radiocarbon ages are given in years B.P. (before present). It should be noted that the fundamental assumption of a constant radiocarbon production rate within the period accessible to radiocarbon dating techniques (about 40,000 years) is not fulfilled exactly (de Vries-effect), as proven by dating dendrochronologically dated wood samples (H. E. Suess, 1970). The presented dates are not corrected for this effect.

The dates are given together with their single standard deviation $\sigma$ originating in statistical fluctuations of the radioactive decay. The results of repeated datings carried out on the same sample should scatter in the way that 68 percent of the results lie within the interval given by $\pm \sigma$, and 32 percent outside it. If the $^{14}C$ content of a sample is lower than the detection limit, the corresponding maximum age that can be measured is given preceded by the sign ">". This maximum age has a standard deviation of about $\pm$ 3,000 years. Concerning principles of the carbon-14 method and measurement see e.g., M. A. Geyh, 1971, W. Drost et al., 1974.

## 2.3. Wadi Ar Rimah

### 2.3.1. The Quaternary Development of the Upper Part of Wadi Ar Rimah

(H. Hötzl, H. Felber, J. G. Zötl)

#### *2.3.1.1. General remarks on wadi investigations*

For the investigation of the Quaternary morphogeny of the Arabian Peninsula the study of the wadis is of utmost importance. The phases of erosion and accumulation are interesting not only for the geomorphology but also for studies of the climate and its changing conditions. Right at its beginning the Quaternary project thus planned to carry out field reconnaissance in the areas of the large wadi systems. As far as possible a correlation of the studies concerning the coastal regions on the one hand, and the investigations of the wadi systems on the other, was envisaged.

The Arabian Shelf shows various wadi systems as a consequence of the intensive surface runoff in humid periods especially of the Upper Tertiary and Lower Quaternary. The direction of the large systems depends on the flat dip of the layers towards the north, east, and southeast, which is very clearly demonstrated by the wide arc of the cuesta landscape of central Saudi Arabia. The depth caused by fluviatile erosion is different for the various wadis.

At the eastern margin of the shield the wadis are, in general, flat valleys with low gradients; however, the influence of the wadis on the relief of the pediplain is insignificant. In contrast, fluviatile erosion has cut steep canyons causing a strong relief intensity in the cuesta landscape, especially in the area of the Tuwayq Mountains. Towards the east, in the area of As Summan Plateau the landscape becomes flatter again. In the course of the writers' first Quaternary studies only a few wadis selected from a large number could be investigated because of the enormous size of the country. The parts of the largest wadis originating in the shield and crossing almost the whole of central and eastern Saudi Arabia appear to be full of promise for research work.

From north to south, the following parts of the major wadis were studied:

The upper part of Wadi Ar Rimah which runs towards the northeast as Wadi Al Batin,

Wadi Atk which probably had a connection with Wadi Al Miyah,

Wadi Hanifah and Wadi Birk with their junction in the basin of Al Kharj and a continuation towards the east as Wadi As Sah'ba,

Wadi Ranyah and Wadi Bishah joining Wadi Ad Dawasir, and its delta fan in Ar Rub' Al Khali.

The following chapter describes the geological and morphological survey in the upper part of Wadi Ar Rimah. The Quaternary studies were carried out over a distance of about 250 kilometers west of Buraydah and 'Unayzah, along about lat 26° N, between long 41° 30′ E and long 44° E. A further field trip was made to the uppermost part of the discharge area at Hulayfah lat 26° N, long 40° 45′ E).

### *2.3.1.2. Geological conditions in the area of Wadi Ar Rimah*

The part of the recharge area of Wadi Ar Rimah described belongs mainly to the eastern margin of the Precambrian shield (Fig. 56). Its continuation towards the east leads into the Mesozoic cuesta landscape, where large areas and parts of the old wadi are covered with dunes. The uppermost tributaries of the wadi reach westwards into the area of Harrat Khaybar, where the Precambrian shield is covered with large basalt flows of Tertiary and Quaternary age (G. F. BROWN et al., 1963; R. A. BRAMKAMP et al., 1963 a).

The eastern border of the study area just reaches the cuesta escarpment of the Permian Khuff Formation, which consists of light limestone with interbedded dolomite, marl, and gypsum. The limestone is underlain by Cambrian-Ordovician Saq sandstones; north of Wadi Ar Rimah there also exist sandstones of the Silurian and Devonian. The outcrop of the sandstones in Wadi Ar Rimah is about 40 kilometers wide, the sandstone lying discordantly above the Precambrian rocks of the shield.

The Precambrian basement of this area consists mainly of acid plutonic rocks of different age and of a heterogeneous sequence of schists, phyllites and epimetamorphic volcanic series (Fig. 56). West of the outcrop of the Cambrian sandstone, near Ar Rass, for a distance of 70 kilometers the surroundings of Wadi Ar Rimah consist of granites. The metamorphic sequence prevails in the adjacent western recharge area. The granites and metamorphites were eroded by intensive Precambrian and Cambrian denudation and thereafter covered by thick clastic sediments. Remnants of the Saq sandstone, e.g., at Jabal Tamiyah, represent this sedimentation after the Precambrian denudation. It may be assumed that parts of the old denudation plain are exhumed and are now part of the recent morphology (G. F. BROWN, 1960).

### *2.3.1.3. Geomorphological features*

A very impressive feature of the landscape in the area of Wadi Ar Rimah is the wide flat pediplain reaching beyond the discharge area and sloping gently to the east-northeast. This landscape obtains its characteristics by broad pediments, narrow pediment-passes, flat trough valleys, inselbergs and thin flat alluvial fans. The average gradient of the wadi is in conformity with the flat pediplain. In the eastern part of the study area, between Buraydah and 'Unayzah, the wadi floor has an altitude of 602 meters; 200 kilometers farther upstream, in the area of Aqlat As Suqur, the altitude reaches 725 meters (Fig. 56). This makes a gradient of about 0.6 per mill for the wadi and, since the distance nearly corresponds to a bee-line, the gradient of the pediplain is roughly the same (Fig. 57).

Since west of this area the pediplain is covered with Late Tertiary basalt, the origin of the pediplain must be older. Plain remnants of higher altitudes are to be found on various inselbergs; however, it is not possible to correlate them. A very impressive remnant of such a higher plain is the top plain of Jabal Tamiyah (lat 25° 31′ N, long 42° 00 E), which consists of Saq sandstone (Cambrian-Ordovician) in its upper part. This Saq sandstone has its present cuesta escarpment about 150 kilometers east of Jabal Tamiyah. With its top surface at an altitude of about 1,300 meters, i.e., 480 meters above the recent pediplain, Jabal Tamiyah towers completely isolated in the

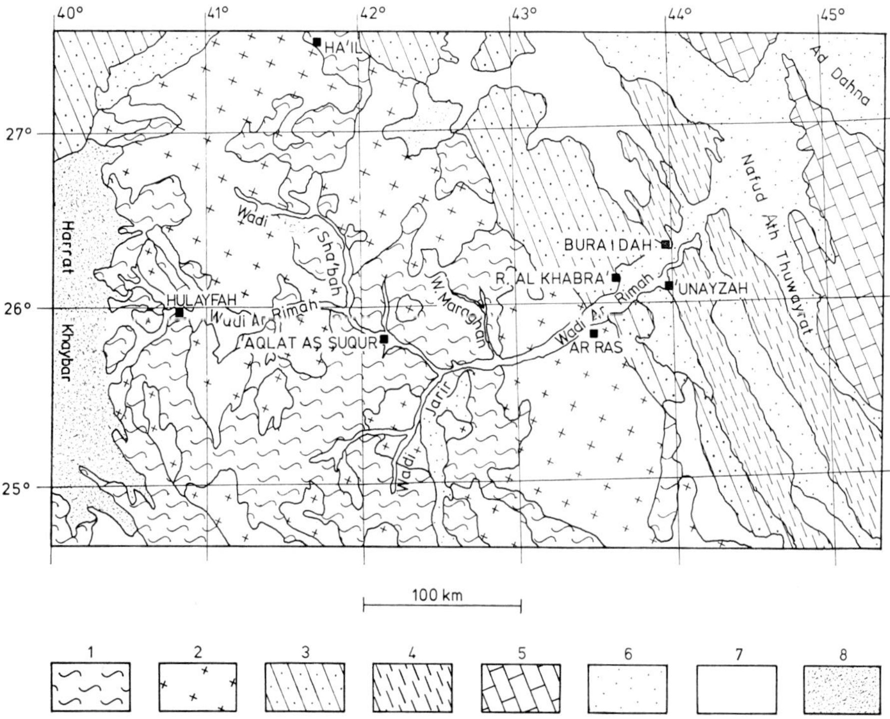


Fig. 56. Geologic map of the discharge area of Wadi Ar Rimah. Generalized after U.S. Geol. Survey and Arabian American Oil Company, 1963. (*1* Precambrian metamorphic rocks, mainly schists; *2* Granite and granite gneiss; *3* Cambrian-Devonian sandstones; *4* Permian and Triassic sandstone, gypsiferous, and carbonate rocks; *5* Jurassic and Cretaceous carbonate rocks; *6* Eolian sand; *7* Alluvium and related surficial deposits; *8* Tertiary and Quaternary basalts

recharge area of the Wadi Ar Rimah. Since it is not possible to date the top plain of the jabal, only the amount of erosion can be demonstrated.

Taking into consideration that, as mentioned above, the surface of the pediplain is older than the Quaternary, the Quaternary morphogeny is primarily limited to the wadi channel, to the cover of eroded pediments with sand, to the fanglomeratic sheets with desert pavement, to alluvial fan and to sand accumulation through wind (dunes).

The nature of alluvial fans depends mainly on the kind of outcropping rocks. Granite, which very easily weathers into fine-grained detritus, predominantly causes sand accumulation through sheetfloods, e.g., in the 70 kilometers of the wadi west of Ar Rass. The various profiles to be seen in wells reveal that the deposits consist almost entirely of sandy granite detritus only sometimes with thin layers of fanglomerates (e.g., well of Qasr Ibn 'Uqayyil, lat 25° 48′ N, long 43° 20′ E, Table 49, sample No. 26/1974). On the contrary in the area of crystalline schists, e.g., in Wadi Sha'bah, a tributary of the Wadi Ar Rimah west of Aqlat As Suqur, coarse-grained alluvial fans cover the flat eroded old pediments.

Fig. 57. Pediplain in the area of Wadi Ar Rimah sloping gently to the east-northeast. In the background Jabal Tamiyah. (Photo: J. G. Zötl, 1974)

The morphogeny of the wadi channel is of Pliocene and Quaternary age. As can be seen from the pediment's margin as well as from various small tributaries of Wadi Ar Rimah, the wadi originally was a broad trough valley with a flat cross-section and slopes of low gradient. The recent wadi channel, about 4 kilometers wide, is the result of a young vertical and lateral erosion. The difference in altitude between the pediment and the recent wadi floor, i.e., the depth of the wadi channel, increases downstream. In Wadi Sha'bah it amounts to 4–6 meters, in Wadi Ar Rimah near Aqlat As Suqur, to 8 meters; east of the mouth of Wadi Maraghan, to 12–15 meters; and near Ar Rass, to 15–20 meters.

Since in recent times accumulation has dominated in Wadi Ar Rimah, the depth of erosion must reach below the recent wadi floor. Unfortunately, there are no data available concerning the thickness of the wadi fill (deposits of sand and gravel) in the central part of Wadi Ar Rimah. In the marginal part of the wadi are some hand dug wells, showing a thickness of the young accumulation of 4 to 8 meters. This agrees with the findings in the tributaries.

In Wadi Maraghan, near its junction with Wadi Ar Rimah, geological well profiles reveal a thickness of 5 meters of the young deposits as well as a minimum of 4 meters in Wadi Sha'bah.

In almost the entire area of Wadi Ar Rimah visited by the writers, the result of the young erosion is a more or less sharp border between the wadi channel and its surroundings. In the area of the crystalline schists, in particular, the scarp sloping down into the wadi is very steep and extends over a distance

Fig. 58. Terraces of Tertiary and/or Quaternary age at the border of the wadi east of Aqlat As Suqur. (Photo: J. G. Zötl, 1974)

of more than 100 kilometers west and east of the village of Aqlat As Suqur. However, in the granite zone downstream towards Ar Rass, the terrace slope is developed as a flat transition zone up to 2.5 kilometers broad.

No doubt, the differences in the morphology of the terrace edges depend on the various stages of weathering which affected the outcropping rocks. In the region of the deeply weathered granites, the fine-grained detritus permits only a low gradient of the slopes. Near the inselbergs, the terrace slopes are locally covered by young sandy alluvial fans (0.5 meter to 3 meters thick). In the areas of crystalline schists the sharp terrace edges are mostly cut into the outcropping rocks, showing a thin cover of autochthonous weathering material. Only in some limited depressions and flat erosional channels of the old pediplain, forming the terrace surface, fanglomeratic or sandy sediments have been accumulated. Cutting from the young wadi channel, they present extensive accumulation terraces. But most of the Quaternary gravel terraces in the area of Wadi Sha'bah mapped by G. F. Brown et al. (1963) are solely rock terraces with a thin cover of autochthonous weathering material.

It is worth mentioning that no remnants of any other terrace system could be found in large areas of Wadi Ar Rimah. However, an exception to this is the area around the mouth of Wadi Maraghan. There exists a distinctly marked step 2 to 4 meters above the present wadi floor (i.e., about 10 meters below the upper edge of the terrace), which is visibly eroded into the crystalline schists. A very similar situation exists in the main wadi immediately east of Aqlat As Suqur (Fig. 58). This small rock terrace seems to correspond to a step of about 2 to 3 meters above the wadi floor in the granite area downstream near the bridge about 25 kilometers west of Ar Rass. These remnants seem to mark a phase of vertical erosion in the wadi channel.

#### *2.3.1.4. Quaternary sediments*

Beginning with the recent conditions mentioned before, the whole area of Wadi Ar Rimah which was surveyed during the field campaign of 1974, discloses a morphogenetic phase of active accumulation. Episodic precipitation during late winter causes an episodic surface runoff, sometimes flowing through all the wadi channels upstream of Ar Rass. This was the situation which was found during the 1974 field trip (Fig. 59).

The quantity and the area of rainfall determine in what part of the wadi seepage and evaporation bring an end to surface runoff, and where the sedimentation of the river load occurs. In general surface runoff occurs only over short distances. The low gradient of the wadi floor and the shrub vegetation that grows especially in these channels favour sedimentation.

The sediments on the wadi floor consist mainly of fine clastics with prevailing silt-size particles. In the granite regions in particular, larger quantities of fine and medium sands can be found in various places. In addition to the clastic components, consisting predominantly of quartz and feldspar with minor quantities of clay minerals (Table 49, sample 15), the content of evaporites (chloride, sulfate) and calcium carbonate may rise to 50 percent and more.

In some areas the sediments were covered by a salt crust. Salty clay (wadi sabkhah soil) was frequently found in the area of Riyadh Al Khabra and 'Unayzah, where the main wadi channel is covered by sand dunes to an ever increasing extent.

Due to the lack of wells in the central part of the wadi it is impossible to give exact information on the thickness of the young wadi deposits and the depth of the base level. The small number of investigated well profiles at the margins of the wadi lead to the conclusion that the base of the recent wadi fill lies at least 5 meters below the wadi floor (i.e., the surface of the deposits). The wadi filling seems to be almost exclusively deposited under the same conditions as do the recent sediments.

Besides these wadi channel sediments, sand, scree material, alluvial veneer and fanglomerate, having their origin mainly in the inselbergs are deposited on the terrace plain and the old pediments. A good example of a recent plain consisting of these materials and of its development can be seen around the area of the granitic Aban Al Asmar massif (50 kilometers west of Ar Rass).

From the recent and subrecent deposits, there are to be distinguished thin gravels, sands and fanglomerate-like sediments, covering the older, smoothly eroded pediments. Their exceptional position results from the fact that they have been cut by the young wadi channel, therefore they must be older than the recent wadi slopes. These deposits were found east of Aqlat As Suqur and in the surroundings of Hulayfah as well as in the area of Wadi Sha'bah.

Fig. 59. Episodic surface runoff in the upper part of Wadi Sha'bah, a tributary of Wadi Ar Rimah, March 1974. (Photo: J. G. Zötl, 1974)

In the latter area, these fanglomerates are widespread at the orographic right hand side of the wadi channel from 30 to 55 kilometers north of the wadi's junction with Wadi Ar Rimah. The accumulation plain there is sloping from various inselbergs towards the south-southwest. The thickness of the fanglomerates covering the smoothly eroded surface of metamorphic rooks depends on the relief of the pediment. This veneer usually averages just a few meters. The recent surface of the fanglomerate plain is at the same time a desert pavement. The angular rock fragments are generally coated by desert varnish and their diameter ranges mainly from 2 up to 15 centimeters. They originate exclusively in the local outcrops of metamorphic and igneous rocks. When two pits were dug to a depth of 60 centimeters, the profiles disclosed terra rossa-like clayey, reddish material with edged pebbles below a surface zone, of a few centimeters thickness, consisting of grey sand between the rock fragments of the pavement. As it is usually the case in desert soil profiles, the number of pebbles directly below the surface was small, but increased the deeper the pits were dug.

While in one of the pits down to a depth of 60 cm only loose earthy material occurred, in the second one calcium-carbonate incrustations coating small crumbs and pebbles were found below a depth of 30 centimeters. On various pebbles the crust had a thickness of up to 2 millimeters. Carbon-14 measurements of this crust (Sample VRI 442) disclosed an age of 28,900 $\pm$ 1,300 years.

The clayey material with crusts of calcite is obviously a fossil red-earth horizon having originated in a more humid period. This result is confirmed by mineralogical investigations in the course of which kaolin, illite, and mixed-layer clay minerals could be found in the red clayey material (Table 49, sample No. 17).

In the area along the right bank of Wadi Sha'bah, the base of the fanglomerate was not visible. In contrast, on the left side of the wadi, very close to the mouth of Wadi Khillah, a layer of gypsum was found below the fanglomerate, at a lower step of the terrace. This gypsum layer, probably an old desert crust and more than a meter thick, became more sandy with increasing depth.

The area of terraces east of Aqlat As Suqur, where sedimentation is similar to that mentioned above, show the same sequence of strata. The most conclusive outcrop can be found there in a young tributary, which partly runs parallel to Wadi Ar Rimah (Fig. 58). In this erosion channel, about 8 meters deep, the lowest part cuts into phyllitic schists with thin quartz layers. Farther downstream the entire terrace consists of this material. In the area of this young tributary the smoothly eroded surface of the schists is covered by sandy and gravelly sediments forming the flat niveau of the terrace. The thickness of this accumulation amounts to about 4 meters.

In the lower part these deposits consist of sand which is interspersed with gypsum partly as crusts. A thin layer of gravel with desert pavement on top follows over them. Just as in Wadi Sha'bah the pebbles are embedded in terra rossa-like material. From samples (Table 49, sample No. 18) collected in this region roentgenographic investigations showed the same composition as those of the samples from Wadi Sha'bah: kaolin, illite, montmorillonite, quartz and feldspar. The red colour can presumably be ascribed to X-ray-amorphous ferric oxide. Up to 20 centimeters below the surface of the terrace incrustations of gypsum occur, but no crusts of calcite were found. In comparison to the desert pavement in the area of Wadi Sha'bah, the pebbles, having diameters of up to 15 centimeters are rounded more smoothly and consist of schists, sandstone, quartzite, granite, and felsite. Most striking was the discovery of well-rounded white quartz pebbles with diameters up to 4 centimeters, which do not exist in the fanglomerate of Wadi Sha'bah.

Similar conditions as in the area of Aqlat as Suqur were found in the surroundings of Hulayfah, 140 kilometers upstream. The terrace slope ascends there only 2–3 meters above the wadi floor. The surface of the terrace, too, acts as a desert pavement. Among the pebbles of granitic and metamorphic rocks the well-rounded small ones of light quartz, probably reworked from

Cambrian-Ordovician Saq sandstone, again attract special attention. Even there terra-rossa-like materials are interspersed among the pebbles below the grey porous sandy soil of the desert pavement; below a depth of 40 centimeters cementation of gypsum occurs. These terrace gravels are accumulated directly above the outcropping series of the Precambrian Halaban Formation at some places, e.g., at the cliff of Jabal Al Hadawi 25 kilometers westsouthwest of Hulayfah. However east of Hulayfah they cover a 0.5–1 meter thick layer of gypsum (J. DELFOUR, 1966).

Of great importance for the stratigraphic classification of the terrace gravels are their superposition by a basalt flow west of Hulayfah. This flow arises with several tributaries encircling Precambrian steptoes from the wide basalt area of Harrat Khaybar. The lava made use for runoff of the old discharge system following the main valley about 40 kilometers in a northeast direction down to the area of Hulayfah. The basalt (its thickness there comes to 5 meters) shows an intensive weathering, with lateritic soil washed out mostly leaving boulders of decomposition and disintegration. Basalt samples were collected for potassium-argon age determination but are not measured until now.

It is worth mentioning that in the granite area of Wadi Ar Rimah, ranging from the mouth of Wadi Maraghan to Ar Rass, no remnants of the older deposits could be mapped. As stated in 2.3.1.3. the flat terrace plain and terrace slopes are covered by sandy detritus of granite. The latter has a composition like fanglomerates in local parts only (Fig. 88, sample 26). It is possible that this material consists of recent detritus as well as reworked older deposits; it may be assumed that the granite detritus cemented by crusts of gypsum near the village of Qasr Ibn 'Uqayyil belongs to such older components.

### *2.3.1.5. The cycle of accumulation and erosion*

From the sedimentological point of view on the one hand and from the geomorphological on the other, the factors concerning the geomorphogeny of the area of Wadi Ar Rimah in the Late Tertiary and in the Quaternary may be summarized as follows:

1. The present landscape is dominated by the wide pediplain of pre-Quaternary origin. In compliance with the theory of D. BUSCHE (1973) this pediplain did not have its origin in arid climate conditions. It is not possible to determine the period of the first flat minor erosion of this pediplain. Such a process may have taken place in the end of the pedimentation phase or immediatly thereafter.

2. The levelling of the slight dissected pediments occurred by accumulation of sand and gypsum at first, as can be seen in the terraces near Aqlat As Suqur, east of Hulayfah and in Wadi Sha'bah. The kind of sediments and especially their high percentage of gypsum disclose that this sedimentation occurred in an arid climate. In the course of following, probably only slightly more humid phases thin layers of conglomerates and fanglomerates were deposited by the surface runoff.

3. The resumption of magmatic activity in the area of Harrat Khaybar led to an intensive extrusion of basaltic lava. One of the flows followed the old Wadi Ar Rimah system down to Hulayfah.

4. During and/or after the time of magmatic activity a humid climate caused an intensive lateritic weathering of the basalts as well as terra-rossa soils on the accumulation plains. The surface runoff, renewed by the increased precipitation, produced the terrace-like dissection of the plains by the formation of the recent wadi channels, which cut back into the upland basaltic areas.

5. The succeeding arid climate has remained prevalent till today, only interrupted by some more humid phases. The filling of the wadi channels by fine clastic material with a varying percentage of evaporites has occurred and occurs recently under the arid conditions. The interposed slightly more humid phases may be assumed on account of the young calcium-carbonate cementation and incrustation, e.g., in terrace soil in Wadi Sha'bah (carbon-14 age determination: 28,900 ± 1,300 years), precipitated in the zone of ground water oscillations. A further indication for that is the morphology of a young tributary east of Aqlat As Suqur, which can only have been eroded after accumulation of the young wadi fill.

## 2.3.2. Hydrogeological and Hydrochemical Investigations in the Upper Part of Wadi Ar Rimah

(H. Hötzl, C. Job, H. Moser, W. Rauert, W. Stichler)

### *2.3.2.1. Hydrogeological conditions in Wadi Ar Rimah between Aqlat As Suqur and 'Unayzah*

(H. Hötzl)

The hydrogeological conditions of this area, a wadi channel with a length of 200 kilometers and a breadth of 5 to 8 kilometers surrounded by plains, depend on the extension of unconsolidated rocks of generally small thickness, and the kinds of hard rock on the plain's surface. Groundwater and groundwater-exchange in the hard rock plains are dependent on rock weathering. Hydrogeologically, there is thus a difference between the schist areas in the west, the granitic rocks in the central part, and the sandstone outcrop in the east (Figs. 56 and 60).

Except for some talus fans and alluvial veneers, unconsolidated rocks are restricted to the main wadi and its tributaries, but even there they are of minor thickness.

Since there are no wells or drillings in the inner part of the wadi channel, no data are available concerning the thickness of these sediments. However, as mentioned in chapter 2.3.1.4., one may conclude from wells at the border of the wadi channel and its tributaries that the alluvial accumulation does not exceed 10 meters in the main wadi, and that in the tributaries accumulation may sometimes be less than 5 meters.

Fanglomerates are dominant at the edge of the main wadi and particularly in the flat terraces, sometimes reaching widths up to 3 kilometers. They also occur in the tributaries. In granite areas coarse sand replaces fanglomerates. Remnants of gravel terraces can be found only in the area of Aqlat As Suqur. The flat floor of the main wadi shows fine sand, silt and sabkhah sediments (Section 2.3.1.4.), which interfinger with fanglomerates and talus fans towards the border of the wadi.

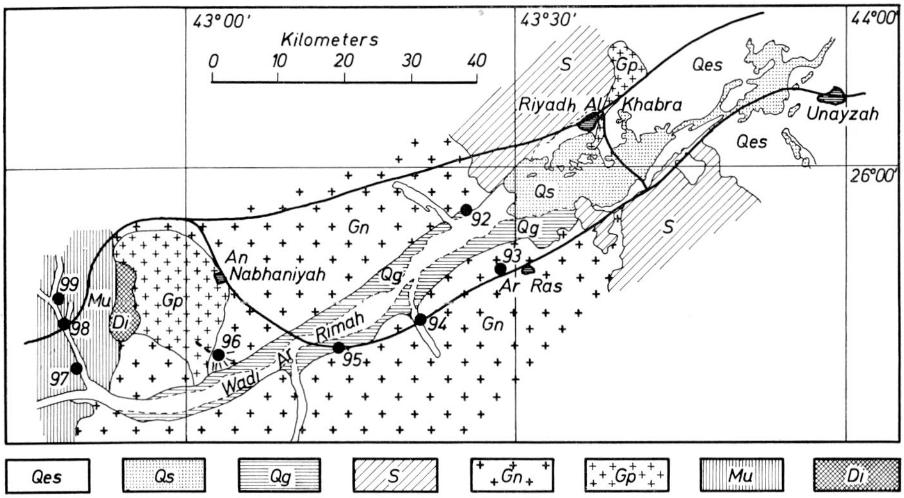


Fig. 60. Generalized geologic map of the area under investigation, with numbers of wells or water samples (see R. A. Bramkamp et al., 1963). *Qes*, *Qs*, *Qg* Quaternary eolian sand, silt and gravel, *S* Saq sandstone, *Gn* Gneissic gray granite, *Gp* red and pink granite, *Mu* Murdama schist, *Di* diorite

The varying permeability of these materials also influences the locations of wells; the wells lie mainly at the border of the main wadi channel, at flat talus fans or in the tributaries, but never in the wadi floor itself. Furthermore, the low permeability of the fine clastic sediments in the wadi floor is the reason for a low infiltration rate and a high percentage of surface runoff even after relatively short precipitation. Such rainfalls produce shallow ponds in larger flat sabkhah pans, as in the area between the villages of Riyadh Al Khabra and Ar Rass, or in many small basins scattered among small sediment hills. The water absorption ability of these fine clastic sediments is high and capillary attraction produces a high soil humidity for many days, but no groundwater enrichment. The high salt concentration of these silty and clayey sediments and surface salt crusts reveal that almost all the water evaporates. In Wadi Sha'bah west of Aqlat As Suqur, floor raising as a consequence of hydration and pressure of salt crystallization caused by evaporation after a humid period could be observed. One of these areas had a size of 50 $\times$ 20 meters and the uplift by pressure of salt crystallization was 70 centimeters. Because there were no wells, no groundwater samples from the wadi channel could be collected.

The local water supply comes from groundwater in the transitional area at the wadi border towards higher terraces and plains and tributaries filled with clastic fanglomerates of medium to coarse sizes. From hand-dug wells a discharge of some litres per second was possible through pumping in the humid March of 1974. Considering the small thickness of the aquifer in dry periods its productivity may rapidly decrease. Recharge and hydrochemical composition of these waters are closely linked with the hydrogeological conditions of the hard rock forming the peneplain. The aquifer is confined to the shallow weathering zone.

In the schists, the main rock type in the western part of the investigated wadi, deep physical disintegration plays a relatively insignificant role. Chemical weathering caused by infiltration water results in decreasing permeability. The consequences are decreasing infiltration as well as increasing surface runoff and evaporation. Thus, only in a few small channels filled with detritus and in the sediment material of the tributaries does shallow groundwater occur sometimes episodically. The groundwater level fluctuates strongly in most cases, and the yield is generally low. In accordance with these hydrogeological conditions the depths of the wells range between 2 and 8 meters.

At the mouth of Wadi Maraghan, the thickness of the poorly sorted scree material (sample 97, Fig. 60) was measured as 5 meters in a well profile. In spite of rainfall the groundwater level was 1.5 meter deeper than the base of the alluvium. Some of the open joints of the unweathered rocks collected the water from the overlying weathered material. Similar conditions could be observed in other wells. It can be assumed that these water-bearing joints in the schists are restricted to near the surface of the unweathered hard rock.

Unlike the schists, in granite and granitic gneisses occupying the central part of the investigated area, weathering of the outcropping rocks in its typical grusy kind reaches very deep. In a recently dug well at the northern edge of Wadi Ar Rimah, west of the village of Riyadh Al Khabra, completely weathered granular granite was found down to a depth of about 20 meters (sample No. 92, Fig. 60). Only gradually does one encounter more compact granite below, transected by open water-bearing joints. The groundwater level was about 22 meters below the surface; from the hand-dug well (about 2 $\times$ 3 meters wide) up to 8 liters of water per second could be pumped for some hours each day.

Owing to the experience with schists of Wadi Maraghan, the first assumption was that this groundwater is enriched by infiltrating rainwater through the mantle of weathered granite. Isotopic measurements carried out in water samples collected from the above mentioned well (samples No. 92, 93, 95; see 2.3.2.2.) showed that the water contains practically no tritium. The conclusion therefore is that no recharge from present precipitations occurs. This is surprising because of the uniformly weathered mantle and a groundwater level which is relatively near to the land surface. The yearly average of precipitation of about 100 millimeters also speaks in favour of an episodic recharge of the groundwater from precipitation. However, that hardly any rainwater infiltrates may be explained by the fact that feldspars weather to kaolin, thus forming impermeable layers in the granitic weathering mantle.

These clay components are washed out in the uppermost zone and on the land surface which is partly covered by a floated grusy detritus to a thickness of 0.5 to 4 meters. Still farther down, the weathered material lies in situ and the weathered feldspars act as a water damming horizon. This implies that two groundwater horizons exist in the more or less homogeneous granitic material. The upper horizon of the two carries groundwater with a recharge from precipitation and has its underground flow to and along the channels of the tributaries for which Wadi Ar Rimah acts as local base level of erosion. As a consequence of the fact that quartz grains prevail in all these unconsolidated materials the shallow groundwater has a low mineralization (sample No. 94 and 96). The yield of this shallow groundwater body depends in this case on the thickness of the accumulated material, the size of the recharge area and the seasonal distribution of rainfall.

The lower groundwater horizon (sample No. 92, 93, 95; see 2.3.2.2.) flowing mainly through open joints in the transition zone of weathered to unweathered granite carries relatively old water. The weathering processes in feldspars, mica and hornblende therefore have a long-term effect on waters which consequently produce higher mineralization. The water-bearing zone, except for areas of stronger tectonics, is generally restricted to the uppermost region of unweathered granitic rocks, since open joints decrease in number rapidly with greater depth.

The hydrogeological situation changes again towards the east, where sandstone of the Cambrian and Ordovician Saq Formation occurs. Together with the overlapping Ordovician and Devonian Tabuk sandstones, north of Wadi Ar Rimah, their thickness reaches 1,000 meters and more. The water in these rocks, circulating through joints as well as through the actual pore volume, is used from numerous wells belonging to villages and oases in the vicinity. In the area around Riyadh Al Khabra on a terrace plain north of Wadi Ar Rimah the groundwater level lies 15 to 20 meters below the land surface. Although an exact geodetic levelling has not been made, it appears that this groundwater level corresponds to the shallow groundwater in the wadi channel.

Farther eastwards, in the area of the towns of Buraydah and 'Unayzah, the sandstone is overlain by limestone, clay and gypsum from the Permian Khuff Formation. C. Voute (discussion in M. Rubin, 1963) mentions a $^{14}C$-age of about 20,000 years B. P. for the water pumped up from sandstone in this area, i.e., from a depth of 300 meters.

Considering the studies of T. Dinçer et al. (1974), one can assume that in the large area occupied by the outcrop of the sandstone—given an average yearly precipitation of about 100 millimeters—a certain amount of groundwater recharge has taken place in recent times. Contrary to this assumption, the water samples collected in the oases of Riyadh Al Khabra contained no tritium. On the other hand, chemical analyses and values gained from stable isotopes indicate the recycling of water pumped from the above mentioned sandstone aquifer and the seepage of this water used for irrigation. Relatively good preconditions concerning infiltration could be assumed from this very fact (see 2.3.2.2.).

Table 36. *Temperature, oxygen-content (mg/l) and chemical composition of water samples collected in the area of Riyadh Al Khabra.* Ion concentration in µe/l, boric acid and silicic acid in mg/l. SK = sum of cations, SA = sum of anions in µe/l. W = type of well (section 2.1.6.1.4.). Samples collected between 18th and 24th March 1974. For sampling locations see Fig. 6C

| No. | W | °C | $O_2$ | $Na^+$ | $K^+$ | $Ca^{++}$ | $Mg^{++}$ | SK | $Cl^-$ | $Br^-$ | $SO_4^{--}$ | $HCO_3^-$ | $NO_3^-$ | $F^-$ | SA | $H_3BO_3$ | $H_2SiO_3$ |
|---|---|---|---|---|---|---|---|---|---|---|---|---|---|---|---|---|---|
| 87 | DP | 27.9 | 7.4 | 18,225 | 208 | 17,200 | 2,300 | 37,933 | 22,857 | 23.1 | 12,000 | 1,727 | 1,572 | 23 | 38,202 | 2.17 | 12.2 |
| 88 | DP | 27.2 | 7.3 | 24,815 | 324 | 20,960 | 3,800 | 49,899 | 26,455 | 24.6 | 19,076 | 2,128 | 2,274 | 24 | 49,982 | 3.34 | 12.2 |
| 89 | DP | 27.9 | 6.8 | 67,247 | 706 | 34,031 | 10,160 | 112,144 | 72,368 | 47.5 | 32,098 | 2,450 | 4,798 | 24 | 111,786 | 8.69 | 13.9 |
| 90 | DP | — | 8.2 | 11,657 | 141 | 10,400 | 1,620 | 23,818 | 12,073 | 15.1 | 8,899 | 1,626 | 1,443 | 29 | 24,085 | 2.11 | 12.5 |
| 91 | DP | 27.4 | 7.3 | 17,747 | 290 | 17,100 | 2,819 | 37,956 | 18,381 | 18.5 | 16,088 | 2,068 | 1,726 | 25 | 38,307 | 2.86 | 13.3 |
| 100 | DP | 27.7 | 9.0 | 11,791 | 146 | 11,200 | 1,330 | 24,467 | 12,195 | 14.1 | 9,144 | 1,616 | 1,564 | 28 | 24,561 | 1.89 | 13.1 |
| 101 | DP | 27.4 | 7.0 | 85,159 | 1,006 | 39,917 | 14,153 | 140,235 | 92,494 | 74.6 | 38,562 | 2,771 | 5,963 | 24 | 139,889 | 10.00 | 16.7 |
| 102 | DP | 27.6 | 7.2 | 17,399 | 287 | 16,800 | 2,760 | 37,246 | 18,154 | 18.6 | 15,562 | 2,008 | 1,703 | 25 | 37,471 | 2.74 | 13.9 |
| 103 | DP | 27.4 | 7.6 | 21,490 | 422 | 21,320 | 3,560 | 46,792 | 21,737 | 22.5 | 20,739 | 2,309 | 1,847 | 26 | 46,681 | 3.54 | 15.6 |

To find an explanation for this contradictory evidence and to clarify the question concerning kind and quantity of groundwater recharge in these sandstone areas, detailed and long-term hydrogeological investigations will be necessary.

RIYADH AL KHABRA

Watertable 16 - 20 m b.s. Borehole 60 - 75 m

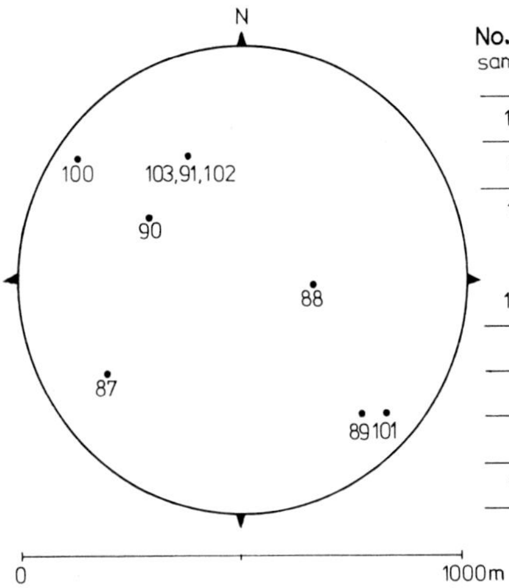


| No. of sample | extraction since | sampled after start | temp °C | $O_2$ mg/l | TDS g/l | δD ‰ | $\delta^{18}O$ ‰ |
|---|---|---|---|---|---|---|---|
| 100 | 1 month | ~8-10 hours | 27,7 | 9,0 | 1.60 | -9,6 | -2,35 |
| 90 | ,, | ,, | ? | 8,2 | 1.56 | -9,3 | -2,24 |
| 103 | 12 months | 3 min | 27,4 | 7,6 | 3,02 | -9,9 | -2,06 |
| 91 | | ~8 hours | ,, | 7,3 | 2,57 | -9,2 | -2,18 |
| 102 | | 12 ,, | 27,6 | 7,2 | 2,42 | -9,6 | -2,11 |
| 87 | 18 ,, | ~8-10 ,, | 27,9 | 7,4 | 2,41 | -10,9 | -2,37 |
| 88 | ? | ,, | 27,2 | 7,3 | 3,19 | -9,7 | -2,20 |
| 89 | 10 years | ~ ,, | 27,9 | 6,8 | 6,96 | -8,5 | -1,73 |
| 101 | ,, | ~1 hour | 27,4 | 7,0 | 8,67 | -8,5 | -1,25 |

Fig. 61. Parts of well area of Riyadh Al Khabra with numbers of wells or water samples. The table shows sample number, the age of the wells (i.e., time since extraction by pumping), the time interval between the start of the pumping and sample collecting (= sampled hours after start), water temperature, oxygen content (mg/l), total mineralization in g/l (T.D.S.) and content of stable isotopes (δ D, $\delta^{18}O$) in per mill

### *2.3.2.2. Hydrochemical investigations and isotope measurements in the areas of Riyadh Al Khabra, Wadi Ar Rimah and Wadi Maraghan*

(C. Job, H. Moser, W. Rauert, W. Stichler)

#### *2.3.2.2.1. Riyadh Al Khabra*

The area of Riyadh Al Khabra includes a few small oases in Wadi Ar Rimah about 30 kilometers west of the town of 'Unayzah (Fig. 60). Water samples were collected from drilled wells (their depths ranged between 60 and 75 meters). Generally, the drillings have a casing of just a few meters below the land surface. The wells cut through minor fanglomerates and reach down into the Saq sandstone. The water level lies 16 to 20 meters below the land surface. The groundwater has a temperature of about 27.5 °C and is saturated with oxygen. The degree of mineralization is highly variable (1.6 to 8.7 g/l) in spite of the small distances between the various wells. According to statements made by farmers in the vicinity the first yield of newly-drilled wells is of good quality, but the degree of mineralization increases in the course of following years. This information agrees with the fact that the salt content of water samples increases in accordance to the age of the wells (Fig. 61).

From the chemical point of view the waters are more or less concentrated Na—Ca—Cl—$SO_4$ solutions containing nitrate up to 370 mg/l and boric acid up to 10 mg/l (Table 36) [1, 2]. Furthermore, it is remarkable that the salt content decreases during pumping every day (Fig. 61). The enrichment of various ions does not increase regularly with the total concentration. Table 37 shows the factors of enrichment with regard to the least concentrated water of sample 90. Naturally, there are some differences in enrichment caused by the limited solubility of $CaF_2$, $CaCO_3$, $CaSO_4$, Ca-borate and Ca-, Mg-silicate. On the other hand, it is most striking that bromide and nitrate are less concentrated than chloride, although both bromide and nitrate salts are very soluble. This phenomenon was also observed in the drainage waters of Al Qatif oases (2.1.6.1., Table 17). The proportional concentration increase of other very soluble salts can be explained by evaporation or solution of an increasing quantity of homogeneous salt mixture or by both of these processes. A closer insight into the causes of salination of these groundwaters is obtained by the measurement of their contents of the environmental isotopes, deuterium, oxygen-18 and tritium (Table 38).

Concerning the definitions of $\delta$ D- and $\delta\, ^{18}O$-values and fundamentals for their hydrological interpretation see section 2.1.8.2. In the present case, however, beside primary evaporation of sea water, subsequent secondary and tertiary evaporation have to be taken into consideration: if a part of the rain-water or surface water evaporates again, an enrichment of heavy isotopes, especially as $H_2\, ^{18}O$, occurs in the liquid phase. In this case $m$ and therefore also $t$ are not constant any more but depend on several climatic factors (quantity and intensity of rainfall, humidity, air temperature, wind velocity, etc.) which determine the extent of this secondary evaporation. Varying isotope contents of precipitation infiltrating into groundwater produce an average by mixing and exchange. Capillary elevation from the subterranean aquifer to the land surface—if it occurs—goes moderately without changing the balance of $m$ and $t$. However, if groundwater again reaches the land surface through wells and springs, kinetically influenced evaporation will become active again (tertiary evaporation). Infiltrating once more these waters have lower $t$ values in the relation $\delta D = 8\,\delta\, ^{18}O + t_8$. (Under the conditions of $m = 8$, $t$ will be written as $t_8$ in the following text.) In this infiltration water the salt-concentration will be higher, too; this is not so much a consequence of evaporation as owing to the high salt content of arid soils. In the case of recycling of these waters into wells, the enrichment of the stable isotope content by tertiary evaporation will be therefore combined with an increasing salt concentration. Similar correlations may also occur by mixing of groundwaters of different isotope and salt content. Examples are described in chapters 2.1.6. and 2.5.4.

---

[1] The samples were collected between March 18 and 24, 1974.

[2] As to international drinking water standards, nitrate concentrations of more than 100 mg/l are detrimental to health (formation of methaemoglobin by reduction of nitrates to nitrites and development of carcinogenic nitrosamines).

Concerning the content of deuterium and oxygen-18, the waters of Riyadh Al Khabra differ strikingly from the isotopically poorer old karst waters of the Umm er Radhuma Formation. The δ D-values are between — 8.5 and — 10.9 per mill, the δ $^{18}$O-values between — 1.25 and — 3.37 per mill (Table 38, Fig. 62). The deuterium content shows no significant relation to the total mineralization ($r_s = 0.33 < 0.58$ for $n = 9$ and $\alpha = 0.05$) [1]. How-

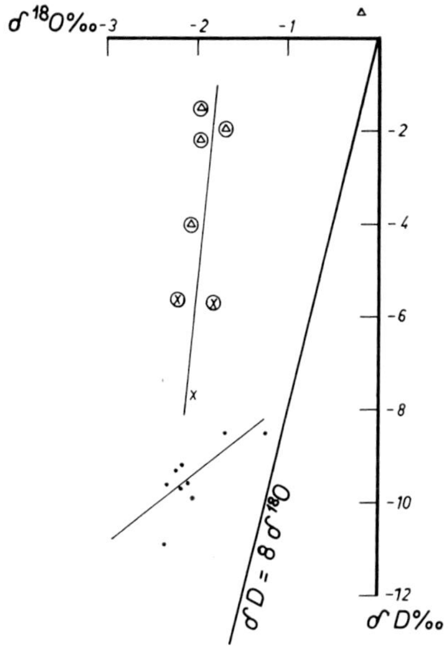


Fig. 62. δ D/δ $^{18}$O relation of recent well waters in the area of Riyadh Al Khabra (dots), in Wadi Ar Rimah (triangles), and in Wadi Maraghan (crosses). Samples having a tritium content of > 4 T.U. are marked by circles. Note the different gradient of the equation lines of Riyadh Al Khabra waters (influenced by evaporation) and the mixed waters of Wadi Ar Rimah

ever, the δ $^{18}$O-values become significantly higher corresponding to an increasing degree of total mineralization ($r_s = 0.73 > 0.68$ for $n = 9$ and $\alpha = 0.025$).

The difference between the δ D- and δ $^{18}$O-values depends on the fact that the δ D-values in relation to measurement accuracy are much less influenced by the kinetics of evaporation than δ $^{18}$O-values. In the relation $\delta D = 8\,\delta\,^{18}O + t_8$ the $t_8$-value is significantly correlated with the salt content ($r_s = -0.83 > -0.81$ for $n = 9$ and $\alpha = 0.0005$). This is typical for waters of tertiary evaporation. The very low tritium content, which could be measured only in two samples taken from the most highly mineralized waters (well No. 89, 101), shows that recent rain water cannot be the main agent washing soil salts into the wells.

---

[1] $r_s$ = Spearman's correlation coefficient; the following number is the check quotient for the number of spot checks ($n$); $\alpha$ = respective mean errors (see section 2.1.6.1.4.).

Table 37. *Factors of enrichment of the various anions and cations of the waters collected in the area of Riyadh Al Khabra in relation to the least mineralized water of well No. 90*

| No. | $Na^+$ | $K^+$ | $Ca^{++}$ | $Mg^{++}$ | $Cl^-$ | $Br^-$ | $SO_4^{--}$ | $HCO_3^-$ | $NO_3^-$ | $F^-$ | $H_3BO_3$ | $H_2SiO_3$ |
|---|---|---|---|---|---|---|---|---|---|---|---|---|
| 90 | 1 | 1 | 1 | 1 | 1 | 1 | 1 | 1 | 1 | 1 | 1 | 1 |
| 100 | 1.01 | 1.04 | 1.08 | 0.82 | 1.01 | 0.93 | 1.03 | 0.99 | 1.08 | 0.97 | 0.90 | 1.05 |
| 102 | 1.49 | 2.04 | 1.62 | 1.70 | 1.50 | 1.23 | 1.75 | 1.23 | 1.18 | 0.86 | 1.30 | 1.11 |
| 87 | 1.56 | 1.48 | 1.65 | 1.42 | 1.89 | 1.53 | 1.35 | 1.06 | 1.09 | 0.79 | 1.03 | 0.98 |
| 91 | 1.52 | 2.06 | 1.64 | 1.74 | 1.52 | 1.23 | 1.81 | 1.27 | 1.20 | 0.86 | 1.36 | 1.06 |
| 103 | 1.84 | 2.99 | 2.05 | 2.20 | 1.80 | 1.49 | 2.33 | 1.42 | 1.28 | 0.90 | 1.68 | 1.25 |
| 88 | 2.13 | 2.30 | 2.02 | 2.35 | 2.19 | 1.63 | 2.14 | 1.31 | 1.58 | 0.83 | 1.58 | 0.98 |
| 89 | 5,76 | 5.01 | 3.27 | 6.27 | 5.99 | 3.15 | 3.61 | 1.51 | 3.33 | 0.83 | 4.12 | 1.11 |
| 101 | 7.31 | 7.13 | 3.84 | 8.74 | 7.66 | 4.94 | 4.33 | 1.70 | 4.13 | 0.83 | 4.74 | 1.34 |

Table 38. *Isotope contents of water samples collected in the areas of Riyadh Al Khabra, Wadi Ar Rimah, Wadi Maraghan and rain water sample from a thunderstorm at 'Unayzah, March 18, 1974 (air temperature 18.5 °C).* Well water samples collected between March 18 and 25, 1974. $\delta$ D and $\delta\,^{18}O$ in per mill deviation of SMOW. 1 T.U. equals 3.2 pCi tritium per litre water. $t_8$ calculated from $\delta D = 8 \cdot \delta\,^{18}O + t_8$. For the tritium content the double standard deviation is given. The respective mean error amounts: for $\delta$ D ± 1.0 per mill, for $\delta\,^{18}O$ ± 0.2 per mill

| No. | $\delta$ D ‰ | $\delta\,^{18}O$ ‰ | $t_8$ | T.U. | |
|---|---|---|---|---|---|
| 87 | — 10.9 | — 2.37 | 8.06 | 0.8 ± 1.7 | |
| 88 | — 9.7 | — 2.20 | 7.90 | 0.2 ± 1.7 | |
| 89 | — 8.5 | — 1.73 | 5.34 | 2.5 ± 1.9 | |
| 90 | — 9.3 | — 2.24 | 8.62 | 0 ± 1.7 | |
| 91 | — 9.2 | — 2.18 | 8.24 | 1.3 ± 1.8 | Riyadh Al Khabra |
| 100 | — 9.6 | — 2.35 | 9.20 | 0.4 ± 2.3 | |
| 101 | — 8.5 | — 1.25 | 1.50 | 2.2 ± 1.9 | |
| 102 | — 9.6 | — 2.11 | 7.28 | 1.9 ± 2.1 | |
| 103 | — 9.9 | — 2.06 | 6.58 | 0.6 ± 2.3 | |
| 92 | + 0.6 | — 0.18 | 2.04 | 2.9 ± 1.8 | |
| 93 | — 1.9 | — 1.66 | 11.38 | 5.4 ± 1.8 | |
| 94 | — 1.5 | — 1.99 | 14.42 | 59.1 ± 3.9 | Wadi Ar Rimah |
| 95 | — 4.0 | — 2.05 | 12.40 | 9.1 ± 1.9 | |
| 96 | — 2.1 | — 1.98 | 13.74 | 40.2 ± 3.1 | |
| 97 | — 5.6 | — 2.21 | 12.08 | 23.0 ± 3.1 | |
| 98 | — 5.7 | — 1.81 | 8.78 | 22.7 ± 3.1 | Wadi Maraghan |
| 99 | — 7.7 | — 2.05 | 8.70 | 1.7 ± 2.3 | |
| 86 | + 9.2 | — 0.02 | 9.36 | 33.5 ± 2.8 | Rain Water |

Table 39. *Chemical composition of water samples collected in the areas of Wadi Ar Rimah (92–96) and Wadi Maraghan (97–99) and from rainfall (86).* Ion concentration in µe/l, $O_2$, boric acid and silicic acid in mg/l. SK = sum of cations, SA = sum of anions in µe/l. W = well type (see section 2.1.6.1.4.). Samples collected on March 19 and 29, 1974. For sampling locations, see Fig. 60

| No. | W | °C | $O_2$ | $Na^+$ | $K^+$ | $Ca^{++}$ | $Mg^{++}$ | SK | $Cl^-$ | $Br^-$ | $SO_4^{--}$ | $HCO_3^-$ | $NO_3^-$ | $F^-$ | SA | $H_3BO_3$ | $H_2SiO_3$ |
|---|---|---|---|---|---|---|---|---|---|---|---|---|---|---|---|---|---|
| 92 | HP | 27.0 | 6.6 | 64,750 | 244 | 56,385 | 5,943 | 127,322 | 70,333 | 107.6 | 38,023 | 1,657 | 17,094 | 145 | 127,360 | 10.57 | 18.9 |
| 93 | HP | 26.9 | 8.0 | 34,189 | 180 | 47,840 | 6,480 | 88,689 | 56,181 | 59.2 | 22,867 | 2,128 | 7,450 | 86 | 88,771 | 8.57 | 23.6 |
| 94 | HP | 26.9 | 9.0 | 2,114 | 31 | 2,450 | 209 | 4,804 | 810 | 0.9 | 1,220 | 2,356 | 339 | 126 | 4,852 | 0.71 | 27.5 |
| 95 | HP | 26.2 | 6.6 | 38,647 | 498 | 34,240 | 4,960 | 78,345 | 42,657 | 62.6 | 32,457 | 2,179 | 1,838 | 135 | 79,329 | 7.20 | 22.8 |
| 96 | HP | 27.9 | 7.8 | 5,763 | 23 | 5,914 | 240 | 11,940 | 3,900 | 9.1 | 3,700 | 2,876 | 1,451 | 100 | 12,036 | 2.14 | 28.9 |
| 97 | HP | 27.7 | 4.2 | 17,833 | 79 | 14,694 | 4,630 | 37,236 | 8,352 | 6.6 | 25,478 | 3,092 | 522 | 51 | 37,502 | 5.23 | 18.9 |
| 98 | HP | 26.7 | 6.2 | 38,060 | 134 | 36,144 | 13,413 | 87,751 | 42,257 | 31.0 | 41,880 | 2,249 | 1,209 | 67 | 87,693 | 6.57 | 20.6 |
| 99 | HP | 26.8 | 1.8 | 3,636 | 196 | 2,185 | 321 | 6,338 | 3,181 | 0.0 | 1,404 | 1,635 | 85 | 5 | 6,310 | 1.03 | 6.7 |
| 86 | Rain | — | — | 287 | 478 | 680 | 388 | 1,833 | 241 | 0.0 | 450 | 938 | 230 | 0 | 1,859 | 0.0 | 4.1 |

Summarizing all observations made in the area of Riyadh Al Khabra the following arguments speak in favour of a reflux of drainage waters into the wells (recycling): The decreasing degree of mineralization of the water during pumping (i.e., the lowering water level) gives evidence that the upper zone of the groundwater has a higher salt concentration than the lower one. Consequently, the salt infiltration must come from above. Furthermore, salination proved to be correlated with isotopic enrichment, which—since it is kinetically influenced—can have taken place only on the land surface. Finally, the fact that the increasing degree of mineralization corresponds to the age of the well supports the theory of recycling; the longer this recycling process lasts, the more surface-evaporites accumulated in the Quaternary sediments are dissolved and carried into the aquifer.

### *2.3.2.2.2. Wadi Ar Rimah, Wadi Maraghan*

For the location of the wells see Fig. 60; the chemical composition of the waters and their content of environmental isotopes are illustrated by Tables 38 and 39 [1].

Well No. 92 is hand-dug in the flat-dipping terrace at the northern border of Wadi Ar Rimah. The water level lies about 22 meters below land surface. The aquifer consists of weathered granite, its water temperature is 27 °C. The water is strongly mineralized and contains more than 1 g/l of nitrate. Its tritium content is low; its stable isotopes are enriched. The wells No. 93–95 lie along the road from the village of Ar Rass to An Nabhaniyah; well No. 96 is dug at the northern part of the wadi, about 10 kilometers west of the road. These wells are hand-dug wells reaching down into the weathered granite which is covered by sand fans and fanglomerate. Wells No. 94 and 96 have relatively slightly mineralized waters which contain tritium from present precipitations (e.g., sample No. 86). Their low mineral content is likely to result from their short subterranean flow through scarcely soluble granite. In spite of the presence of the same rock material the water of well No. 93 is much more strongly mineralized. The tritium content is lower and the water therefore older than that of wells No. 94 and 96. The same is true of well No. 95.

The wells of Wadi Maraghan (No. 97–99) reach 2–5 meters into the weathered schists of the Murdama Formation. The water is partly collected from the alluvial cover and partly from open joints in the schists. In wells No. 97 and 98 the water has a moderate to strong degree of mineralization and tritium content.

The correlation tests of the waters 92–98 show a strong relation between the tritium content and the total mineralization ($r_s = -0.96 = -0.96$ for $n = 7$ and $\alpha = 0.001$), i.e., the mineralization decreases in correspondence to the increasing tritium content of the water. Only the water of well No. 99 does not show the same tendency; the mineralization is low, but the water has no tritium. This is rather difficult to explain; it may be that there exists

[1] The samples were collected on March 19 and 20, 1974.

a direct infiltration into outcropping open joints not covered by weathered material.

The $t_8$-values of waters of the wells No. 92–98 decrease significantly with their total mineralization ($r_s = -0.93 > -0.89$ for $n = 7$ and $\alpha = 0.005$). In comparison with the water samples collected in the area of Riyadh Al Khabra it seems that the waters of higher mineralization may have undergone a stronger evaporation process.

However, as can be seen from Fig. 62, in the samples 93–98 decreasing $t_8$-value is not linked with an increasing $^{18}O$-content as it is in the Riyadh Al Khabra waters when evaporation is involved, but mainly with a decreasing deuterium content. In this case it seems that the change of $t_8$-value—and therefore also the differences in mineralization—depends on the fact that older waters, free from tritium, poor in deuterium and more strongly mineralized, become diluted by younger waters of a lower mineralization, carrying tritium and an enriched deuterium content. Therefore, a decreasing $t_8$-value combined with an increasing mineralization is not always a proof of tertiary evaporation, but can also occur in mixed waters. The gradient of the equation line shows which of these two possibilities is effective. For well No. 92 the high $^{18}O$-content again proves a strong tertiary evaporation and a recycling similar to the Riyadh Al Khabra waters. An argument against recycling seems to be the fact that this well with high mineralization and a stable isotope composition typical of recycling, has practically no tritium in its water in spite of the fairly good conditions for infiltration. The question arises where the tritium from the present episodic rainfalls disappears. This can be explained by the fact that the recycling process is restricted to the area around the well. In comparison to the large quantities of irrigation water infiltrating into the underground the amount of rain water is very small, and the tritium value, therefore, remains below traceability. Thus, in this case the lack of tritium is no argument against recycling. It should be mentioned once more that the drainage waters of Al Qatif oases include no tritium either.

On the other hand, recycling is not always the only reason for a high degree of total mineralization of the groundwater. Thus, the high salinity of the wells in Wadi Maraghan (samples No. 97, 98) is probably attributable to a contamination by an infiltration of river water, which infiltrates into the well through salinated Quarternary deposits (see 2.3.2.1.). The inflow of waters of high salt content from Quaternary deposits explains the occurrence of highly mineralized water in less soluble rocks, e.g., granite, schists, and sandstone.

The geological finding on some well-profiles in Wadi Ar Rimah suggests that clayey weathered feldspars divide the groundwater in the weathered granites into two horizons (see 2.3.3.1.). Such an impermeable layer would not only prevent the influx of tritium-containing rain but also the reflux of irrigation water into the well. There, the different conclusions drawn from hydrochemical investigations on the one side and from hydrogeological conditions on the other make special local investigations, and the application of artificial groundwater tracing techniques necessary. General conclusions from hydrochemical investigations carried out in waters from the areas of Riyadh Al

Khabra and of Wadi Ar Rimah make it clear that for the agricultural development of arid regions not only the drilling of the new wells but also the selection of areas is decisive for a successful outcome. Not only does the water supply have importance for irrigation but also a well-controlled drainage system. From this point of view cultivation should generally not use wadi floors and basins but higher located areas and flat slopes where there is a free runoff of irrigation water, and no infiltration of salt water from Quaternary accumulations may occur. Shallow basins should be used as evaporation pans. To calculate the depth of boreholes and their casing, the possibilities of mineralization of the groundwater through reflux or inflow of highly mineralized waters from Quaternary wadi sediments should be taken into consideration.

On the other hand there are large quantities of groundwater of good quality and enough recharge by recent precipitation, e.g., Wadi Nisah, Wadi Hanifah (water supply for the capital of Ar Riyadh) et al. Such wadis should become strictly protected areas without agriculture and settlement.

The studies carried out in the area of Riyadh Al Khabra demonstrate that drilling of new wells for irrigation without planning a well-working drainage system may endanger an oasis that has been viable so far.

## 2.4. Cuesta Region of the Tuwayq Mountains

### 2.4.1. General Geology and Stratigraphy

(E. Schyfsma)

By the cuesta region of the Tuwayq Mountains is meant the area where the landscape is controlled by asymmetric ridges with one face (dip slope, cuesta back slope) long and gentle, conforming with the dip of the resistant Mesozoic beds, rarely exceeding 1°, and the opposite face (scarp slope, escarpment) steep and clifflike, formed by the outcrop of these resistant rocks, mostly limestones. It is a vast area occurring in a great curved belt parallel to the eastern margin of the Arabian Shield. In this belt relatively low hills protrude above flat plains. The gentle slopes of these hills conform with the dip of the beds, and the steep slopes descend abruptly from their crests, forming steep escarpments facing west and capped by resistant limestones. The west-facing scarps offer excellent opportunity for studying the rock sequences; some of the rock units can thus be traced over hundreds of kilometers. The softer formations between cuestas consist mostly of sandstones and shales which are less resistant against erosion and denudation. They occur in the intervening valleys and plains, but although they are often only partially masked by gravels and sands, these units are fairly well known. The cuesta region of the Tuwayq Mountains constitutes one of the most spectacular features of the central Arabian landscape.

| Time units: Periods | | Epochs | Rock formations | Thickness meters |
|---|---|---|---|---|
| Tertiary | | Paleocene | Umm er Radhuma | |
| | | | disconformity | |
| Cretaceous | Upper | Maestrichtian Campanian | Aruma | 140 |
| | | | disconformity | |
| | | Turonian? Cenomanian | Wasia | 40 |
| | | | unconformity | |
| | Lower | Aptian Barremian | Biyadh | 425 |
| | | | unconformity | |
| | | Hauterivian | Buwaib | 45 |
| | | | unconformity | |
| | | Valanginian | Yamama | 45.5 |
| | | | unconformity | |
| | | Berriasian | Sulaiy | 170 |
| Jurassic | Upper | Tithonian | Hith | 90 |
| | | | Arab | 127.5 |
| | | Kimmeridgian | Jubaila | 130 |
| | | | Hanifa | 101 |
| | | Oxfordian | Tuwaiq Mountain | 215 |
| | Middle | Callovian Bathonian Bajocian | Dhruma | 375 |
| | Lower | Toarcian | Marrat | 111 |
| | | | unconformity | |
| Triassic | Upper | | Minjur | |

The geological information for this region is entirely derived from R. W. POWERS et al. (1966) [1].

[1] For locations of places mentioned in this text see R. A. BRAMKAMP, L. F. RAMIREZ, 1958.

The cuestas are shaped in Jurassic and Cretaceous rocks, mostly in Upper Jurassic and early Lower Cretaceous. These periods are subdivided into the following epochs, to which belong the following formations:

**Marrat Formation** (Lower Jurassic, Toarcian). In the Tuwayq Mountains, the oldest formation to display cuestas with west-facing escarpments is the Marrat Formation of Lower Jurassic (Toarcian) age.

The base of the formation is at the unconformable contact between tan, thin-bedded limestone above and gray, friable, angular, crossbedded sandstone below (Minjur Sandstone). The top is at the change from Marrat gray (golden-brown weathering) limestone to olive-green, occasionally gypsiferous shale of the overlying Dhruma. This contact is conformable at the type section but unconformable further south.

The formation consists of three natural topographic-lithologic members:

1. A lower limestone member forming the lower scarp of the Maraht cuesta (lower Marrat limestone), 40 meters thick.
2. The Marrat red shale member, 55 meters thick.
3. The upper limestone member forming the upper Marrat limestone, 16 meters thick.

The formation can be traced over a distance of more than 650 kilometers. The outcrop band reaches its maximum width of about 15 kilometers in the vicinity of Maraht.

From Khashm Adh Dhibi south, the formation loses topographic distinctiveness as it dips under the protective ledge of the Tuwayq Escarpment. Further south, up to Khashm Al Jufayr, the three members can still be traced, although significant facies changes have taken place. Beyond this point, unit boundaries become increasingly uncertain.

North of Shaqra, the escarpments above and below the red shale member gradually break down and the three-fold subdivision of the Marrat Formation is no longer found north of lat 25° 30′ N.

**Dhruma Formation** (Middle Jurassic). The type section was measured at Khashm Adh Dhibi. The top of the formation is placed at the contact between olive-green shale of the upper Dhruma and light-brown, soft, chalky limestone of the Tuwayq Mountain Limestone. This contact represents a conformable transition from shale to limestone in the central outcrop area.

The major subdivisions of the Dhruma Formation are, from bottom to top:

1. Lower Dhruma (Bajocian): Shale with some limestone, containing a thin layer of gypsum at the base, and an upper bed of massive, cliff-forming limestone (Dhibi limestone).
2. Middle Dhruma (Bathonian): Limestone containing beds of clean, current-washed calcarenite at various levels, and an upper oolite bed that forms a prominent cliff.
3. Upper Dhruma, at the type section divided into two members on lithologic and faunal evidence, the Atash member (Bathonian) below, consisting of limestones, and the Hisyan member (Callovian?) above, consisting of shales.

Beds of the Dhruma Formation extend from lat 19° 20′ N to lat

27° 05′ N, a distance of more than 900 kilometers. The outcrop belt reaches a maximum width of 20–25 kilometers at Wadi Birk and Khashm Adh Dhibi and narrows irregularly to the north and south.

Dhruma topography in the vicinity of Khashm Adh Dhibi is dominated by three prominent limestone benches. The lowermost of these is capped by the Dhibi limestone, i.e., the top of the lower Dhruma Formation. The middle is formed by the upper oolite bed of the middle Dhruma, and the third by the Atash member of the upper Dhruma.

From Khashm Adh Dhibi south, the Dhruma is remarkably well exposed in a steplike series of high, limestone-capped scarps and broad dip slopes. Towards the south, the limestones gradually change to shales and sandstones. As a consequence, massive scarps fade, and the Dhruma retreats under a protective cap of Tuwayq Mountain Limestone.

North of Khashm Adh Dhibi, the Dhruma is generally poorly exposed and consists mainly of intensely weathered shales under capping ledges of limestone.

**Tuwayq Mountain Limestone** (Middle-Upper Jurassic). The Tuwayq Mountain Limestone itself constitutes the backbone of Jebal Tuwayq.

The top of the formation is at the contact between the massive coral-bearing limestone (top of the Tuwayq Mountain Limestone) and the overlying marl and soft limestone of the Hanifa Formation.

The Tuwayq Mountain Limestone has been mapped from lat 17° 30′ N to lat 27° 20′ N, a distance of more than 1,200 kilometers. The width of outcrop in the central area is 20 kilometers and narrows irregularly to about 10 kilometers at the northern and southern extremities. Although strike of the beds shifts gradually along the outcrop belt, dip remains relatively constant, averaging about 0° 40′ E. The Tuwayq Mountain Limestone dominates both the stratigraphy and the topography of central Arabia. Over most of its extent the limestone forms a precipitous west-facing escarpment, gently curved in the center and concave toward the west. The scarp in front of the central part of the outcrop is impressive. Away from this area, the height of the escarpment diminishes gradually until a low, weathered bench of a few meters remains in the extreme south and north.

The western boundary of the formation is marked by a vertical limestone cliff which in many places exposes some 100 meters of lower Tuwayq Mountain Limestone. An additional 30 to 40 meters of Tuwayq Mountain rocks occur below the cliff-forming section, but this unit is simply a weathered as an extension of the gentle slope established in the underlying Dhruma beds. The upper part of the formation forms a strong plateau surface tilted gently eastward. This surface is not a true dip slope but is slightly beveled so that successively younger beds are present down dip. The eastern margin of the formation is usually marked by a wide, flat valley eroded in front of the much softer Hanifa Formation.

The marked lithologic continuity of the Tuwayq Mountain Limestone over a great distance is one of the most spectacular features of the geology of Arabia. In the central outcrop area, the formation is a massively-bedded aphanitic limestone with some calcarenitic limestone and calcarenite in the upper part. The lower 25 to 40 meters is well-bedded, highly fossiliferous, soft, chalky limestone with some soft calcarenitic limestone, calcarenite, and shale. The lower chalky unit is gradational from Dhruma shale below to massive limestone at the top and forms an even talus slope at the foot of the vertical cliff. The base of the cliff-forming limestone is not a stratigraphic horizon. Tracing the "notch" weathered at the junction of the hard and soft beds confirms this observation. When the basal beds of the limestone become more argillaceous and calcarenitic, the "notch" rises in the section, and vice versa. The typical lithologic-topographic pattern of a shale slope at the base, a limestone cliff above, and a broad, bench-ridden back slope cut in the upper limestone beds maintains itself throughout the central outcrop belt.

North of the central outcrop area, the upper cliff-forming, massive limestones gradually change into softer rocks such as chalky limestones and shales, resulting in the loss of the sharp contact with the overlying Hanifa. In its northern extremity the formation has changed to an argillaceous and arenaceous facies.

South of the central outcrop area, the first lithologic change in Tuwayq Mountain rocks occurs in the lower part where the basal 10 meters are replaced by sandstone. The upper part of the formation remains calcareous throughout with some dolomite-calcarenite lenses. In its southern extremity the formation consists of upper aphanitic limestone overlying a unit of sandstone with interbeds of calcareous sandstone and sandy limestone containing numerous fossils, mostly *Nerinea.*

The Dhruma-Tuwayq contact is conformable in the central outcrop area and north of this. South of the central outcrop region, at lat 19° 24′ N, the Dhruma pinches out and the Tuwayq Mountain Limestone rests with low angular discordance on the Sudair Shale (Lower Triassic). Continued transgression southward carries the Tuwayq Mountain across the Sudair and Khuff Formations (Upper Permian), and finally brings the formation at its southern extremity into unconformable contact with the Wajid Formation (pre-Khuff, Lower Permian?).

**Hanifa Formation** (Upper Jurassic). The top of the Hanifa Formation is at the break between clean-washed, tightly cemented, oolite-pellet calcarenite of the Hanifa below and slabby weathering, aphanitic and calcarenitic limestone of the Jubaila Limestone above.

The Hanifa Formation crops out in a narrow, gently curving belt from lat 18° 00′ N to lat 26° 00′ N, a distance of about 1,100 kilometers. The width of outcrop varies from 0 to 25 kilometers but averages about 10 kilometers.

The Hanifa Formation is a unit of relatively soft rocks situated on the hard and resistant Jebal Tuwayq Plateau and protected above by harder and more resistant limestones of the Jubaila Formation.

Lithologically, the Hanifa Formation is somewhat more variable than the underlying Tuwayq Mountain Limestone; basically it is another nearly pure carbonate unit. It is made up of alternating aphanitic and calcarenitic limestone, much of it oolitic, especially in the upper part of the formation. In fact, the top of the Hanifa is marked by a massive bed of oolite-pellet calcarenite that can be traced without interruption almost the entire length of the outcrop. Argillaceous limestone units and some shale are present, particularly in the northern area; dolomite is common in the south. A few coral-bearing beds occur in the central area and become increasingly prominent southward.

The conformable contact between the Hanifa Formation and the overlying Jubaila Formation is somewhat more difficult to place than the lower limit, especially at the far ends of the outcrop where dolomite replacement obscures the break. In the type area, the upper contact is obvious: the distinctive brown colour of the Hanifa rocks, the top unit of which is a characteristic gray-brown, tightly cemented, pellet-oolite calcarenite, contrasts sharply with the light-coloured, usually cliff-forming, somewhat more thinly bedded limestone of the overlying Jubaila Limestone.

**Jubaila Limestone** (Upper Jurassic). The top of the Jubaila is at the change from brown, ledge-forming, tight limestone below to gray, softer, rubbly-weathering dolomite and clean-washed calcarenite of the Arab Formation above. The formation has been mapped from lat 18° 00′ N to 26° 00′ N, a distance of nearly 1,100 kilometers.

The Jubaila Limestone, an extremely resistant limestone, sandwiched between two softer units, typically weathers to a series of high steps. Lower beds form a prominent scarp, often more than 75 meters high, capping the Hanifa. The upper Jubaila breaks away to a gently eastward-tilted plateau which is a slightly beveled upland surface rather than a true dip slope. Consequently, the western margin of the formation is a sharp scarp; the eastern boundary is less regularly defined by patches of the overlying Arab Formation erratically scattered on the resistant Jubaila slope.

North of the type locality, the strong western scarp soon changes into two or three lower benches which lose relief gradually farther to the northwest. In the south, the Jubaila forms low benches with intervening wide, slightly eastward-dipping slopes; scarps become relatively subdued, usually less than 5 meters high.

The Jubaila Limestone, like the underlying Tuwayq Mountain and Hanifa Formations, is basically a shallow-water carbonate unit. Though it is mostly aphanitic and calcarenitic limestone, some highly persistent layers of clean-washed lime sand (calcarenite) are present, particularly in the central and northern areas. Corals and stromatoporoids, few in the type area, become abundant to the north and south. Limestones change to dolomite and sandstone both to the north and to the south.

**Arab Formation** (Upper Jurassic). The top of the formation falls at the conformable change from tan aphanitic limestone below to massive anhydrite of the Hith Formation above.

The Arab Formation consists of four marine carbonate units, separated by laterally persistent beds of anhydrite. The carbonates, together with their overlying anhydrite beds, represent four main cycles of deposition. These four members (A, B, C, and D) can easily be recognized in boreholes. This is not the case in the field, however, where the area of Arab exposure is mainly a broad, gently undulating plain. A few low, isolated hills stand above the general ground level. Effects of slumping are everywhere evident due to the solution of anhydrite beds and the subsequent collapse of the carbonate units left behind.

The formation has been mapped from about lat 20° 00′ N to about lat 26° 00′ N, a distance of nearly 700 kilometers. The width of outcrop over this distance averages about 10 kilometers.

**Hith Anhydrite** (Upper Jurassic). The Hith Formation is mainly massive anhydrite. Well sections usually contain some interbedded dolomite and limestone, particularly in the upper part.

The top of the Hith is at the conformable contact of massive anhydrite and a limestone breccia, with evenly-bedded oolite calcarenite of the overlying Sulaiy Formation.

The area of outcrop where the Hith should occur is a rather broad zone of badly slumped beds. This belt of collapsed rocks can be traced over a distance of some 550 kilometers and involves mostly Sulaiy strata.

**Sulaiy Formation** (Lower Cretaceous). The lithology of the Sulaiy is remarkably uniform. The main rock type is tan, chalky, massively bedded, aphanitic limestone. Some thin beds of skeletal and pelletoid calcarenitic limestone and calcarenite occur throughout but become particularly prominent in the lower 40 meters. The basal beds also include moderate amounts of fine-grained quartz sand.

The Hith-Sulaiy contact is visible only at Dahl Heet where it is gradational and conformable. The Sulaiy is overlain conformably by the Yamama Formation. The contact is marked by a change from tan, chalky aphanitic limestone below to massive beds of clean-washed Yamama calcarenite above.

The Sulaiy crops out in a gently curving arc from lat 22° 38′ N to lat 25° 30′ N, a distance of 350 kilometers. The most prominent feature of the Sulaiy landscape is the 50–100 meters thick, west-facing slump escarpment in the upper part of the formation. This escarpment, the Hith escarpment, marks the boundary between two areas of quite different topographic expression, and can be followed for over 200 kilometers. To the east is a gently tilted, steplike, upland surface formed by the weathering of tight, well-bedded limestone. To the west is an undulating surface of contorted limestone beds and low, complexly slumped hills, characterizing lower Sulaiy outcrops on dissolving Hith Anhydrite.

**Yamama Formation** (Lower Cretaceous). The Yamama Formation, like the Sulaiy, is basically a nearly pure carbonate unit. Calcarenite, calcarenitic limestone, and aphanitic limestone occur in varying proportions along strike. Clean-washed calcarenite and calcirudite are the main rock types at the type locality; away from this area progressively muddier sediments, aphanitic and calcarenitic limestone, and even very thin layers of shale come in.

The Yamama Formation rests conformably on the Sulaiy, i.e., limestone on limestone. Generally, however, the break is marked by a change from tan, moderately porous, aphanitic limestone below to brown, tightly cemented pellet calcarenite above. In addition, Sulaiy beds tend to be thicker and more resistant to erosion, and on weathering, these beds form a step-like series of 2–10 meters sheer cliffs separated by short intervals of dip slope. In the

Yamama, on the other hand, short intervals of soft beds form low-angle slopes each capped by a thin 0.5–2 meters resistant ledge.

The contact between the Yamama and overlying Buwaib Formation is unconformable. It is also a limestone on limestone contact and quite inconspicuous. Exact placement must await detailed faunal control.

**Buwaib Formation** (Lower Cretaceous). The Buwaib Formation is essentially a carbonate unit. Aphanitic limestone and calcarenitic limestone are most common, but the sequence is broken with thin intervals of calcarenite, sandstone, and shale. Beds of clean quartzose sandstone are confined to the upper part of the formation.

The top of the Buwaib is at the conformable and gradational contact between the highest thin marine limestone below and the sandstone of the overlying Biyadh Formation. Outcrops of the Buwaib Formation have been mapped in a narrow, irregular, slightly curved band from lat 20° 38′ N to lat 25° 24′ N, a distance of nearly 600 kilometers.

The formation weathers to a relatively low-relief plain. The generally flat surface is broken, however, by several small discontinuous benches that appear at various levels.

**Biyadh Sandstone** (Lower Cretaceous). The Biyadh Sandstone is essentially an arenaceous unit. Its top is at the unconformable contact of coarse-grained quartz sandstone with quartz pebbles and the overlying furruginous silty sandstone of the Wasia Formation, i.e., sandstone on sandstone.

Sandstone of the Biyadh can be traced over a distance of about 650 kilometers, from lat 20° 30′ N to lat 25° 40′ N. From Al Kharj southward, weathering of the Biyadh results in a characteristic landscape made up of numerous small dark-coloured to black hills between which are gently-rolling white, gravel-covered plains.

**Wasia Formation** (Upper Cretaceous). The Wasia Formation is a relatively thin sandstone and shale unit which is at least partly marine in the central and northern areas and probably nonmarine in the south. It contains, in the central and northern areas, lenses of dolomite and soft nodular limestone with marine fossils about 25 meters below the top of the formation.

The Wasia Formation is everywhere in disconformable contact with the overlying Aruma. The break is clearly marked by a change from sandstone below to reddish-brown, coarsely crystalline dolomite and dolomitic limestone above. At the base of the Wasia is a profound unconformity that cuts into progressively older beds from south to north around the arc of outcrop.

Wasia outcrops form a great arc from lat 20° 30′ N to lat 30° 00′ N, a distance of nearly 1,450 kilometers.

**Aruma Formation** (Upper Cretaceous). The upper limit of the Aruma Formation is taken at the change from yellow-brown dolomitic shale below to gray crystalline *Lockhartia*-bearing dolomite of the Umm er Radhuma Formation (Paleocene) above. This contact represents a disconformity on the basis of paleontologic evidence.

Rocks of the Aruma Formation crop out from lat 20° 30′ N to beyond the Iraq–Saudi Arabia border, a distance of more than 1,600 kilometers. Over this distance the strike of the outcrop belt changes gradually from northeast to north to northwest, describing a great arc wrapped around the Central Arabian Arch. As the belt approaches Iraq it again swings gently north.

The western edge of Aruma limestone is characteristically a steep west-facing escarpment in the central and northern area of outcrop. Generally this scarp represents the western limit of the formation, but about 40 kilometers south of the type area it becomes progressively lower and moves eastward diagonally across the outcrop band as the limestone is replaced from the bottom up by less resistant sandstone in a southward direction.

The Aruma is the youngest Mesozoic formation showing the typical cuesta shape in the curved belt of sedimentary rocks bordering the Arabian shield. East of this cuesta region is a broad expense of relatively low-relief terrain in which Tertiary and Quaternary deposits effectively mask older units.

The central Arabian graben and trough system, a great arc of tensional structures, upsets the otherwise continuous cuesta formed in the Mesozoic rocks of central Arabia.

## 2.4.2. Accumulation Terraces of Wadi Hanifah and Wadi Al Luhy

(H. Hötzl, H. Felber, V. Maurin, J. G. Zötl)

### *2.4.2.1. Morphological and sedimentological features*

The Central Tuwayq Mountains west and southwest of Ar Riyadh are intensively dissected by several wadi systems and tectonic grabens (Fig. 63).

The main wadis are Wadi Hanifah, Wadi Al Luhy, Wadi Bu'ajya and Wadi Al 'Ayn; the two grabens which also act as important young discharge systems are the grabens of Wadi Nisah and Wadi Al Awsat. All of them drain into the basin of Al Kharj (R. A. Bramkamp, L. F. Ramirez, 1958).

R. Wolfart (1965) described the development of this drainage pattern originated at first by erosion with meandering valleys and then partly transformed by the reactivation of the graben tectonic during the Pliocene and the Quaternary. R. Wolfart believed to have indications, from the occurrence of accumulation terraces in the wadis and their nonoccurrence in the grabens, that the tectonic movements have continued till recently. This is the first mention of accumulation terraces in the literature.

These young terraces seemed to us of great importance not only for the tectonic events, but also for the recognition of the sedimentological and climatological processes of the Quaternary Period, so the investigation was started with the wadi filling in Wadi Hanifah and its tributary, Wadi Al Luhy [1].

Wadi Hanifah has its headwaters in the area of the northern Tuwayq Mountains. From its upper part north of Al Jubaylah down to Al Ha'ir, a small village at the junction of the Wadi Hanifah and Wadi Al Luhy (see Fig. 63), i.e., a beeline of about 75 kilometers, the wadi channel runs from northwest to southeast and remains in the Jubaila Limestone (Upper Jurassic) to the approximate latitude of Ar Riyadh. From this area towards the south-

[1] To T. Dinçer, UNDP hydrologist at the Ministry of Agriculture and Water, Ar Riyadh we have to extend our sincere thanks for his ready assistance during our first excursions in this area.

east the eastern border of the wadi partly consists of limestone and dolomite dating back to the Arab Formation (strata overlying the Jubaila Limestone). Some 10 kilometers southeast of Al Ha'ir the wadi changes its direction towards the east, crossing the Arab Formation on its way into the basin of Al Kharj.

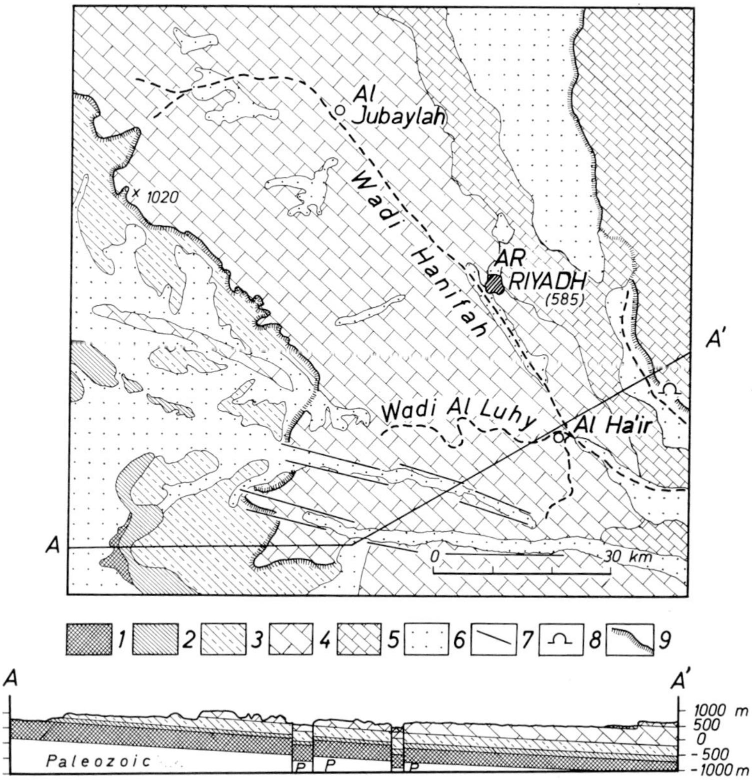


Fig. 63. Generalized geological map of Tuwayq Mountains in the surroundings of Wadi Hanifah. *1* Triassic Minjur Sandstone, *2* Lower Jurassic Marrat Formation, *3* Middle Jurassic Dhruma Formation, *4* Upper Jurassic Tuwayq Mountain Limestone, Hanifa Formation, Jubaila Limestone, and Arab Formation, *5* Lower Cretaceous formations, *6* Upper Pliocene and Quaternary sediments, *7* fault, *8* cave, *9* dominant escarpment

The first pattern of small recent erosion channels can be found in older wadi deposits in the area of Al Jubaylah. In the area west of Ar Riyadh, the remnants of the accumulation terrace have a thickness of more than 3 meters. The erosion caused by recent episodic floods has formed a vertical edge of this terrace, which has a flat surface without any vegetation (Fig. 64). The height of the terrace edge increases down stream from Ar Riyadh to Al Ha'ir. In general it amounts to 4–6 meters and ranges up to 8 meters at particularly narrow sections with active erosion. The same differences are to be observed in the lower part of Wadi Al Luhy. Downstream from Al Ha'ir, Wadi Hanifah at present shows active accumulation, and the wadi channel becomes flat, so that the terrace corresponds to the highwater bed.

Even though the thickness of the beds varies, the geologic sequence of the wadi sediments is rather constant. The uppermost 2.5–4 meters are occupied by wadi sabkhah sediments. These are mainly clayey silt and fine sands resulting from eolian deposits and fluviatile redeposition. The internal structures such as lamination and wavy bedding are sometimes disturbed by plant roots giving partly the impression of nonstratified deposits. At several places they contain fossil snail shells.

Fig. 64. Accumulation terrace in Wadi Hanifah southwest of Ar Riyadh. (Photo: J. G. Zötl, 1973)

Gravel layers with thicknesses up to 4 meters (in parts to 6 meters) underlie the wadi sabkhah sediments. The pebbles consist primarily of carbonate rocks of local origin. A new hand-dug well, under construction in 1974, 8 kilometers upstream of Al Ha'ir shows again clayey silts below the gravels. The whole wadi filling may have a thickness of 50 to 70 meters [1]. Concerning grain-size distribution and mineral composition of these sediments see chapter 3.1.2., Tables 48, 49 (samples No. 20, 22, 24, 31).

The terraces in Wadi Hanifah and Wadi Al Luhy show the same uniform morphological and sedimentological features. Only a slightly different sequence of deposition exists in the lower part of Wadi Al Luhy. Above the gravel there are interbedded stillwater sediments of grey clayey silts with coaly layers (sample No. 22) partly of lacustrine character. These deposits

[1] Personal communication, Mr. Schubarth, Riyadh Water Supply.

were formed by an episodic damming of the tributary by the river in Wadi Hanifah or by more rapid sedimentation there. Near the mouth into Wadi Hanifah these grey silts in Wadi Al Luhy reach a thickness of several meters; upstream they thin out. About 5 kilometers upstream from the mouth their thickness is already reduced to 0.5 meter. The 3.5-meter-high terrace wall there is formed from top to bottom by a 2.5-meters wadi sabkhah sequence (samples No. 263, 264), then 0.5 meter of grey calcareous stillwater silts

Fig. 65. Terrace in Wadi Al Luhy. Layer of stillwater sediments with abundant shells of gastropods (see Fig. 66) 2.5 meters below terrace surface underlain by gravels. Research group collecting shells of gastropods. (Photo: J. G. Zötl, 1974)

(sample No. 20) and at the base a 0.5-meter-thick gravel horizon. The stillwater sediments abound with shells of gastropods, which were collected for carbon-14 dating (Fig. 65). As a result of the carbon-14 measurement (sample VRI-384), an age of 8,400 ± 140 years B.P. was disclosed.

### 2.4.2.2. *Paleontological description of Gastropoda*

A paleontological determination was carried out on 170 shells of gastropods collected from the stillwater sediments in Wadi Al Luhy 5 kilometers upstream from the mouth near the village of Al Ha'ir. They can be divided into four species (Fig. 66):

| | Number of collected shells |
|---|---|
| *Biomphalaria pfeifferi ruppellii* (Dunker) | 11 |
| *Bulinus truncatus* (Audoin) | 2 |
| *Lymnea natalensis* (Krauss) | 21 |
| *Melanoides tuberculata* (Müller) | 138 |

A more detailed description of these species is given below. For special information refer to G. MANDAHL-BARTH (1957), H. A. PILSBRY, J. BEQUAERT (1927), B. W. SPARKS, A. T. GROVE (1961). With the exception of *Melanoides*, which belong to the subdivision of the Prosobranches, all collected gastropods can be assigned to the subdivision of Pulmonates.

*Biomphalaria pfeiferi ruppellii* (DUNKER), Fig. 66 a

The dextral shell is spiral-shaped and flat. It has 4.5 to 5 involute biconcave whorls. The sutures are deeply notched, the cross-section of the whorls

Fig. 66. Shells of gastropods from the grey calcareous stillwater silts of the Wadi Al Luhy terraces. Sample location: Wadi Al Luhy, 5 kilometers upstream from the village Al Ha'ir; 2.5 meters below terrace surface. Graphic scale in millimeters. (Photo: H. HÖTZL, 1976.) a) *Biomphalaria pfeifferi ruppellii* (DUNKER); b) *Bulinus truncatus* (AUDOIN); c) *Lymnea natalensis* (KRAUSS); d) *Melanoides tuberculata* (MÜLLER)

and the opening well-rounded in the upper part; however, they are slightly, yet noticeably, angular in the lower part. More or less regular growth lines form a feebly-developed sculpture.

| Sizes: | | |
|---|---|---|
| | largest diameter | 9 millimeters |
| | diameter of the umbilicus | 2.5 millimeters |
| | height of the shell | 2.9 millimeters |
| | height of the last whorl | 2.9 millimeters |

According to G. Mandahl-Barth (1957), the decisive criteria for the subdivision of the species of *Biomphalaria* are, firstly, the relations between the diameter of its umbilicus as well as, secondly, the relation of the shell diameter to its height. As to the investigated shells the gradient and the parallel straight lines refer to the criteria for *Biomphalaria pfeifferi ruppellii* described by G. Mandahl-Barth (1957; see Fig. 67). In recent times this species has been found in Yemen, Eritrea, Sudan, Tanganyika and Zaire.

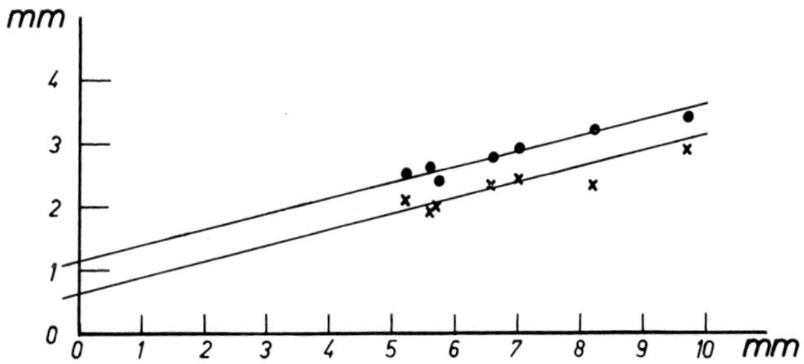


Fig. 67. Graphs of diameters of *Biomphalaria* sp. plotted against height of last whorl (×) and against diameter of umbilicus (●)

*Bulinus truncatus* (Audoin), Fig. 66 b

The evolute form of the shell has a slightly stretched spire. Four sinistral whorls clearly show steps and the suture is clearly developed. The oval mouth emphasizes the stretched form of the whorls. The outer margin of the aperture has a sharp edge, just as does the acute-angled parietal margin. The slightly reflexed columellar margin forms a narrow umbilical opening. Irregular growth lines cause a wavy sculpture.

| Sizes: | height of the shell | 8.1 millimeters and 2.8 millimeters |
|---|---|---|
| | diameter of the shell | 4.8 millimeters and 1.8 millimeters |
| | height of the aperture | 4.4 millimeters and 2.2 millimeters |

Both shells obviously belong to the subdivision of "*truncatus*"; their habit agrees with that in the description made by G. Mandahl-Barth (1957). This subdivision has recently been found in Egypt, Sudan and Uganda.

*Lymnea natalensis (*Krauss), Fig. 66 c

The relatively thin walls of the shell have sharp, conically stretched spires. The whorls, in most instances three or four, are clearly divided by steps. The concavely vaulted dextral whorls are symmetrically arranged and furnished with regular growth lines. The upper part of the aperture is rather acute-angled, the lower part is broadly curved with sharp edges. Its columellar margin is slightly reflexed.

| Sizes: | height of the shell | 6 millimeters |
|---|---|---|
| | height of the aperture | 4.1 millimeters |
| | diameter of the shell | 3.2 millimeters |

Recently, this species has been found in the East African lakes, the region of the White Nile and South Africa. Descriptions of fossil shells of this kind from the Tibesti Mountains and the Sudan are available.

*Melanoides tuberculata* (MÜLLER), Fig. 66 d

The slender, tower-like shell has 9 or 10 dextral, very regular, conic whorls, which are divided by clearly developed sutures. The whorls are straight and slightly inclined except the last one, which is concave; their heights only gradually increase. The aperture is approximately ovate; its outer margin has a sharp edge, its columellar margin is reflexed and lies directly on the last whorls. The basal margin shows partly a siphonoidal protrusion. The species is characterized by spiral and axial sculpture forming nodules and ribs.

| Sizes: | height of the shell | 15 millimeters (max. 25 millimeters) |
|---|---|---|
| | height of the aperture | 5 millimeters (max. 8 millimeters) |
| | diameter of the shell | 5 millimeters (max. 8.5 millimeters) |

In recent times, *Melanoides tuberculata* has been found in North and East Africa, on the Arabian Peninsula, in India and Southeast Asia. Descriptions of fossil and subfossil shells exist from the Tibesti Mountains, and from the area of El Jafr (Jordan) and the East African lakes.

### *2.4.2.3. Biogeographic and ecological aspects*

The four types of gastropods collected from the stillwater sediments of the terrace accumulation in Wadi Al Luhy belong to the Afro-Ethiopian fauna. This, however, is not surprising because the area of Wadi Al Luhy lies at a latitude of 23° N. B. W. SPARKS, A. T. GROVE (1961), and K. KAISER et al. (1973) assume that the boundary between the shell species mentioned above and the paleoarctic fauna elements was, even during the northern (glacial) pluvials, at a latitude of just 25° N; in recent times, this boundary lies more than 1,000 kilometers farther north, near the African coast of the Mediterranean Sea.

There is a striking conformity with fossils of the same age collected at the same geographical latitude in Africa. J. D. TOTHILL (1946) describes gastropods collected from clay at the Red Sea coast in the Sudan, e.g., six subdivisions including *Biomphalaria pfeifferi ruppellii, Bulinus truncatus,* and *Melanoides tuberculata.* Among the Mollusca from lake sediments in the Tibesti area all the species were found which, in recent times, are well-known from Central Arabia. Carbon-14 measurements in fossils of the Tibesti region revealed ages of 6,000 to 10,000 years before present.

In contrast, there is no conformity with fossil gastropods of the same age or of the Würm pluvial, collected in the basin of Damascus where paleoarctic species were found (H. SCHÜTT, 1973). The same is true of the area of El Jafr west of Wadi Araba (Jordan), except for *Melanoides tuberculata*, which was found there.

Moreover, it should be stated that from the ecological point of view all kinds of gastropods described from Wadi Al Luhy are fresh water fossils. However they are very adaptable; thus *Melanoides tuberculata* may also be found in brackish and thermal waters at present. Shallow calm lakes which at the same time act as a place of deposits for clayey muds are the habitat for such species living in recent times. Concerning the conditions prevailing in Wadi Al Luhy, all these factors require a more humid climate for the period of the age of the collected fossils.

## 2.4.3. Wadi Birk

(H. Hötzl, V. Maurin)

Wadi Birk is a large old river system crossing the southern Tuwayq Mountains. These mountains, a part of the cuesta landscape of central Saudi Arabia, are a sedimentary complex of Mesozoic limestones and sandstones. In the area of Wadi Birk the Tuwayq Mountains rise to 1,100 meters above sea level; the strata dip towards the east to an altitude of 600 meters.

Wadi Birk crosses the mountains from west to east at latitude 23° 20′ N. The western part of the breaching is 2.5 to 4 kilometers wide; eastwards the wadi becomes narrower, finally forming a canyon one kilometer wide. At the eastern margin of the Tuwayq Mountains, i.e., at the escarpment of the Arab Formation, the wadi turns northward, collecting a number of tributaries in its meandering valley. The most significant tributary is Wadi Al Hawtah, which enters Wadi Birk near the oases of Al Hawtah. Wadi Birk itself at present turns into the plain of Al Kharj.

### *2.4.3.1. Wadi deposits*

Except for some parts which will be described later, Wadi Birk is an accumulation area. The thickness of the wadi fill reaches about 50 meters as can be seen in its wells. In the oases of Al Hawtah wadi deposits to a depth of 40 meters below surface could be investigated in hand-dug wells.

The wadi fill mainly consists of silt with a varying percentage of clay and sand. Often marly, the sediments show a remarkably high content of gypsum, sometimes in the form of a crust, but also as deposits in pores. As these are sediments in very flat parts of the wadi where high evaporation occurs after episodic rainfalls, the deposits resemble those found in sabkhahs and are typical of sediments accumulated under arid climatic conditions.

Beside thick layers of silt, there are minor strata of eolian and fluviatile sands with gravels. These gravels—as far as they were investigated down to 40 meters—consist of material from strata of the Tuwayq Mountains. In smaller tributaries local scree material is dominant as can be observed in hand-dug wells. Borehole profiles were not available. In some of the hand-dug wells in the area of Al Hawtah it can be clearly seen that the groundwater discharges from a gravel layer at a depth of 40 meters. When the water level becomes lower through pumping, the water flow from this gravel layer can be clearly seen. The gravel layer is overlain by wadi sabkhah sediments, as well as eolian and fluviatile sands. Other gravel layers follow at depths ranging between 8 and 15 meters below surface. These gravel strata are likewise overlain by fluviatile sand and sabkhah sediments, covered by eolian sand towards the surface in the uppermost 4 to 6 meters.

In the area of Al Hawtah the uppermost parts of the wadi deposits can be seen in a young erosion channel. Remnants of charcoal could be collected from a "fire-horizon" 6 meters below the terrace surface on the left side of Wadi Birk about 700 meters downstream from the mouth of Wadi Al Hawtah (Fig. 68). Samples were collected for carbon-14 determination, the measurements of which revealed an age of about 2,000 years B.P. (Table 40). The terrace profile to 10 meters depth may be described as follows:

| Depth in cm from the land surface | Description of sediments |
|---|---|
| 0–490 | Uniform silty fine sand, not clearly stratified; pattern of small tubes as a consequence of weathered roots (sample No. 131; Tab. 48). |
| 90 | Cultural level, possibly using a natural erosion terrace (about 25 meters wide). |
| 135 | Another cultural level, possibly using a natural small erosion terrace (about 4 meters wide). |
| 270 | Charcoal remnants (Fig. 68). |
| 490–554 | Alternating strata of sand and silt; silt layers partly covered by clay film; strata forming waves of an amplitude of up to 50 centimeters. Pattern of tubes as a consequence of weathered roots, filled with sand. At a depth of 505 centimeters, horizon with curled clay flakes. |
| 554–570 | Silt with gypsum, narrow pattern of root molds. Small horizon of charcoal remnants at a depth of 560 centimeters. |
| 570–603 | Fine sand, cross-bedded. Traces of ripple marks. Sample No. 130 (Tab. 48). |
| 603–612 | Silt with gypsum; narrow pattern of small plant root moulds. |
| 612–614 | Charcoal horizon of reddish to black-greyish colour (VRI-503; Tab. 40). |
| 614–625 | Silty sand, reddish towards the charcoal horizon. |
| 625–638 | Sandy silt, crystals of gypsum up to 2 millimeters long (No. 129; Tab. 48). |
| 638–689 | Fine- and medium-grained sand, horizontal lamination. |
| 689–697 | Silt with sand and fine-grained gravels. |
| 697–752 | Irregular alternation of brown sand and grey-brown silt. |
| 752–784 | Changing strata of fine sand and gravely coarse sand; cross-bedding. |
| 784–812 | Sand with clay galls and limestone gravels up to diameters of 2 centimeters. |
| 812–818 | Lens of fine and coarse pebbles. |
| 818–856 | Shifted medium-grained sand with clay lenses and limestone gravels of diameters up to 2 centimeters. |
| 856–926 | Sand with fine- and coarse-grained gravels, various gravels of diameters up to 15 centimeters. |
| 926–941 | Medium- to coarse-grained sand (sample No. 128; Tab. 48). |
| 941–966 | Sand and gravels of diameters up to 15 centimeters. |

Fig. 68. Remnants of charcoal (dark horizon above the hammer) interbedded into the accumulation terrace of Wadi Birk. (Photo: V. MAURIN, 1975)

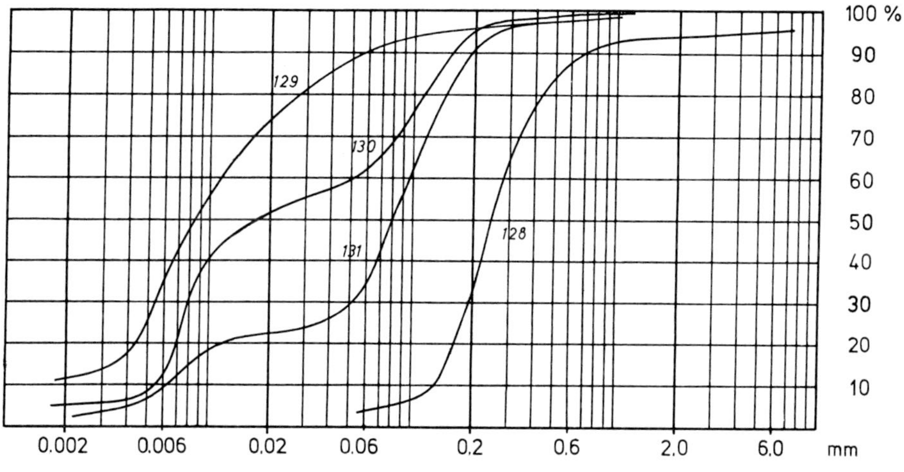


Fig. 69. Cumulative grain size distributions of sediments from the terrace profile near Al Hawtah, Wadi Birk: *131* terrace profile 0–490 centimeters, *130* terrace profile 570–603 centimeters, *129* terrace profile 625—638 centimeters, *128* terrace profile 926–941 centimeters. Detailed description see p. 210; the grain size parameters are given in chapter 3.1.2., Table 48

In summary the following conclusions can be drawn:

1. The stratum from 0–490 centimeters is mainly an eolian sedimentary complex and in some parts still has the structure of dunes. This structure was partly destroyed by roots and shifting.

2. The deposits ranging from a depth of 490 down to 752 centimeters are products of a period showing an alternation of runoff and dry seasons; flood and channel sediments are alternating with wadi sabkhahs and eolian accumulation.

3. The sedimentary complex ranging from 752 down to 966 centimeters consists of fluviatile, accumulated clastic sediments.

The curve of grain size distribution of the samples (for sample numbers see profile) is shown in Fig. 69; their sedimentological parameters are given in Table 48.

On the terrace surface outside of the recent oasis of Al Hawtah a lot of destroyed well hills, remnants of a formerly more extended cultivation are to be found. On two of these hills, which were heaped up by the raised material from the depth, sinter pebbles were collected. One sample (Table 40, Lab. No. 4,240) derives from immediately downstream Al Hawtah, the other (Lab. No. 4,241) is from a ruined village 5 km wadi downward. Carbon-14 measurements revealed the results listed in Table 40. For the lack of more information about the origin and exact profile position of these sinter pebbles, the age values give us only a vague reference of more humid phases about 7,000 and 27,000 years B.P.

Table 40. *Results of carbon-14 measurements of charcoal and calcite sinter collected in the area of Wadi Birk*

| Lab. | Reg.-No. | Material | $^{14}$C-Content (% modern) | $^{14}$C-Age (Years B.P.) |
|---|---|---|---|---|
| IRKW[1] | VRI-503 | Charcoal | — | 2,170 ± 130 |
| IRM[2] | Lab.-No. 4,240 | gravel of calcite sinter | 2.9 ± 0.7 | 27,140 ± 1,940 |
| IRM | Lab.-No. 4,241 | gravel of calcite sinter | 36.1 ± 1.3 | 6,880 ± 290 |

[1] Institut für Radiumforschung und Kernphysik, Vienna.
[2] Institut für Radiohydrometrie, Munich.

The described wadi fill corresponds almost fully to recent sedimentation in Wadi Birk. At present wadi sabkhahs can for example be found in the meandering centre of the breach valley as well as in eastern parts of the wadi south of Al Hilwah, and also in the wadi section (about 25 kilometers long) north of the village of Al Hawtah, near its mouth into the plain of Sha'ib As Sawt. The same is true of dunes which extend around the 'Ayn Al Bakrah south of Al Hilwah, and into the wadi about 15 kilometers north of Al Hawtah as well as in many parts exposed to the main wind direction, as can be especially noticed in Wadi Al Hawtah. Accumulation of sand and fine-grained gravels, too, now occurs downstream from the short erosion channels immediately north of Al Hawtah.

In recent deposits only small layers of coarse gravels were discovered. The transport of coarse material requires larger water quantities than the recent runoff. It seems that a slight increase in humidity is just enough to change conditions as observed in the area of Sha'ib 'Uthaylan joining Wadi Birk north of Al Hawtah. The floor of this tributary consists solely of fine- to medium-grained gravels which have been transported and shifted in recent times; this becomes evident in the cone formed by accumulation at its mouth into Wadi Birk.

Fig. 70. Erosion channel and the almost vertical terrace wall north of Al Hawtah, Wadi Birk. (Photo: V. MAURIN, 1975)

### *2.4.3.2. Accumulation terrace in the area of Al Hawtah*

In contrast to the largest part of Wadi Birk with a homogeneous flat wadi floor, there is an erosion channel 30 to 60 meters wide and reaching a maximum depth of 10 meters in the area of Al Hawtah. The channel walls are almost vertical; consequently, the terrace has formed a sharp edge (Fig. 70).

The terrace begins with a small step of 1 to 3 meters near the village of Al Hilwah, just south of the oasis of Al Hawtah. Upstream from the mouth of Wadi Al Hawtah the walls are as high as 8 meters. Downstream from the oasis the terrace becomes rapidly lower; after 5 kilometers the wall rises only 1 to 1.5 meters, and after another 5 kilometers the terrace has completely disappeared.

In the tributary Wadi Al Hawtah near the discharge into Wadi Birk the terrace is even more impressive reaching an altitude of 10 meters. The erosion there is still active as can be seen from the scour marks. However, even there the erosional channel flattens in a short distance upstream the oasis. A broad flat discharge section—sometimes occupying the whole wadi floor—remains. Further terraces are to be found wadi-upward only in the oases of Na'am and Al Hariq. In some places a polyphase terrace development is presented by lateral-eroded and intercut dune areas, high water marks as well as by collapsed well hills. From geomorphological features, agricultural cultivation, and carbon-14 measurements it can be concluded that the age of the upper part of these terraces is just a few thousand years (Table 40, VRI-503).

### *2.4.3.3. Morphogeny of Wadi Birk*

Wadi Birk, in large areas now covered with dunes and divided into various independent runoff basins, was originally a river system crossing through the entire Arabian Shelf, with its headwaters in the top of the Arabian Shield and its mouth into the Arabian Gulf. The main upper courses of the system were Wadi Ar Rayn, Wadi Al Jifr and Wadi Al Badriyah, which now end up in the central sand plain of Nefud Adh Dhai. Downstream, Wadi Birk joined the rivers of Wadi Nisah, Wadi Al Luhy, Wadi Hanifah and Wadi As Seley in the area of Al Kharj; the joint runoff went over to the recent Wadi As Sah'ba towards the east into the Arabian Gulf. With regard to the area of Tertiary deposits with their gravels from the crystalline Shield it must be assumed that the river systems such as Wadi Ar Rimah, Wadi Atk, Wadi Birk and Wadi Ad Dawasir, arising in the Arabian Shield and subsequently crossing the Tuwayq Mountains, must have been in function since Middle Tertiary (Miocene).

Considering the morphology it can be recognized that the river system of Wadi Birk originally crossed the outcrop of the Arab formation in a wide curve to the south, discharging into the accumulation basin of the Kharj Formation which is of Miocene to Pliocene age.

The course of this old Wadi Birk system towards the sedimentary basin of the Upper Tertiary is a key to the morphogeny of this system. However, the age of the Kharj Formation has not yet been exactly determined. According to R. W. Powers et al. (1966), these deposits should be compared to sediments of the Hofuf Formation (Upper Miocene to Lower Pliocene), although a Lower Miocene age, too, has to be taken into consideration.

This old Wadi Birk system followed a course about 20 kilometers south of Al Hawtah presumably 20 meters above the recent wadi floor. The view towards the west clearly shows that the valley must have been wider and more rectilinear.

After sedimentation of the Kharj Formation a lower erosional base level was the reason for developing a joint drainage system of the Tuwayq Mountains through Wadi As Sah'ba towards the east. Another reason for such a development may be found in a tectonic movement and the formation of Wadi Nisah—Wadi As Sah'ba graben (E. Burek, 1973). The result of such developments was an intensive headward erosion upstream from the Al

Kharj plain, forcing Wadi Birk to turn from south of Al Hawtah towards the Al Kharj basin. This turn-away was combined with a strong vertical erosion which reached a depth of more than 70 meters in the meanders of the breaching valley. The strong vertical erosion in the main valley transformed a number of tributaries to hanging valleys, and also massive landslides occurred in Wadi Birk itself (Fig. 71).

Fig. 71. Huge landslide caused by strong vertical erosion at the valley slope (central part of the picture) of lower Wadi Birk. (Photo: J. G. Zötl, 1975)

This strong vertical erosion in Wadi Birk required a heavy surface runoff. The wadi arcs of valley meanders, too, needed large quantities of river water. The hanging valleys of the local tributaries as well as the steep slope of the entire breaching valley through the Tuwayq Mountains reveal that these enormous water quantities arose from crystalline rocks of the Arabian Shield. Obviously, the amount of precipitation was remarkably higher in the Shield area than in the region of the Tuwayq Mountains.

The whole situation must have required more humid climate conditions for the period of this strong vertical erosion. Considering the large erosional extent an equivalent massive accumulation must have taken place in the foreland during this period. Indeed, vast accumulations can be found in the delta fan east of Wadi As Sah'ba (see 2.1.3.4.). For both the gravel composition of this delta cone and the gravel size, a large amount of river water was necessary, available only in a very humid climate. In agreement with chapter 2.1.3.4. these conditions prevailed during the Late Pliocene and Early

Pleistocene. This phase was followed by a period of arid climate. The fact that no gravels that originated in the crystalline Shield are to be found in recent wadi deposits makes it clear that during the younger Quaternary the runoff coming from the Shield was stopped in the sand regions of nefud. Consequently, Wadi Birk became an area of sedimentation by local tributaries and eolian sands. This doubtless must have been the situation during the younger Quaternary Period. During the humid phases of this time only materials from local tributaries of the Tuwayq Mountains were transported into Wadi Birk. A more detailed examination of these deposits should provide more information about the humid phases of the younger Quaternary.

## 2.4.4. Chemistry and Isotope Content of Some Wadi Groundwaters in the Central Parts of the Tuwayq Mountains

(C. JOB, H. MOSER, W. RAUERT, W. STICHLER)

This chapter is devoted to the groundwaters of the Wadis Hanifah, Nisah, Al Luhy, Al Hawtah and Birk. The Quaternary clastic sediments of these valleys carry waters that are mineralized to various degrees and that, according to their content of heavy isotopes and occasionally tritium, originate from rainfalls of different ages.

### *2.4.4.1. Wadi Hanifah, Wadi Nisah, Wadi Al Luhy* [1]

Wadi Hanifah cuts in an eastward direction through the Tuwayq Mountains some 50 kilometers northwest of Ar Riyadh, turns southeast and finally flows into the wide basin of Al Kharj after another 100 kilometers. Wadi Al Luhy flows into Wadi Hanifah as a small tributary near Al Ha'ir, some 30 kilometers south of Ar Riyadh. In the vicinity of Al Ha'ir there is the central plant of a big waterworks supplying the capital, Ar Riyadh, from wells of both Wadis Hanifah and Nisah. The latter follows a large fault crossing the Tuwayq chain in a west-east direction at lat 24° 10′ N and opens into the basin of Al Kharj. The thickness of the fluviatile accumulation reaches approximately 60–80 meters in Wadi Hanifah and some 300 meters in Wadi Nisah.

The water samples No. 49 (1973) and 112 (1974) originate from a 60 meters deep well in Wadi Hanifah near the village of Al Ha'ir (Fig. 72).

The temperatures of these waters are about 27 °C; they are saturated with oxygen and contained some 9 me/l of dissolved salts in 1973 and 9.4 me/l in 1974 (Table 41, 42). This slight increase in concentration in the course of one year is worth mentioning, as the deuterium and the $^{18}O$-content (Table 43) experienced a pronounced rise as well. Both samples No. 48 (1973) and 111 (1974) suggest such a connection, too. These somewhat stronger

---

[1] Wadi Al Luhy is the correct designation of the wadi named Sha'ib Ha in the geologic map (R. A. BRAMKAMP, L. F. RAMIREZ, 1958) and earlier publications.

mineralized waters come from a drilled well in a small oasis, some 2.5 kilometers upstream from the waterworks. The high tritium content of the Hanifah waters indicates that this aquifer receives short-term affluents from recent rainfalls. Consequently, it is natural to assume that the fluctuations observed in the salt content depend on the yearly quantity of precipitation (amounting to about 20 to 160 millimeters in this area), thus producing a solution volume that fluctuates widely. Since the enrichment of stable isotopes is likewise dependent on the quantity of rainfall—scanty precipitation is iso-

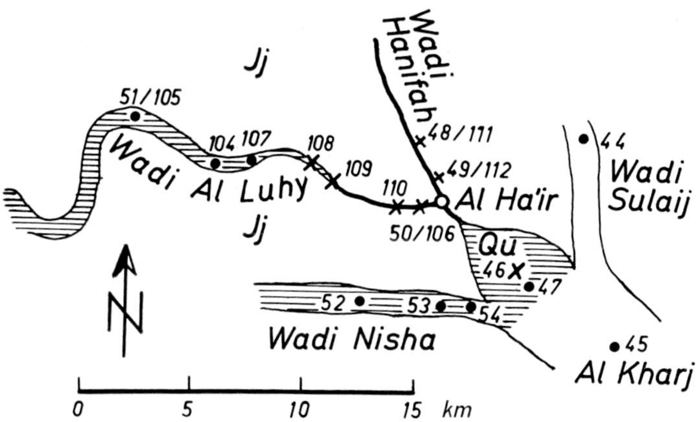


Fig. 72. Sketch of the area under investigation and sampling spots in Wadi Hanifah, Wadi Al Luhy, Wadi Nisah and Al Kharj basin. *Jj* Jurassic Jubaila Formation, *Qu* Quaternary; samples with values above 4 T.U. are marked with a cross

topically heavier than large amounts of rainfall provided that the climatic conditions are the same—it can be assumed that both the salt and the isotope content depend on the quantity of rainfall.

In order to supply the city of Ar Riyadh with water, several shallow drillings were made in Wadi Hanifah and also one deep drilling down to 1,200 meters into the Minjur Sandstone. Sample 115 taken from this well clearly shows the difference in the content of stable isotopes between the young, shallow wadi groundwater and the old storage in deeper formations. In comparison to the isotope data which make a differentiation of these two kinds of waters possible at once, the chemical differences are not so striking. Neither the total concentration nor the ion distribution displays the different provenance and history of these waters. If there were not any higher temperature and the lack of oxygen as well as the reduction of nitrates, one might be tempted to conclude that the Minjur water would be nothing else than wadi groundwater enriched with a lower degree of sodium chloride and gypsum.

From all the waters of Wadi Nisah, sample 52 was collected from the drilling "DP 5" of the Riyadh waterworks, which reaches 300 meters down into the accumulation and supplies poorly mineralized water (some 6.1 me/l).

Table 41. *Temperature, oxygen content (mg/l) and chemical composition of the waters from Wadi Hanifah, Wadi Nisah, and Al Kharj basin.* Ion concentrations in µe/l, boric acid and silicic acid in mg/l. SK and SA represent the sums of cations and anions. W stands for the type of well (cf., section 2.1.6.1.4.). The sample numbers with the same index stand for waters from the same wells. Samples 44–54 were collected between April 8 and 10, 1973, the remaining samples between March 28 and 30, 1974. Sample 113 was taken from a thunderstorm on March 28, 1974, and 114 from the following wadi flood. Sample 115 was collected from drilling H-1 near Al Ha'ir

| No. | W | °C | $O_2$ | $Na^+$ | $K^+$ | $Ca^{++}$ | $Mg^{++}$ | SK | $Cl^-$ | $Br^-$ | $SO_4^{--}$ | $HCO_3^-$ | $NO_3^-$ | $F^-$ | SA | $H_3BO_3$ | $H_2SiO_3$ | |
|---|---|---|---|---|---|---|---|---|---|---|---|---|---|---|---|---|---|---|
| 48[1] | DP | 27.0 | 6.5 | 2,784 | 123 | 4,440 | 3,094 | 10,441 | 1,425 | 1.3 | 3,992 | 4,723 | 181 | 37 | 10,359 | 1.01 | 15.3 | Wadi Hanifah |
| 49[1] | DP | 27.0 | 7.4 | 2,405 | 77 | 3,989 | 2,565 | 9,036 | 1,339 | 1.7 | 3,281 | 4,178 | 230 | 44 | 9,074 | 0.74 | 14.7 | |
| 111[1] | DP | 27.1 | 7.6 | 4,350 | 163 | 5,600 | 4,520 | 14,633 | 2,650 | 0.5 | 6,618 | 5,291 | 123 | 27 | 14,710 | 1.77 | 16.7 | |
| 112[1] | DP | 27.2 | 8.3 | 2,610 | 84 | 4,200 | 2,527 | 9,421 | 2,299 | 2.4 | 3,357 | 3,363 | 288 | 30 | 9,339 | 0.66 | 13.3 | |
| 113 | Rain | — | — | 148 | 51 | 160 | 33 | 392 | 64 | 0.0 | 80 | 203 | 52 | 0 | 399 | 0.0 | 0.2 | |
| 114 | Runoff | — | — | 191 | 73 | 2,680 | 268 | 3,212 | 145 | — | 500 | 2,392 | 153 | 6 | 3,196 | 0.01 | 9.2 | |
| 115 | DP | 50.5 | 0.0 | 6,090 | 253 | 8,440 | 3,360 | 18,143 | 5,491 | 8.5 | 9,237 | 3,203 | 0 | 19 | 17,959 | 1.40 | 19.5 | |
| 52 | DP | 30.5 | 6.2 | 1,182 | 94 | 3,508 | 1,416 | 6,200 | 970 | 2.0 | 1,928 | 2,947 | 194 | 30 | 6,071 | 0.61 | 15.4 | Nisah |
| 53 | DP | 29.2 | 5.0 | 5,437 | 94 | 8,560 | 3,448 | 17,539 | 4,900 | 7.7 | 8,233 | 3,380 | 790 | 40 | 17,351 | 1.75 | 14.8 | |
| 54 | DP | 29.0 | 5.6 | 3,306 | 111 | 6,655 | 3,085 | 13,157 | 2,814 | 5.0 | 5,124 | 3,814 | 1,307 | 35 | 13,099 | 1.25 | 15.3 | |
| 44 | — | 28.0 | 6.0 | 6,087 | 147 | 31,920 | 6,320 | 44,474 | 5,755 | 5.3 | 35,702 | 2,452 | 391 | 76 | 44,381 | 2.05 | 15.9 | Al Kharj |
| 45 | — | 30.7 | 5.4 | 5,652 | 169 | 32,840 | 6,480 | 45,141 | 4,436 | 3.8 | 37,205 | 3,088 | 165 | 90 | 44,988 | 2.40 | 16.9 | |
| 46 | DP | 29.7 | 6.0 | 12,814 | 248 | 14,440 | 6,520 | 34,022 | 14,776 | 23.7 | 13,397 | 2,750 | 3,000 | 41 | 33,988 | 3.22 | 14.6 | |
| 47 | DP | 30.1 | 5.5 | 5,872 | 188 | 7,016 | 3,308 | 16,384 | 5,126 | 5.7 | 7,092 | 3,320 | 758 | 35 | 16,337 | 1.76 | 14.9 | |

[1] Samples collected from the same wells 1973 and 1974 resp.

Table 42. *Temperature, oxygen content (mg/l) and chemical composition of the waters from Wadi Al Luhy, Wadi Al Hawtah and Wadi Birk.* Symbols are the same as used in Table 41. Sampling date: 50 and 51 April 9, 1973, 104–110 March 27 and 28, 1974, 173–182 between March 4 and 8, 1975

| No. | W | °C | $O_2$ | $Na^+$ | $K^+$ | $Ca^{++}$ | $Mg^{++}$ | SK | $Cl^-$ | $Br^-$ | $SO_4^{--}$ | $HCO_3^-$ | $NO_3^-$ | $F^-$ | SA | $H_3BO_3$ | $H_2SiO_3$ | |
|---|---|---|---|---|---|---|---|---|---|---|---|---|---|---|---|---|---|---|
| 50 [1] | DP | 28.5 | 5.4 | 15,842 | 394 | 11,360 | 9,120 | 36,716 | 13,789 | 11.3 | 18,343 | 3,678 | 908 | 47 | 36,776 | 3.39 | 15.4 | Wadi Al Luhy |
| 51 [1] | DP | 29.5 | 6.4 | 18,908 | 506 | 11,680 | 9,809 | 40,903 | 18,222 | 17.8 | 17,566 | 2,609 | 2,490 | 66 | 40,971 | 3.41 | 15.4 | |
| 104 | DP | 29.5 | 13.0 | 15,963 | 430 | 9,560 | 8,000 | 33,953 | 14,486 | 13.5 | 15,518 | 2,851 | 1,306 | 58 | 34,233 | 3.20 | 16.3 | |
| 105 [1] | DP | 29.2 | 8.0 | 20,530 | 549 | 13,360 | 11,680 | 46,119 | 21,301 | 19.1 | 18,763 | 2,610 | 3,568 | 60 | 46,321 | 3.71 | 17.8 | |
| 106 [1] | DP | 28.4 | 7.5 | 17,109 | 422 | 12,600 | 10,040 | 40,171 | 15,129 | 10.9 | 20,233 | 4,016 | 948 | 46 | 40,383 | 4.11 | 16.1 | |
| 107 | DP | 29.6 | 8.6 | 15,876 | 422 | 9,440 | 7,880 | 33,618 | 13,748 | 11.9 | 15,401 | 2,912 | 1,393 | 62 | 33,528 | 3.54 | 15.0 | |
| 108 | DP | 29.6 | 8.4 | 12,527 | 352 | 8,160 | 6,520 | 27,559 | 10,951 | 8.5 | 12,462 | 3,193 | 876 | 53 | 27,544 | 2.83 | 15.9 | |
| 109 | DP | 29.1 | 7.6 | 13,919 | 394 | 9,280 | 7,320 | 30,913 | 12,113 | 7.1 | 14,361 | 3,354 | 989 | 53 | 30,877 | 3.06 | 15.0 | |
| 110 | DP | — | 7.6 | 18,703 | 464 | 14,200 | 11,760 | 45,127 | 17,227 | 13.1 | 23,004 | 3,885 | 992 | 46 | 45,167 | 4.46 | 16.1 | |
| 174 | HP | 26.5 | — | 1,411 | 86 | 2,966 | 1,749 | 6,212 | 958 | 2 | 1,948 | 2,775 | 677 | 17 | 6,377 | 1.09 | 16.1 | Wadi Hawtah |
| 175 | HP | 26.2 | — | 1,196 | 84 | 3,494 | 1,470 | 6,244 | 243 | 2 | 2,300 | 3,169 | 237 | 22 | 5,973 | 0.74 | 15.0 | |
| 176 | HP | 26.3 | — | 1,278 | 96 | 4,240 | 2,016 | 7,630 | 1,579 | 3 | 2,640 | 3,078 | 242 | 20 | 7,562 | 0.69 | 15.0 | |
| 177 | HP | 26.8 | — | 2,349 | 113 | 4,900 | 2,700 | 10,062 | 2,791 | 4 | 4,385 | 2,550 | 363 | 19 | 10,112 | 0.97 | 15.6 | |
| 178 | HP | 27.0 | — | 2,958 | 123 | 6,020 | 4,720 | 13,821 | 2,189 | 4 | 8,474 | 3,000 | 212 | 22 | 13,901 | 1.71 | 16.7 | |
| 179 | HP | 28.5 | — | 1,459 | 92 | 3,700 | 1,505 | 6,756 | 1,165 | 0 | 2,692 | 2,629 | 246 | 24 | 6,756 | 0.86 | 15.6 | |
| 180 | HP | 29.6 | — | 1,435 | 86 | 3,540 | 1,749 | 6,810 | 1,060 | 0 | 2,712 | 2,856 | 185 | 32 | 6,845 | 0.70 | 8.6 | |
| 181 | DP | 29.2 | — | 6,633 | 152 | 7,100 | 7,940 | 21,825 | 6,334 | 6 | 12,168 | 3,210 | 266 | 38 | 22,022 | 2.34 | 17.5 | Wadi Birk |
| 182 | HP | 28.8 | — | 2,871 | 118 | 5,520 | 4,096 | 12,605 | 3,389 | 3 | 6,514 | 2,680 | 167 | 29 | 12,782 | 1.03 | 17.5 | |
| 173 | HP | 28.8 | 5.9 | 3,371 | 135 | 6,340 | 5,620 | 15,466 | 4,841 | 4 | 8,715 | 1,640 | 484 | 25 | 15,709 | 1.20 | 16.7 | |

[1] Samples collected from the same wells 1973 and 1974 resp.

The two other samples (53, 54) collected in the lower part of Wadi Nisah near its mouth into Al Kharj basin from a 60-meter well showed stronger mineralization (17.5 and 13.1 me/l). The Nisah waters display lower isotope contents than the Hanifah waters and are free from tritium. This might imply that the Nisah aquifer primarily carries water from heavy precipitation falling under different climatic conditions in a possibly more distant recharge area situated higher up, whereas the Hanifah aquifer is alimented by poorer rainfalls with a younger infiltration.

Table 43. *Deuterium, oxygen-18 and tritium contents of the waters from the central parts of the Tuwayq area.* For sampling dates refer to Tables 41 and 42. Deuterium (δ D) and oxygen-18 values (δ $^{18}$O) have double standard deviations of ± 1 per mill and ± 0.2 per mill, respectively. 1 T.U. corresponds to 3.2 pCi tritium/l water. For the tritium values a double standard deviation is given

| | δ D (‰) | δ $^{18}$O (‰) | $t_8$ | Tritium content (T.U.) | |
|---|---|---|---|---|---|
| 48 [1] | — 2.8 | — 2.30 | 15.60 | 74 ± 10 | |
| 49 [1] | — 3.1 | — 2.40 | 16.10 | 67 ± 8 | |
| 111 [1] | — 1.2 | — 1.96 | 14.48 | 72.1 ± 5.0 | |
| 112 [1] | — 2.5 | — 2.11 | 14.38 | 72.6 ± 5.0 | Wadi Hanifah |
| 113 | + 11.7 | — 0.45 | 15.30 | 53.7 ± 4.1 | |
| 114 | + 13.2 | — 0.07 | 13.76 | 54.4 ± 4.1 | |
| 115 | — 38.9 | — 5.46 | 4.78 | 0 ± 2.3 | |
| 52 | — 8.9 | — 2.95 | 14.70 | 0.5 ± 0.4 | |
| 53 | — 11.2 | — 3.30 | 15.20 | 0 ± 3.8 | Wadi Nisah |
| 54 | — 6.7 | — 2.80 | 15.70 | 0.7 ± 3.4 | |
| 44 | — 6.9 | — 2.67 | 14.46 | 0 ± 2 | |
| 45 | — 8.1 | — 2.68 | 13.34 | 0 ± 4 | Al Kharj |
| 46 | — 5.3 | — 2.31 | 13.18 | 7 ± 2 | |
| 47 | — 7.1 | — 2.96 | 16.58 | 1.7 ± 3.8 | |
| 50 [1] | — 6.4 | — 2.74 | 15.52 | 16 ± 2 | |
| 51 [1] | — 6.4 | — 2.56 | 14.08 | 1.6 ± 4.0 | |
| 104 | — 5.8 | — 2.43 | 13.64 | 2.3 ± 1.9 | |
| 105 [1] | — 5.9 | — 2.29 | 12.42 | 3.1 ± 2.2 | |
| 106 [1] | — 5.6 | — 2.20 | 12.00 | 13.7 ± 2.4 | Wadi Al Luhy |
| 107 | — 6.2 | — 2.31 | 12.28 | 3.4 ± 2.7 | |
| 108 | — 5.8 | — 2.34 | 12.92 | 10.2 ± 2.4 | |
| 109 | — 5.7 | — 2.43 | 13.74 | 11.4 ± 2.4 | |
| 110 | — 3.9 | — 2.25 | 14.10 | 18.2 ± 2.2 | |
| 174 | — 3.4 | — 2.85 | 19.40 | 28 ± 2.3 | |
| 175 | — 2.8 | — 2.84 | 19.92 | 30 ± 2.2 | |
| 176 | — 3.9 | — 2.89 | 19.22 | 28 ± 2.2 | |
| 177 | — 4.1 | — 2.88 | 18.94 | 24 ± 1.7 | Wadi Al Hawtah |
| 178 | — 4.0 | — 2.98 | 19.84 | 14.1 ± 1.7 | |
| 179 | — 5.1 | — 2.89 | 18.02 | 20 ± 2.0 | |
| 180 | — 6.9 | — 3.31 | 19.58 | 2.7 ± 1.4 | |
| 181 | — 7.2 | — 3.21 | 18.48 | 0.7 ± 1.5 | |
| 182 | — 6.6 | — 3.12 | 18.38 | 3.7 ± 1.1 | Wadi Birk |
| 173 | — 5.8 | — 3.21 | 19.88 | 3.9 ± 1.4 | |

[1] Samples collected from the same wells 1973 and 1974 resp.

The water samples 46 and 47 were taken from drillings near Umm Musha'al in the bordering area of Al Kharj basin into which Wadi Hanifah flows from the northwest and Wadi Nisah from the west. Chemically, sample 47 is very similar to sample 53 collected in the lower course of Wadi Nisah. Sample 46, which is twice as strongly mineralized and isotopically enriched as sample 47, contains a little tritium (7 T.U.). Both mineralization and isotope enrichment as well as the high content of nitrates argue that mineralization is caused by infiltration through the wadi surface. Sample 45 was collected from the big 'Ayn Dola which apparently came into existence from

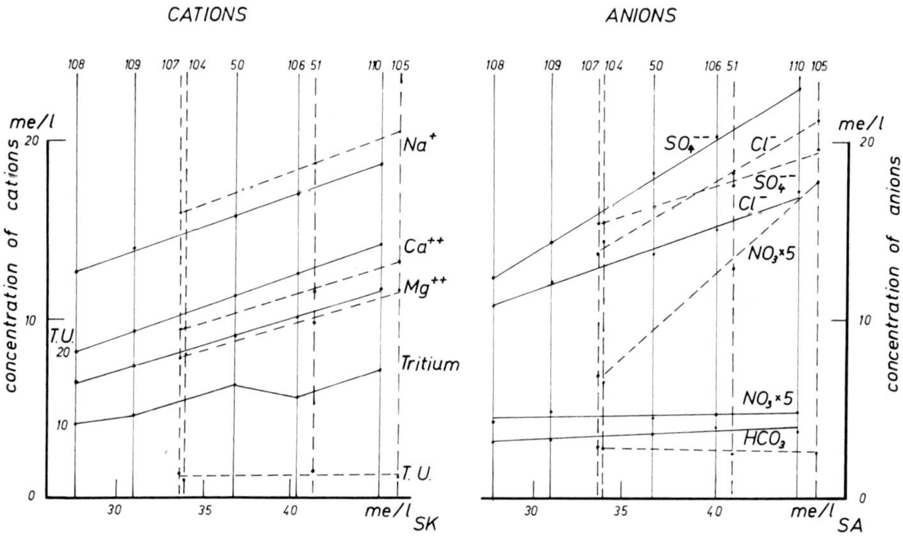


Fig. 73. Content of cations and anions (ordinate) of the waters from Wadi Al Luhy dependent on the total concentration. Broken line = water group with a low tritium content

a dolina of gypsum in the central parts of the basin. The water is saturated with gypsum; chemically, it is related to samples 46 and 53. The same is true of sample 44 from a cave (Dahl Heet) in Wadi As Seley that follows a northwest course towards Al Kharj basin, parallel to Wadi Hanifah [1].

Concerning the content of isotopes, the waters of Wadi Al Luhy range between those of Wadi Hanifah and those of Nisah-Al Kharj. In samples 51 (and 105/1974), 104 and 107, only traces of tritium were found, whereas samples 50 (and 106/1974), 108, 109, and 110 contained T.U. values between 10.2 and 18.2. There is no significant difference as to the total mineralization of these two groups; in their ion distribution and their content of nitrate (Fig. 73), however, a major difference becomes apparent. In both groups,

[1] Above the cave entrance there is an outcrop of gypsum belonging to the Hith Anhydrite. The analysis of this gypsum resulted in $\delta\,^{34}S = 16.2 \pm 2$ per mill. This value corresponds to the ratio of isotopes of Jurassic gypsums (16–19 per mill); measurement E. PAK, 1974.

ion contents increase with the total concentration in a linear fashion. Obviously, these waters belong to different water groups. It cannot be said with certainty whether the formation of groups may be ascribed to a mixing or dissolving relationship (i.e., by dissolution of different quantities of a mixture of substances made up homogeneously). In the group producing higher tritium values the content of tritium seems to rise together with that of salt. This phenomenon is an argument in favour of mixed waters. However, the correlation ($r_s = 0.85$) does not exceed the critical value (0.8 for $n = 5$ and $\alpha = 0.005$) convincingly enough. In 1973 and 1974 samples were collected from a well of each group. Just as in the two Hanifah wells, there too a stronger mineralization together with an enrichment of deuterium and $^{18}O$ as well as a decrease in the $t_8$-value was noted in 1974. In well 50/106 (the tritium content of which suggests a recent infiltration) the rise in concentration might be attributed to the fact that minor quantities of rainfall caused more strongly mineralized waters to infiltrate into the well (cf., Hanifah well). This hypothesis is supported by the somewhat lower, but hardly significantly different, tritium content in 1974. However, this explanation fails to apply in well 51/105, since its water is almost free of tritium. It may be that the recycling process of irrigation water plays a considerable part in this case. These conclusions are corroborated by a stronger share of nitrates in the process of concentration (a 43 percent rise in nitrate while the total concentration rose by only 13 percent). If such a recycling occurs as was observed in Riyadh Al Khabra (section 2.3.2.2.), then it must be taken into consideration that concentrations are dependent on the time of sampling after having started to work the pump.

### 2.4.4.2. *Wadi Al Hawtah, Wadi Birk*

Wadi Al Hawtah (Fig. 74) cuts through the Upper Jurassic strata of the Tuwayq Mountains in a west-northwest direction and flows into Wadi Birk at about lat 23° 30′ N. The wadi lies some 550 meters above sea level in the breaching area. In the west a widened entrance touching the edges of the Tuwayq Escarpment (about 1,000 meters above sea level) collects the precipitation through numerous tributaries in the massive outcrop of the Dhruma Formation. Wadi Birk, farther south, crosses the entire range of the Tuwayq Mountains. The canyon-like breaching turns sharply northwards in the lower part and receives Wadi Al Hawtah near Al Hilwah. Both valleys are filled with thick Quaternary accumulations. Major oases can be found in Wadi Al Hawtah at the beginning of the breaching near Al Hariq, as well as farther east near Na'am and at the wadi's mouth into Wadi Birk. The latter area is cultivated in the lower parts only, in the vicinity of Al Hilwah. All hand-dug wells, which in some cases are very old, are pumped by means of diesel engines. The waters of Wadi Al Hawtah contain 6.2 to 14 me/l of dissolved salts; those of Wadi Birk, 12.6 to 21.8 me/l (Table 42). The ion distribution fluctuates widely and suggests no dependence on the total mineralization (Fig. 75). The $\delta$ D-values show a marked relationship to the tritium content as well as the water temperature, particularly in waters of Wadi

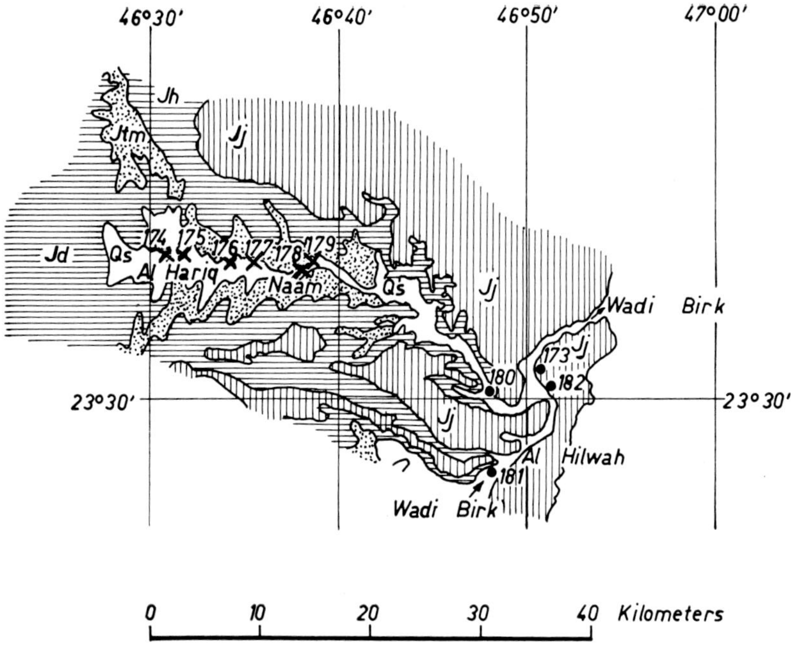


Fig. 74. Very simplistic geological sketch of the area under investigation with sampling spots in Wadi Al Hawtah, and in the lower part of Wadi Birk. Samples with values above 4 T.U. are marked with a cross. Jurassic Formations: *Jd* Dhruma, *Jtm* Tuwayq, *Jh* Hanifah, *Jj* Jubaila, *Qs* Quaternary (silts predominantly)

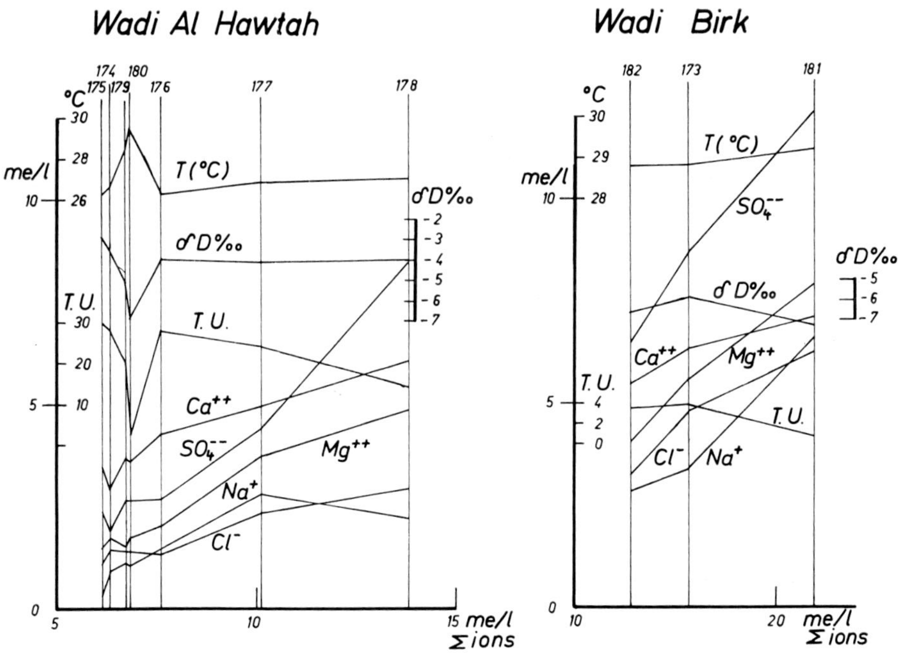


Fig. 75. Temperature (*T*), deuterium (*δ D*), tritium (*T.U.*) and ion concentrations of the waters from Wadi Al Hawtah and Wadi Birk plotted against the total concentration

Al Hawtah (174 to 180). Both δ D and δ [18]O increase together with the tritium content ($r_s = 0.87 > 0.86$ for $n = 7$ in $\alpha = 0.01$), whereas the water temperature falls with the rising content of tritium and of stable isotopes ($r_s = -0.96$ and $-0.93, > 0.89$ for $n = 7$ in $\alpha = 0.005$). It may thus be concluded that warmer waters that are free from tritium and contain less D and [18]O become mixed with cool affluents enriched with tritium and stable isotopes. Presumably, these tritium-containing affluents have a relatively short infiltration time, which points to a nearby recharge area. They probably flow in from the rather steep and short tributaries in the valley flanks of the breaching. Their recharge area of the older, tritium-free component may be the area west of the breaching.

A similar situation may prevail in Wadi Birk as well. In this case, only traces of tritium could be detected. Because these samples (173, 181, 182) were collected from the lower parts of the wadi, the longer subterranean flow might be responsible for it. Sample 180 taken at the end of Wadi Al Hawtah has also very little tritium, although waters with a relatively high tritium content join the middle section of the aquifer.

In the δ D–δ [18]O diagram (Fig. 76) the waters of Wadi Al Hawtah and Wadi Birk form a homogeneous group which is differentiated by a remarkable higher $t_8$-value (~ 19) and a steeper gradient of the equation line ($\delta D = 7.94 \cdot \delta\, ^{18}O + 18.99$) [1] from the Hanifah-Al Luhy and Nisah waters ($t_8$-value 14, $\delta D = 6.08 \cdot \delta\, ^{18}O + 9.41$) [2]. Both indicate that the Hawtah-Birk waters experience less evaporation than those of the Hanifah-Al Luhy-Nisah group. Considering the fact that the processes of mixture and evaporation may produce reverse changes in the isotope content (cf., section 2.5.4.), the course of the equation line speaks in favour of the theory that the recent affluents into Wadi Al Hawtah are exposed to less evaporation than the other waters examined in the central parts of the Tuwayq Mountains. The causes may simply be a greater infiltration velocity into the coarse gravels of the tributaries, the much steeper relief and the jointed rocks in the Al Hawtah-Birk area.

### *2.4.4.3. The δD–δ [18]O relation of waters in the central parts of the Tuwayq Mountains*

Any interpretation of the isotope content in groundwaters or a judgment of the extent of evaporation, in particular, is made difficult by lack of long-term measurements of isotope contents in rainfalls in the area under investigation. The nearest rain gauge supplying isotope values is 450 kilometers northeast, at the island of Bahrain in the Arabian Gulf. There, monthly measurements of the precipitation depth and the isotope content have been made by the International Atomic Energy Agency ever since 1961. The data are completely available for the period between November 1961 and Novem-

[1] $r = 0.87 > 0.82$ for $n = 10$ in $\alpha$ 0.001.

[2] $r = 0.84 > 0.82$ for $n = 10$ in $\alpha$ 0.001.

ber 1967 [1]. Assorted according to the amount of falling rain, the average values display a very distinct amount effect (Fig. 76). The values of isotopes of groundwaters mostly lie on the left side of the "Bahrain line". This does not reflect the surely strong secondary evaporation to which rainfalls are exposed above land surface before they infiltrate into the subsoil. Samples 113 and 114 furnish an example of the isotope content from one rainfall

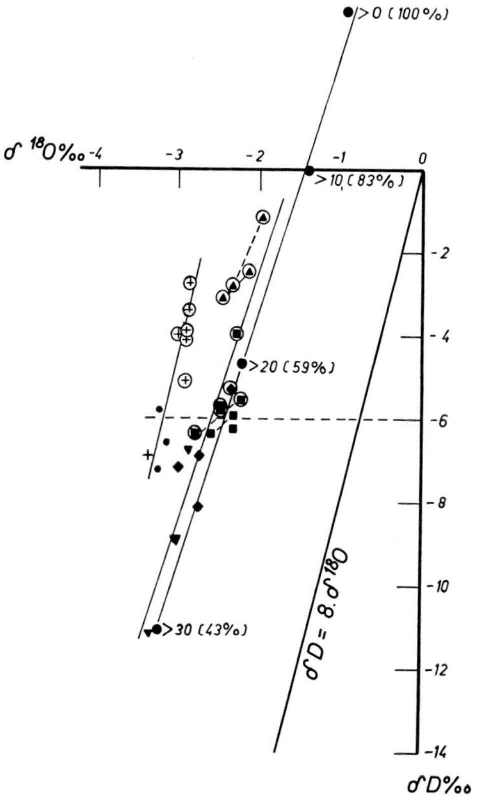


Fig. 76. $\delta D$–$\delta^{18}O$ relation in the waters of the central areas of the Tuwayq Mountains. Symbols used for the provenance of samples: Wadi Al Hawtah +, Wadi Birk ■, Wadi Hanifah ▲, Wadi Al Luhy ■, Wadi Nisah ▼, Al Kharj basin ◆, ● rainfalls in Bahrain between November 1961 and November 1967 (mean values from individual measurements in monthly samples of precipitation selected according to the respective precipitation depth in millimeters). In parentheses: proportional share of the stated precipitation quantity in the total precipitation during the test period. Symbols with a circle: samples with a tritium content $> 4$ T.U.

in the examined area. The first was taken from a thunderstorm breaking on March 28, 1974, near Al Ha'ir in Wadi Hanifah; the latter, from the immediately rising wadi flood. The $t_8$-values of both samples correspond approximately to those of the groundwater of this area.

[1] The values were taken from the report compiled by T. Dinçer and his colleagues (1974).

The displacement of the $\delta\ ^{18}O$-values of groundwaters towards the left on the diagram as compared to the Bahrain rainfalls is in support of the theory that the rain feeding these wadi groundwaters condenses from a water vapour that originated from sea water by a kinetically stronger influenced primary evaporation than the vapour from which Bahrain rainfalls condense. This difference is probably due to the fact that the transport of humid air from the Gulf onto the Arabian Peninsula is done by the wind that intensifies the kinetic fractionation of isotopes while the sea water evaporates.

As to a further interpretation of the $\delta$ D-$\delta\ ^{18}O$ diagram (Fig. 76), it is presumed that the quantitative ratio in the rain falling in the area under investigation is more or less the same as that in Bahrain rainfalls. Under these assumptions it may be gathered from the position of the deuterium values that aquifers are generally recharged by rainfalls of more than 15 millimeters. Two reasons might be decisive for this: firstly, the percentage of rain below 15 millimeters is just about 20 percent of the total rain recharging these aquifers, and secondly, smaller amounts of rain do not flow quickly enough into the wadi from the just slightly dipping plateaus. Consequently, they are largely subject to evaporation.

The values listed in Table 43 suggest that the tritium contents normally lie either below 4 or above 10 T.U. It can be seen from Fig. 76, where the samples above 4 T.U. are marked with a circle, that waters of a higher tritium content nearly always have a higher content of deuterium as well, as was observed in the waters of Wadi Ar Rimah (cf., section 2.3.2.2.). From among 14 samples with a tritium content below 4 T.U. the content of deuterium exceeds — 6.0 per mill thrice just by a hair-breadth, and from 15 samples with a tritium content of more than 10 T. U. $\delta$ D is once very slightly below — 6.0 per mill. All waters having tritium values of more than 10 T.U. were collected from upper and central parts of valleys provided with both a nearby and a remote recharge area, whereas waters with less than 4 T.U. leave the lower parts of broad valleys or else come out of Al Kharj basin far away from the recharge area. The generally low content of deuterium in larger aquifers suggests that these are preferentially recharged by heavy and intense rainfalls. It may be that the major aquifers have stored isotope-poor waters from an earlier rainy period (Neolithic Pluvial?), the remnants of which have mixed time and again with affluents containing more isotopes ever since the transition to an arid climate. This would likewise explain why among no less than 180 samples no deuterium content is detectable, which in fact fills the gap between the ancient waters of sediment sequences ($\delta$ D $<$ — 23 per mill) and the wadi groundwaters ($\delta$ D $>$ — 11 per mill). It would be advisable to make radiocarbon datings of wadi groundwaters free from tritium combined with measurements of the stable isotopes; this might eventually give an answer to this topical issue.

## 2.5. Wadi Ad Dawasir and Its Hinterland

Wadi Ad Dawasir is the only old river system crossing the cuesta landscape of central Saudi Arabia in its southern part. The breaching channel breaks its way through the complex of Paleozoic and Mesozoic strata for about 200 kilometers. The tributaries of the upper part of this old river system collected the surface runoff from almost the entire Al Hijaz highlands between the present towns of At Taif and Abha. The mouth of the river was to be found at the western slope of the large continental basin of the present Rub' Al Khali. The recharge area of this river system covered more than 150,000 square kilometers. The main tributaries were the present Wadi Tathlith, Wadi Bishah, Wadi Ranyah and Wadi Subay' (see Fig. 77).

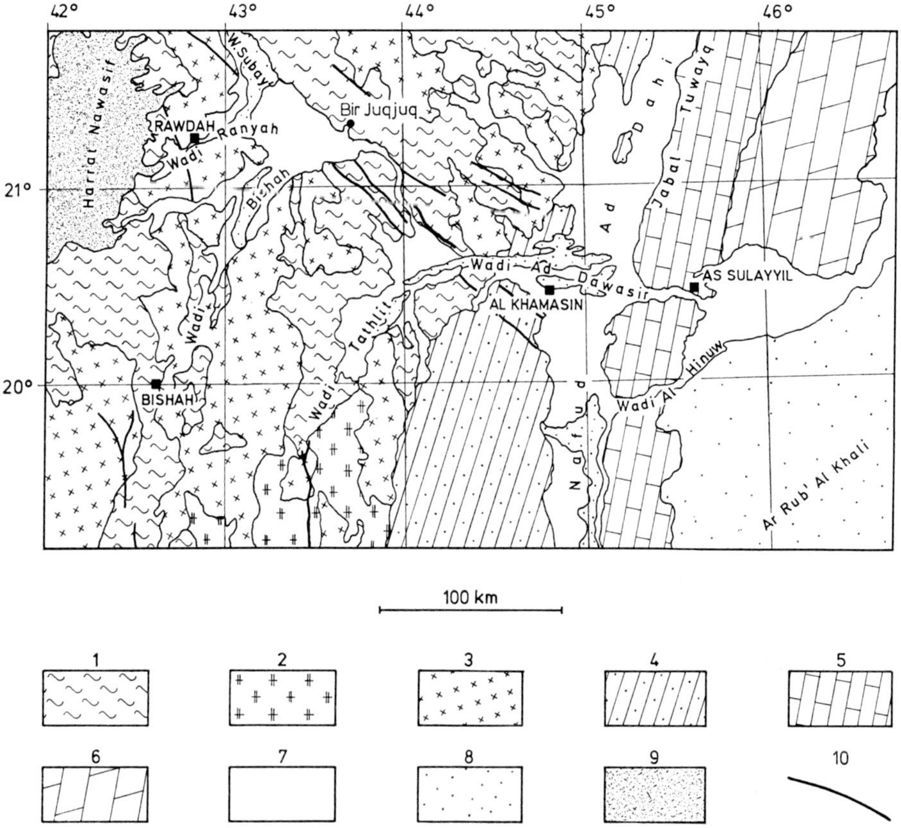


Fig. 77. Geologic map of the recharge area of Wadi Ad Dawasir. Generalized after U.S. Geol. Survey and Arabian American Oil Company, 1963. *1* Precambrian metamorphic rocks, mainly schists; *2* Basic igneous rocks, partly metamorphic; *3* Granite and granite gneiss; *4* Permian (Wajid sandstone and Khuff limestone) and Triassic (mainly silt and sandstone); *5* Jurassic (mainly limestone); *6* Cretaceous (mainly sandstone); *7* Alluvium and related surficial deposits; *8* Eolian sand; *9* Tertiary and Quaternary basalts; *10* Fault

At present no continuous river system exists, and wadis end in large accumulation basins. Towards the east Wadi Ad Dawasir is covered by sand dunes in many places. Only in its eastern part, where all wadis cross the Tuwayq Mountains does a channel still exist, collecting the surface runoff from smaller wadis of the cuesta landscape. This channel, too, finds its end in an accumulation plain west of Ar Rub' Al Khali.

### 2.5.1. General Geology

(H. Hötzl)

The recharge area of the old river system of Wadi Ad Dawasir occupied parts of both the Arabian Shield and the Arabian Shelf. This fact causes great differences in the geological situation (see Fig. 77). The western part consists of metamorphic and igneous rocks of Precambrian orogenic series, locally covered with Cenozoic plateau basalts. The eastern part lies in a complex of Paleozoic, Mesozoic and Cenozoic strata.

The area in the crystalline shield includes two orogens of different age and different structure, the Hijaz and the Najd orogens (G. F. Brown, R. O. Jackson, 1960), separated by the Madinah-Muwayh fault system west-northwest of Wadi Ad Dawasir. This fault system is developed parallel to the sinistral Najd wrench fault farther northeast.

The southern block belonging to the older Hijaz orogen (800–600 million years) underwent intense deformation. Migmatitic gneisses and metamorphic rocks of the greenschist and amphibolite facies, developed from clastic sediments and basic volcanic rocks (e.g., Hali and Baish complexes), indicate a relatively deep tectonic level. Intrusive into these rocks are pre-, syn-, and post-tectonic granites, granodiorites and diorites, more or less gneissic in structure. Owing to the following uplift and extensive denudation, these acidic intrusive rocks can be found over large areas of the present land surface.

The tectonic structures of the Hijaz block have north and northeast directions, and the same is true of the main channels of Wadi Tathlith and Wadi Bishah. These channels meet the northern block at an acute angle.

The northern block is younger (Najd orogen 600–450 million years) and its structure pattern lies parallel to the fault system northwest-southeast. Besides various orthogneisses, slightly metamorphosed rocks like the series of andesite and felsite (Halaban Formation) are dominant, with corresponding intercalations of breccia, conglomerate, quartzite, phyllite and marble. The runoff from this area goes to the southeast-trending lower course of Wadi Bishah and its tributary, Wadi Subay'.

In the northwestern part of the recharge area, the Precambrian rocks are covered with plateau basalts of Tertiary and Quaternary age (Harrat Nawasif). Numerous well-developed volcanic cones as well as lava flows reaching far into the foreland emphasize the young age of these volcanic rocks.

The complex of sedimentary strata discordantly overlying the crystalline rocks of the shield consists mainly of sandstones and limestones. The lowest strata of this flat, eastward-dipping series are the Wajid Sandstone of terrestrial origin, with a thickness of several hundred meters and an assumed age of Devonian or Lower Permian (R. W. Powers, 1968). South of Wadi Ad Dawasir these sandstones extend widely, forming the vast, slightly rising area west of Nefud Adh Dhai; in contrast, they are not to be found north of Wadi Ad Dawasir, where the crystalline rocks are directly overlain by the Permian Khuff Formation. It can be assumed that the base of these sandstones is formed by a Paleozoic peneplain on the crystalline shield, which may have been partly exhumed by younger denudation.

In contrast to the carbonate facies in the north, the Khuff Formation in the area of Wadi Ad Dawasir is more sandy and clayey. Together with the fine clastic sediments of the Sudair Formation, these strata have their outcrops in the flat depression of Nefud Adh Dhai. Regretably, almost the entire area is covered by young eolian and fluviatile sediments.

The cuesta landscape (Jebal Tuwayq in the north, Al 'Aaridh in the south) comprises the Jurassic strata of the Dhruma, Tuwayq, Hanifah, Jubaila, and Arab Formations (see 2.4.1.). Outcrops at the bases of Al 'Aaridh Mountains, south of Wadi Ad Dawasir, show sandstones and schists of Permian and Triassic age overlain by the Jurassic Dhruma Formation, which, in contrast to its characteristic carbonate facies north of lat 22° N, is sandy and of terrestrial origin in the area of Wadi Ad Dawasir.

With the Tuwayq Mountain Formation in the area of Wadi Ad Dawasir the first layers of marine carbonate rocks can be found. Carbonate strata have been developed from the Middle Jurassic to the Lower Cretaceous (Buwait Formation). With a thickness of 700 meters, this limestone complex forms the characteristic cuesta landscape of the Jebal Tuwayq and Jebal Al 'Aaridh.

North of Wadi Ad Dawasir the outcrops of the Biyadh and Wasia Sandstones (Aptian, Turonian) occupy a wide area (see Fig. 77). Southwards, these sandstones are covered by an accumulation plain consisting of sands and gravels from Wadi Ad Dawasir and Wadi Al Hinuw. Farther southeast lie the dunes of Ar Rub' Al Khali, which, in the south, overlap the flat surface of the Jurassic cuesta layers dipping gently towards the east.

Size and flow directions of the old river system of Wadi Ad Dawasir strongly depended on the geologic structure and the tectonic. The recharge area described above is practically the western surface continuation of the large east-dipping basin of Ar Rub' Al Khali. This synclinal structure finds its expression in the outcrops of the sedimentary rocks forming an arc concave towards the west (R. W. Powers et al. 1966). According to P. Burek (1974) this syncline, together with the Hadhramawt anticlinorium in the south and the Arabic homocline in the north, is considered to be an immediate consequence of the origin of the east-west-trending Aden graben system. The Arabian plate was warped parallel to this Upper Cretaceous taphrogeny.

The Tertiary uplift of the Arabian Shield's west part, with the resulting origin of the Red Sea graben, caused a steeper eastward dipping of the Arabian platform. The basin of Ar Rub' Al Khali and its western border were decisive for the development of the drainage systems. In the flatter parts of the shield's crystalline rocks the old structures and faults remained relevant. Thus at present, both Wadi Tathlith and Wadi Bishah follow north- and north-northwest-trending structures over more than 250 kilometers to the Madinah-Muwayh fault, forming in the northeast the border of the Hijaz block. The same is true of parts of Wadi Ranyah and the upper part of Wadi

Subay' west of Harrat Nawasif. Towards the bordering Madinah-Muwayh fault, large accumulation plains are the result of wadi sedimentation. The fault itself is the reason for the change in the drainage direction, now following the trough axis southeast into the basin of Ar Rub' Al Khali. The lower parts of Wadi Bishah and Wadi Subay' follow this fault. In contrast, Wadi Ad Dawasir itself is not strictly bound to this fault; its east-southeast direction follows mainly the trough axis of Ar Rub' Al Khali basin. The development of Wadi Ad Dawasir and its breaching through the complex of strata seems to be connected with the outcrops of different rock series. Thus it is clearly apparent that the great Devonian-Permian Wajid Sandstone formation forms the southern flank of Wadi Ad Dawasir. This formation is missing immediately north of the wadi, where the crystalline rocks are directly overlain by the Middle Permian Khuff Formation.

## 2.5.2. Quaternary Studies on the Recharge Area Situated in Crystalline Rock Regions

(H. Hötzl, H. J. Lippolt, V. Maurin, H. Moser, W. Rauert)

### *2.5.2.1. Wadi Ranyah between the villages of Al Amlah and Rawdhah*

The main tributaries in the northwestern part of the recharge area of Wadi Ad Dawasir are Wadi Ranyah and Wadi Subay'. The area of Wadi Ranyah, investigated during the 1975 field reconnaissance, runs along the southeastern edge of Harrat Nawasif basalt plateau. The wadi collects the episodic runoff from the eastern basalt plateau; the water flows in a narrow breaching through the Kawr Majami-Kawr Barahah Mountains into a wide accumulation plain, which also receives episodic inflows from the northwest (mouth of Wadi Subay') and south (Wadi Bishah).

Wadi Ranyah was selected for field studies because of its function as a drainage system of the Tertiary and Quaternary plateau basalts. Guidelines as to both age and erosional processes were expected to be found from the orographic features of the young basalts.

Special field studies were carried out in the wadi region between the villages of Al Amlah and Rawdhah (long 42° 30′–43° 00′ E). Going from west to east, the wadi crosses the plain west of the village of Al Jirthamiyah, runs through the breaching through the Kawr Majami-Kawr Barahah Mountains, and crosses the basin of Muqabil-Rawdhah and its continuation, eventually leading to the Ranyah-Bishah accumulation plain (cf. R. O. Jackson et al., 1963).

The plain in the western region is a flat peneplain having a low gradient towards the wadi. The plain has a few isolated inselbergs, and its bedrock comprises old granites (granite gneisses), granodiorites and greenschists, intruded by younger granites belonging to the Hijaz orogeny. On the peneplain these rocks form the land surface, and only in a few places do they have a thin alluvial veneer. In a distance of just 3 to 8 kilometers from the wadi, sheetflood sediments become thicker to various degrees and finally produce an accumulation plain partly covered with eolian sand.

The eastern border of this plain is formed by the relatively steep flanks of the Kawr Majami Mountains. These flanks follow a north-northwest-trending fault, where the younger granites are uplifted in relation to the gneisses and greenschists of the plain. Towards the west of the peneplain's border are the flatly-rising plateau basalts of Harrat Nawasif. A few isolated basalt flows penetrate into the plain up to 25 kilometers, following flat erosion channels. This can be very well seen from the basalt flow of Khashm Shayil which follows the flat channel of Wadi Ranyah. Most impressive are the nearly unweathered basalt tongues of Sha'ib Hathag showing distinct contrast of the dark basalt against the light color of the weathered material of the peneplain.

The recent runoff channel of the wadi is 80 to 150 meters wide and, in the region of the villages of Al Amlah and Al Jirthamiyah, some meters lower than the accompanying fluviatile terraces, eolian accumulations and weathered basalt flows. The wadi bed lies in sand and gravels, and the thickness of these young accumulations as well as the depth of the erosion channel is unknown. In wells at the edge of the oasis of Al Jirthamiyah a thickness of 6 to 8 meters can be observed for the eolian deposits, in agreement with the height of the terrace above the wadi floor. The thickness of the underlying coarse gravels is likewise unknown.

East of Al Jirthamiyah, just before the wadi enters the breaching, the wadi channel cuts through intensively weathered basalts. Owing to soil weathering and denudation the basalt surface is uneven, and the wadi channel with a depth of 12–15 meters displays erosional terraces. The land surface is covered with blocks, up to a diameter of 0.5 meters, which originated *in situ* and have been exhumed by erosion and deflation of fine-grained weathering material. These blocks cover red clay, of a thickness of a few meters, formed by autochthonous weathering of the basalts.

The basalts, which are not separated on the map of R. O. JACKSON et al. (1963), belong to a lava flow striking westwards parallel to the wadi channel. Whether this lava flow is isolated or linked with the basalt flow of Khashm Shayil could not be investigated. The intensive weathering of these basalts in contrast to the almost unweathered basalt tongue only 20 kilometers farther northeast in the area of Sha'ib Hathag is remarkable.

Six kilometers farther east Wadi Ranyah crosses the ridge of the Kawr Majami-Kawr Barahah Mountains with their elevation of about 100 meters above the plain. These mountains consist of granite and diorite. Although of small width, these mountains form one unit, a NNW-SSE-striking ridge with a length of more than 50 kilometers. The steep scarp of their western flank follows a fault, thus substantiating the impression of homogeneity.

Concerning the morphogeny, it is of interest that Wadi Ranyah crosses perpendicular to this ridge in its center and at its highest elevation. Together with the local drainage pattern these facts speak in favor of an old system independent of the present geomorphologic features. This old drainage system must have been developed when the present relief did not exist (Lower Tertiary?). During the following phase of denudation the more resistant rocks of the Kawr Majami-Kawr Barahah Mountains formed a ridge now crossed by the epigenetic breaching of Wadi Ranyah. Since the younger peneplain in the west is covered by Miocene basalts, it can be assumed that the epigenetic breaching must be of old origin, too.

Table 44. *Results of Carbon-14 measurements in the areas of Wadi Ad Dawasir and Wadi Ranyah (1975)*

| | Locality | Geograph. position | Lab. | Reg.-No. | Material | $^{14}$C-Age (years B.P.) |
|---|---|---|---|---|---|---|
| 1 | Bir Juqjuq | lat 21° 18′ N<br>long 43° 42′ E | IRM [1] | Lab.-No. 4,231 | Weathered limy crust, soil horizon in terrace | > 35,000 |
| 2 | Bir Juqjuq | lat 21° 18′ N<br>long 43° 42′ E | IRM | Lab.-No. 4,232 | Limy crust in terrace | 10,820 ± 320 [2] |
| 3 | Loam pit Muqabil Wadi Ranyah | lat 21° 14′ N<br>long 42° 46′ E | IRM | Lab.-No. 4,233 | Limy weathered material overlying basalt; sample 0.7 m below land surface | 26,400 ± 1,970 |
| 4 | Loam pit Muqabil Wadi Ranyah | lat 21° 14′ N<br>long 42° 46′ E | IRM | Lab.-No. 4,234 | Limy weathered material overlying basalt; sample 1.0 m below land surface | 29,840 ± 2,600 |
| 5 | Sha'ib Hathag NW of Al Jirthamiyah | lat 21° 15′ N<br>long 42° 44′ E | IRM | Lab.-No. 4,235 | Calcite sinter | 6,700 ± 280 |
| 6 | Sha'ib Hathag NW of Al Jirthamiyah | lat 21° 15′ N<br>long 42° 44′ E | IRM | Lab.-No. 4,236 | Calcite sinter | 6,110 ± 250 |

[1] Institut für Radiohydrometrie, Munich.
[2] Twofold standard deviation.

Along the breaching, Wadi Ranyah is just 70 to 120 meters wide. The flanks, partly convex rock walls formed by physical weathering, provide these parts of the wadi with the character of a gorge. Flood water occupies the whole wadi floor, and only on wider parts do sand and gravel terraces exist, up to 4 meters higher than the actual wadi floor. Marks of different water levels show that these terraces are likewise overflowed. The wadi floor consists of coarse sands and gravels. Local changes of erosion and accumulation occur depending on the strength of the episodic surface runoff.

East of the breaching lies the wide valley of Muqabil-Rawdhah which gradually turns into the Ranyah-Bishah accumulation plain over a distance of about 25 kilometers.

Directly after breaching the wadi channel, deflected by a lava flow, turns north, running along the edge of the mountains for a distance of two kilometers. Here again the wadi has cut its way through a basalt flow. Near the oasis of Al Harf the wadi changes its direction again to east-northeast. The wadi floor is 5 to 8 meters lower in this place than the plain of weathered basalts between the villages of Muqabil and Rawdhah. Since the wadi is partly filled with sand and gravels, the real depth of the erosion channel remains unknown.

The above-mentioned basalt flow shows the same intensive weathering as the basalt rocks of Al Jirthamiyah. Large areas of the accumulation plain, especially south of the wadi channel, consist of autochthonous basalt detritus and eolian sand of a small thickness. The underlying clayey weathered soil also reaches a thickness of about 4 meters. This material is used for building, and

Fig. 78. Lateritic soil complex developed on a basalt flow of 3.5 million years age. This lateritic weathering took place during the Late Pliocene/Early Pleistocene humid phase. During humid phases of the Pleistocene solution of calcite in the upper horizon and its precipitation in a lower horizon lead to a Ca-horizon about 1 meter below the land surface. C-14 determinations yielded about 29,000 years B.P. (Photo: H. Hötzl, 1975)

pits can be found south of the village of Rawdhah and near Muqabil. In these pits several limy soil horizons, signs of former groundwater levels are detectable. These old soil B horizons together with the clayey layers emphasize the intensive chemical weathering process that followed the eruption of the basalts. Samples for Carbon-14 measurements were collected from the limy crusts at depths of 0.7 meter and 1.0 meter below the land surface. The measurements revealed ages of 26,400 and 29,840 years B.P. for the origin of these calcite horizons (see Table 44). In contrast to our expectations, these results lead to the conclusion that there was enough precipitation during the Würm interstadial to produce such soil caliche for this period (Fig. 78).

Various flat hills south of the village of Rawdhah seem to have acted as eruption centers for these lava flows. East of the Kawr Majami-Kawr Barahah Mountains no basalt rocks were known until recently (R. O. JACKSON et al., 1963).

Whereas the plain between the villages of Muqabil and Rawdhah shows a certain relief caused by basalt flows and pediments with outcropping rocks and different detritus covers, the flat basin east of the oasis of Rawdhah takes on more and more the character of an accumulation plain towards the east. The alluvial fans from the south still show a northward flow direction, i.e., towards the old east-west-trending wadi. North of the wadi the accumulation plain displays more and more a homogeneous eastward dip. At present the old wadi channel is accompanied by dunes on both sides. These eolian accumulations demonstrate the wind transport of fine materials blown out from the upper areas of the accumulation plain.

### *2.5.2.2. Age determination of basalt rocks and its significance in the reconstruction of climate conditions during the Quaternary*

The basalt rocks in the recharge area of Wadi Ranyah belong to the plateau basalts of Harrat Nawasif. These rocks occur at the easternmost point of the plateau basalts which cover large areas of the Arabian Shield in the north as well as in the west. Their origin is attributable to a period of magmatic eruptions that began in the Miocene and continued until historic times (R. J. FLECK et al., 1973, 1976; F. BENDER, 1968).

Harrat Nawasif plateau, known as Harrat Al Buqum in its southern parts, trends north-northeast with a maximum length of 170 kilometers and a lateral extent of approximately 50 kilometers. At the plateau's edges the uniform basaltic layer breaks up into various lava flows which may extend as much as 25 kilometers into the plain. Its origin can be ascribed to a joint system which lies parallel to the main axis of the plateau, as this becomes evident from a north-northeast trending chain of volcanic cones in the plateau's center (R. O. JACKSON et al., 1963). Primarily titaniferous olivine basalts are involved, in places interlayered with pyroclastics. There has been lack of exact age dating of these basalts until recently. However, in analogy to the remaining plateau basalts, they can be classified as belonging to the Upper Tertiary and Quaternary.

In the section of Wadi Ranyah between Al Amlah and Rawdhah which was subject to closer investigations, the basalts must be of two different periods of eruption as indicated by their different degrees of weathering. Some of the younger, slightly weathered basalts are three lava flows some 29 kilometers northwest of Al Jirthamiyah that push forward tonguelike from the uniform basaltic layer on the granitic peneplain of the Sha'ib Hathag towards the east. The surfaces of the basalt flows are very fresh, showing polygonal cooling patterns and slaglike structures such as may be observed in recent lava flows. Generally, these basalts are very compact; only towards the edges does one observe a blocky break-up. Despite their breadth of a few kilometers, these basalt flows rise above the granitic peneplain by just 8–15 meters. Their thickness increases in areas of the homogeneous basaltic plateau only.

Beneath the two southern of the three above-mentioned lava flows, the contact with the subvolcanic land surface was examined. The good correspondence of the contact surface with the exposed granitic land surface as well as their height makes it clear that no major erosive activities have taken place since the eruption of the basalts. The detritus of the granitic surface reveals a strong red vitrification in the contact area. In the back regions of the bay formed by the two lava flows mentioned above, sinter formations are visible, in which various kinds of sinter troughs and channels are still in a relatively good state of preservation. Carbon-14 ages of two samples are 6,700 and 6,110 years B.P. (see Table 44, No. 5, 6).

The conclusion that the granitic peneplain, ever since the formation of the lava flows, has undergone neither any remarkable linear erosion nor a denudation of the peneplain authorizes the assumption that arid conditions have prevailed in this period, interrupted by short phases of more humid climate which reactivated sinter and soil development. At the same times, the slight weathering and the well maintained shape of the lava flows thus find a satisfactory explanation. In the area of the homogeneous basaltic plateau itself, a few relatively straight-lined, flat and narrow erosion channels have been formed. They predominantly border adjacent lava flows and may be attributed to erosion by the surface runoff combined with episodic precipitation. No channel formations, however, are detectable in their continuation on the peneplain.

In contrast to these slightly weathered basalts other basalts occur, in the area of Al Jirthamiyah as well as east of the Kawr Majami-Kawr Barahah plateau between Muqabil and Rawdhah, which have experienced intensive lateritic weathering. Both the thick mantle of weathered material and the linear erosion of basalts, which are not considered in the map drawn by R. O. Jackson et al. (1963) were discussed in the previous chapter (2.5.2.1.).

The kind and intensity of this weathering process presupposes humid tropic conditions, a climate that had obviously prevailed before the young, unweathered basalts were exposed. Basalt samples were collected in order to conduct a radiometric age determination of this humid phase and its subsequent arid period. Older basalts were sampled directly at the eastern wadi border of the basalt flow, 1.5 kilometers north of Muqabil (Table 45, sample No. 109). Samples of the younger basalts were collected in the area of Sha'ib Hathag 20 kilometers northwest of Al Jirthamiyah at the end of the bay between the two basalt flows (sample No. 107). Both preparation and investigation of the samples were carried out at the Laboratory of Geochronology at Heidelberg University.

Potassium-argon age determinations were performed following techniques described by P. Horn, H. J. Lippolt and W. Todt (1972). Argon was measured by mass spectrometric isotope dilution; potassium, by flame photometry.

Both basalt samples are fresh rocks without any evidence of contamination by older materials. They are alkali-olivine basalts consisting predominantly of plagioclase, pyroxene and olivine in a groundmass of glass, pyroxene, olivine, calcite, opaques and apatite. The olivine crystals of sample No. 107 are altered at the rims; those of sample No. 109 are more or less completely altered.

The results of the age determinations are tabulated in Table 45. The analyses confirm the age difference between the two basalts which has been assumed from field evidence. Sample No. 109 yielded an age of 3.5 ± 0.3 million years, sample No. 107 has an age of 1.1 ± 0.3 million years, and therefore is younger by about two million years.

Table 45. *Potassium/Argon ages of Harrat-Nawasif basalts*

| Sample No. | Potassium % | Run | Radiogenetic Argon $cm^3/g$ STP | $^{40}Ar$ (atm) % | Age $10^6$ years |
|---|---|---|---|---|---|
| 107 | 0.78 | 2,150<br>2,221 | $0.34 \cdot 10^{-7}$ | 95 | 1.1 ± 0.3 |
| 109 | 0.69 | 2,152<br>2,224 | $0.97 \cdot 10^{-7}$ | 85 | 3.5 ± 0.3 |

The different age values of both kinds of basalts justify a marked time separation of two climatically differing periods, which are most significant for the area of South Arabia, considering the different degrees of weathering of these rocks. Accordingly, a humid climate must have prevailed in the Upper Pliocene and/or the Early Pleistocene—the precise duration cannot be stated with exactitude yet, but surely in the period between 3 million and one million years ago—that in the southern part of Arabia facilitated tropical humid weathering and a pronounced linear erosion together with accumulation in the eastern sediment basin. In the past one million years, maybe even before, arid climatic conditions were prevalent. Less precipitation was neither sufficient to bring about an intensive chemical weathering process nor did it result in appreciable linear erosion combined with denudation.

### *2.5.2.3. Ranyah-Bishah accumulation plain*

Wadi Ranyah and the upper reach of Wadi Bishah unite east of the Rawdhah oasis in a vast accumulation plain (lat 21° 00′–21° 30′ N, long 43° 00′–43° 45′ E). Both the location and the NW-SE trending axis of this accumulation plain (including an area of some 70 × 30 kilometers) are characterized by the Madinah-Muwayh fault system crossing the plain (see chapter 2.5.1. and Fig. 77). The accumulation plain dips in general to the northeast, transverse to the main extension. Its line of surface drainage is situated at the northeast border; its continuation is the narrow lower reach of Wadi Bishah. Together with the lower part of Wadi Subay' discharging from the northwest, the afore-mentioned feature is pointed up by the direction of the drainage system which runs parallel to the fault system. Both Wadi Ranyah and Wadi Bishah (coming from the southwest and south-southwest respectively) lie almost perpendicularly across this geologically formed drainage channel. The course and the form of this entire drainage system, with the angular turn of the main direction as well as the narrow gap of Wadi Bishah's lower reach towards the broad and flat plain in the northwest suggest that the drainage originally must have taken place farther north and northwest, having followed the path of upper Wadi Bishah and Wadi Tathlith. The Red Sea taphrogenesis and the eastward-tilted Arabian Peninsula presumably caused the change in the drainage direction. By a corresponding headward erosion from Wadi Ad Dawasir, which acted most swiftly along the Madinah-Muwayh fault system (today's lower part of Wadi Bishah), the system of Wadi Ranyah as well as Wadi Bishah was obviously captured and diverted towards Ar Rub' Al Khali.

Primarily, the accumulation plain owes its origin to intensive erosion in a far more humid period (Late Pliocene–Early Pleistocene?). This erosive activity was favored by the confluence of many runoff channels and the breaking-up rocks caused by faults. Various wells at the northeast border of the plain south of Jabal Umm Matirah suggest that the original local base level of erosion lay at least 30 meters lower. The relatively quick transition to arid climatic conditions still prevailing today largely caused the rivers to run dry. In the area of sporadically occurring runoff, erosion channels were filled up with fine pebbles, sands and mud. Of some significance were eolian accumulations as becomes evident from the present extension of dune areas. On the geologic map by R. O. JACKSON et al. (1963) various gravel terraces dating back to the Quaternary are shown in the area of the accumulation plain. This was one of the primary reasons why we decided upon making a trip through the northwest part of the vast recharge area of Wadi Ad Dawasir in the course of our excursion. We expected to gain more information on alternating erosion and accumulation processes from these terrace sediments during the most recent geological past. However, these terraces proved to be remnants of an old peneplain built up by the outcropping greenschists and metadiabases. On the higher parts of the terrace, i.e., 5 to 10 meters above the recent accumulation plain, not one nonautochthonous gravel specimen could be secured.

The north-south direction of these erosion remnants suggests that they form the continuation of the Umm Shat ridge located in the south. The Umm Shat ridge accompanies Wadi Bishah in the south, forming its natural border. The relief of its northern continuation, together with scattered remnants and inselbergs caused by west-to-east running channels indicates that the plain must be of relatively old age; this becomes apparent in the originally northern direction of discharge of Wadi Bishah.

If one applies the findings gained from the age determination of basalts of Wadi Ranyah to the vast alluvial fan, the peneplain found in the erosional remnants must consequently be of Pliocene or even older age. This, however, would imply that during the humid phase of the Upper Pliocene–Early Pleistocene, the erosional channels as well as basins dissecting the old peneplain must have been formed. The extensive accumulation into these erosional forms may thus be assumed to have taken place under arid conditions since the beginning of Quaternary time.

### *2.5.2.4. Wadi Bishah's lower part*

Wadi Bishah's lower part acts as the connecting runoff channel from the Ranyah-Bishah accumulation plain to Wadi Tathlith, thus producing Wadi Ad Dawasir. This lower part (approximately 80 kilometers long) of Wadi Bishah forms a relatively narrow valley crossing basaltic and rhyolitic series of the Halaban Formation along the faults of the Madinah-Muwayh system. Similar to the vast accumulation plain, sedimentary material has also accumulated in this wadi. The sedimentary material is derived from the surroundings, from bordering slopes and ridges and from tributaries. In exposed places can be found dunes which in some cases fill the entire wadi width, acting as dams.

The wadi floor itself is even and consists mostly of sandy and clayey sediments, although its flanks are rather steep. Large blocks and coarse debris material occur only in the marginal parts of the wadi. Flat scree fans may have developed at the mouth of various tributaries. The widening southeastern part of the wadi floor is formed by a reg. The present accumulation activity is characterized by the present runoff conditions. There are no deep channels. Apart from the lateral eolian accumulation, the entire wadi width is homogeneously developed. No terrace formations may be observed. There are practically no through-going runoffs. Periodically rising floods may be ascribed to the flank runoff and to various inlets from tributaries. In the area of the well of Abu Harays, wadi sabkhahs can be found. With their clayey material and their salt crusts they convey the impression of damming as well as evaporation of runoffs. The tributaries display clearly formed runoff channels where frequently terraces 2–2.5 meters high are developed beside individual marks of the average and high water levels. In places the sediment body is very inhomogeneous. In a relatively small tributary on the right (some 8 kilometers above the Abu Harays well) the various strata of the accumulation terraces could be examined in a small erosion channel.

The terrace's surface lies 2 to 3 meters above the present river bed. Poorly rounded pebbles can be found most frequently. They overlie sand dunes reaching heights of up to 2 meters at the wadi mouth. Coarse lateral scree fans with a thickness of one meter partly cover the pebble layer. In some instances, the pebbles are slightly cemented by calcite. The whole deposit, including the recent accumulation material exclusively consists of rocks exposed in the wadi. With regard to the area that comprises just a few square kilometers, no conclusions whatsoever may be drawn concerning the climatic changes, owing to a very strong dependence of erosion as well as accumulation on episodic phases of precipitation.

A similar two-meter terrace could be made out in a tributary of Bir Juqjuq on the northern side. The tributary opens with a flatly sloping scree fan into the upper part of the lower reach of Wadi Bishah. In the back area of this tributary (where a small and winding wadi exists) both basalts and gravel accumulations are eroded terrace-like by the recent meandering runoff channel.

Only at Bir Juqjuq is the terrace built up of porphyroides with huge labradorite phenocrysts. These rocks are intensively weathered and are coated with a limy crust. This limy crust horizon may be considered to be a remnant of a soil formation, above the porphyroides, which formed under far more humid conditions. Through processes of a secondary nature it crumbles into fine dust. Samples were taken for a radiocarbon age determination (Table 44).

The floor of the meandering runoff channel is covered with pebbles and gravels.

## 2.5.3. Studies of the Quaternary Development of the Eastern Part of the Recharge Area of Wadi Ad Dawasir

(H. Hötzl, V. Maurin, J. G. Zötl)

### *2.5.3.1. The area between the oases of Al Khamasin and Al Lidam*

In the area between the oases of Al Khamasin and Al Lidam, at the eastern border of the Arabian Shield, where the crystalline rocks are overlain by Wajid Sandstone, Wadi Ad Dawasir flows into the wide plain of an accumulation area west of the escarpment of the Tuwayq Mountains. The bedrock of the present north-south desert area consists of less resistant Upper Permian and Triassic sediments originally crossed by a consistently flowing river system. Both the old wadi system and the eroded old sediments along the recent outcrop of the Tuwayq escarpment are at present a unique accumulation area bearing large inhomogeneous dunes of Nefud Adh Dhai. In the west, the east-west-trending small, longitudinal dunes reach far into the accumulation plain of Wadi Tathlith and Wadi Bishah which are the upper reaches of the Wadi Ad Dawasir [1]. The surface morphology in the region covered by the dunes never indicates the old river system. This is discernible only from a channel of about 10 kilometers width which is accompanied by crystalline inselbergs and mountain ridges in its western part and shows canyon-like breaching through the Tuwayq Mountains in the east. As mentioned above, the system is largely covered by dunes in the areas in between.

On the geologic maps of R. O. Jackson et al. (1963) and R. A. Bramkamp et al. (1956), in the southern part of the wide wadi traces of the main channel can be seen as silty and clayey sediments, now partly covered by dunes, which carried surface runoff in the younger part. An oasis area extends about 18 kilometers along this channel, the main villages of which are Al Khamasin and Al Lidam. North of Al Khamasin lie vast stretches of dunes reaching altitudes of 20 to 30 meters above the level of the oases. The southern border of the oases is shaped by east-west-trending erosional remnants of Permian sandstone and schists.

[1] The name Wadi Ad Dawasir is generally used downstream from the junction of Wadi Tathlith and Wadi Bishah.

One erosional remnant south of the village of Al Khamasin has a size of 350 by 100 meters, and an altitude about 11 meters higher than the oasis level. Its top is a small plain with ruins of a small fort. The lower part of the hill shows up to 4 (partly 7) meters of red sandstone of the Wajid Formation (Fig. 79). This sandstone is disconformably overlain by a grey-yellow shale of the Khuff Formation with thin purple-red layers and some lenses of gypsum having a thickness of a few centimeters. At the northern edge of the hill this strata complex has an uppermost stratum of heavily weathered sandstone about 1.2 meters thick.

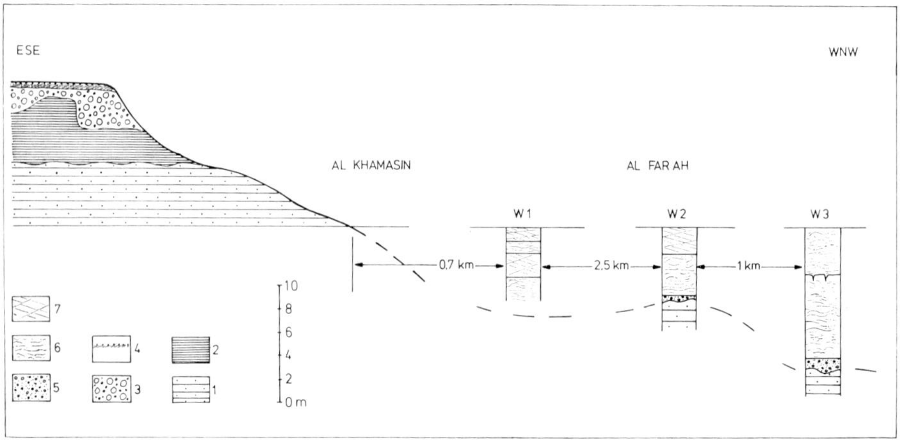


Fig. 79. Schematic geologic section of the area of Al Khamasin—Al Farah, Wadi Ad Dawasir. (*1* Wajid Sandstone; *2* Khuff shale; *3* Tertiary conglomerate; *4* desert pavement with calcareous soil horizon; *5* gravel; *6* wadi sabkhah sediments; *7* eolian sand)

This erosional remnant is covered with younger, just slightly cemented conglomerates in its upper parts. In 1975 these coarse clastic sediments were being used for road construction. The pit revealed that the conglomerate covers a strongly eroded old relief surface with vertical walls in the Permian sediments. The impression of high relief intensity at the time of deposition is emphasized by the boulder-like material of the conglomerate, from which a torrent-like accumulation may be concluded for this period. It could be assumed that this material might be traced back to a pre-Pliocene age, but the lack of further outcrops makes a more exact dating impossible. Some boulders from this conglomerate have diameters of 50 centimeters but are well rounded. The size of the pebbles is not always the same. The matrix material consists of sandy clay of reddish and red-brown colour. The pebble components comprise andesite, rhyolite, felsite, amphibolite, schists, sandstone, limestone and dolomite. The origin of the limestone and dolomite may probably be attributed to the Wajid and Khuff Formations. At the time of this accumulation process the Khuff Formation must have reached much farther west than at present.

The surface of the top plain displays a desert pavement with small, well-rounded quartz pebbles less than 4 centimeters, and angular pieces of crystalline rocks. Below the pavement lies a fossil limy soil horizon (50–70 centimeters thick) of a crumbly consistency, quickly disintegrating to dust. In this carbonate matrix, which makes up more than 50 percent of all the soil material, small quartz grains are intercalated, but very seldom are there gravels with diameters of more than 5 millimeters.

At the northern flank of this erosional remnant mainly eluvial sediments accumulate. There is a dry hand-dug well about 700 meters northwest of the hill, providing the opportunity to study the accumulation profile throughout about 6 meters. One can see a change in bedding of eolian sand (partly with clayey films), cross-bedded dune sand and clayey sabkhah sediments (Fig. 79, W 1).

This alternation of typical eolian and fluviatile sediments can also be observed in two other wells (Fig. 79). Well W 2 is located 2.5 kilometers upstream in the oasis of Al Farah. The water level was 8 meters below land surface, and the downward profile makes the following strata visible (meters below surface):

| | |
|---|---|
| 0–2.5 meters | Fluviatile redeposited dune sand with small limy concretions ($<$ 5 millimeters), sporadic small pebbles, and young, well preserved shells of *Melanoides tuberculata;* |
| 2.5–6.0 meters | Wadi sabkhah sediments with sandy, clayey silt and numerous salt and gypsum crystals; |
| 6.0–6.3 meters | Fluviatile sand with crystalline pebbles, very loose; causing a concave notch in the wall of the well; |
| from 6.3 meters | Weathered Wajid Sandstone, yellow and brown, originally reddish in less weathered parts. |

Well W 3, 1 kilometer farther west, showed the following strata (meters below surface):

| | |
|---|---|
| 0–4.0 meters | Sabkhah sediments of varying percentages of sand, silt and clay; a wavelike bedding; remarkable layers of gypsum up to a thickness of 3 centimeters; |
| 4.0–11.0 meters | Rather homogeneous silty sabkhah clays; at 4.0 meters below surface sand dikes down to a depth of 30 centimeters in the clay can be found. |
| 11.0–12.0 meters | Grey fluviatile sand and pebbles, partly from crystalline rocks; |
| from 12 meters | Yellow Wajid Sandstone. |

Samples were collected from wadi sabkhah sediments 500 meters east of Well W 3 at a depth of 20 centimeters as well as from dune sand. The grain size frequency is exemplified by Fig. 95. For the sabkhah sediments only the curves of the clastic components are drawn. Particles of more than 0.063 millimeters of grain size are mostly mica and quartz. The content of evaporites amounts to 62 percent. The dune sand showed the expected distribution of fine and medium sized sand grains. The carbonate content of this sample was only 5 percent.

The different thicknesses of the young wadi fill in the two wells on the one side and the erosional remnants of Permian sandstone rising above the surface of the plain on the other show an erosional relief covered by young sediments. However, it has to be taken into account that the wells are obviously not located in the actual deep channel. A borehole (owner Mr. Fayez A. Al Hagbani) drilled in 1975 down to a depth of 300 meters in the oasis of Kumdah situated 25 kilometers east of the village of Al Khamasin, shows a young accumulation of about 65 meters. Samples collected from the drilling (every third meter) give an overall view of the various layers:

| m below surface | Rock type |
|---|---|
| 3 | sand, probably from dunes |
| 6 | sand and fine pebbles |
| 9, 12, 15 | sabkhah clay |
| 18, 21, 24 | sabkhah clay |
| 27, 30 | sand with fine pebbles |
| 33, 36 | sabkhah clay |
| 39 | coarse sand and fine pebbles |
| 42 | fine pebbles and gravels, partly dark rock material |
| 45, 48 | sabkhah clay |
| 51, 54, 57 | coarse sand and pebbles, mainly quartz |
| 60, 63 | pebbles; quartz, granite and dark rock particles |
| 72 | Wajid Sandstone, pink and yellow |

It is difficult to interpret the genesis of the accumulation from the profile described above. For one thing the samples were collected from a swash drilling and for another thing samples were collected every third meter only. However, there is no doubt that there exists an alternation of fluviatile sand and gravels with sabkhah clays which leads to the conclusion that an alternation of dry and wet phases (with precipitation and surface runoff) must have taken place.

Provided that the drilling reached the maximum depth of the wadi channel, the complex of coarse sand, pebbles and gravels between the depths of 51 and 63 meters, overlying the bedrock of Wajid Sandstone, could be regarded to have accumulated at the end of the humid phase during which the channel was eroded. In such a case the yield of the surface runoff would decrease and the zone of accumulation in the wadi would move headward. The layers of sabkhah clay correspond with arid conditions in a period in which no continuous surface runoff occurred. This kind of accumulation becomes dominant especially in the upper part of the profile (9–24 meters depth). A wetter phase can be concluded from the sediments at depths between 39 and 42 meters. Concerning the sand in higher horizons, it is not quite clear whether it is of fluviatile or eolian origin, as the samples were collected from a swash drilling.

Determination of the origin of the erosional channel would be a very important accomplishment. With regard to the situation in the upper part of the recharge area of Wadi Ad Dawasir, i.e., Wadi Ranyah, and the dating of basalt rocks of this region (see 2.5.2.2.), it was assumed that the humid phase in the Upper Pliocene/Early Pleistocene was the last pluvial with surface runoff yield necessary for erosion of deep channels as this exists below the young accumulation plain.

From the thick layers of sabkhah clay, arid climatic conditions can be assumed for the Quaternary period. Layers of fluviatile sand and pebbles are indications of intermediate, more humid phases. Both their duration and their precise chronological determination require extremely detailed investigations of the strata in this area.

### *2.5.3.2. The break-through in the Tuwayq Mountains*

Wadi Ad Dawasir crosses the cuesta landscape of Jebal Tuwayq from west to east as a continuous slightly meandering valley about 50 kilometers long, cutting through the complex of Jurassic limestones at lat 20° 28′ N. The steep westward facing escarpment towers above the accumulation plain, situated farther west, by some 300 meters. The wadi floor has an altitude of 585 meters above sea level. The various layers of the strata complex dip slightly towards the east. There the cuesta backslope descends below the alluvial plain into which the wadi also flows.

At the beginning of the break-through area the wadi has a breadth of almost 30 kilometers and becomes ever narrower towards the east, having a funnel-like shape of about 15 kilometers. The actual break-through channel begins right there with a width of 2.5 to 3.5 kilometers, and the wadi becomes even narrower (just 1.5 kilometers) eastward of the village of As Sulayil. The morphology of this complex shows that the last 15 kilometers of the wadi must have developed in a late erosional phase as a straight-line break-through in the Jubaila Formation. When the base level of erosion was located at a higher altitude than now, the old river system had its bed north of As Sulayil, running farther towards the northeast.

The wadi channel's width decreasing downstream, i.e., the funnel-like opening towards the west, which can also be observed in various other wadis of the central Arabian cuesta landscape, induced D. A. Holmes (1960) to assume that the drainage system originally flowed from east to west. But there is no obvious fact corroborating this assumption, neither with regard to the paleogeography of the Tertiary nor concerning the morphogeny of the "funnel-like openings".

At present, the wadi floor is an accumulation plain over its entire breadth, covered by eolian sand and redeposited sediments derived from episodic floods. The existing runoff systems are formed due to episodic heavy floods. The surface runoff from the steep walls dries up at the borders of the accumulation plains where local sabkhahs have developed between dunes and scree fans. Only in the central part of the wadi does a surface channel carry water over a longer distance after heavy rainfalls. The reason for this episodic local drainage system lies in the alimentation of the runoff from tributaries which have large recharge areas on the flatly-dipping cuesta backslope. However, this channel too runs dry towards the accumulation plain in the east. Normally, even the lowest channel of the wadi is dessicated. The wadi floor is eroded down 1 to 2.5 meters over a width of 150 to 300 meters. In these lowest wadi channels one can find transported sand of fluviatile origin; in small basins silt and clay prevail with patterns of dried cracked mud. At the eastern outlet of the wadi the channel splits up into numerous smaller channels, finally petering out in the plain.

These observations reveal that under arid climatic conditions such as prevail today, the entire wadi system breaking through the Tuwayq Mountains is an accumulation area. The sediments are of eolian or episodically fluviatile origin with an appropriate assortment of eolian sand covers, dunes, transported sands of fluviatile origin, silt and sabkhah sediments, including evaporites. These accumulations are intermingled with coarse clastic sediments from the tributaries as well as poorly sorted scree material. Profiles showing the alternation of these sediments can be observed in two wells dug in the oasis of Tamrah at the southern border of the wadi.

In view of an apparent alternation of accumulation and erosion phases particular attention was paid to uncovering remnants of old accumulation terraces. However, no remnants of old wadi fills of a greater thickness could be detected. A gravel accumulation with an altitude four to six meters higher than the wadi floor of the tributary Wadi Tamrah proved to be an eroded scree fan from a small canyon. The terraces which can be found in the wadi break-through at the Tuwayq Mountains are only erosional steps along the resistant, eastward-dipping layers at the entire Jurassic complex.

The various kinds of strata as well as the thickness of the accumulation body below the wadi floor are unknown since there are neither wells nor drilling profiles available. In analogy to the conditions described for the oasis of Al Khamasin about 80 kilometers upstream (see 2.5.3.1.), an accumulation thickness of about 65 meters can also be assumed for the wadi in the Tuwayq area. The alternation of strata may likewise be similar there.

### *2.5.3.3. The wadi's transition into Ar Rub' Al Khali*

After breaching through the Tuwayq Mountains, Wadi Ad Dawasir enters the northwest border of the larger basin of Ar Rub' Al Khali. This basin is an enormous trough of sedimentation. The young sediments begin at the edge of Jebal Tuwayq and extend over more than a thousand kilometers east-northeast to the Arabian Gulf. Towards its eastern part the width of this desert reaches about 800 kilometers. The thickness of the Cenozoic sediments is more than 1,500 meters (G. F. Brown, 1972). The land surface of this gigantic accumulation trough is the world's largest continuous area of sand dunes, some rising up to 300 meters. The borders of the dune area are alluvial fans of numerous radial wadis running towards the trough axis. Most of these wadis discharge from bordering mountains consisting mainly of Mesozoic and Tertiary rocks. There are just a few wadis—only in the western regions—which have recharge areas reaching far back into the hinterland. One of them is Wadi Ad Dawasir, a river system of numerous branches in the southern Hijaz highland in ancient humid periods and a continuous runoff channel down to the basin of Ar Rub' Al Khali.

Under the present arid conditions the part of Wadi Ad Dawasir which breaches the Jebal Tuwayq has solely a local surface runoff (see 2.5.3.2.), fed by episodic floods from the cuesta backslope. Only occasionally does this runoff reach the area of Farshat Al Atwa, depositing silt, clay and evaporites. In the eastern part of this accumulation plain eolian processes are dominant features. Therefore, east of Jebal Tuwayq, the large alluvial fan of Wadi Ad Dawasir is covered by desert pavement, eolian sand and dunes reaching towards the north to the outcrop of Cretaceous and Tertiary strata. There, the Tertiary rocks dip southwards, i.e., towards the trough axis covered by dunes of Ar Rub' Al Khali. This desert reaches far towards the west, where it is in direct contact with the cuesta backslope of the Jurassic strata.

According to R. A. Bramkamp et al. (1956), Wadi Ad Dawasir—in this part also known as Wadi Al Atwa—continues for about 70 kilometers towards the east into the accumulation plain as a flat channel of 10 kilometers breadth. In its western part a slightly rising terrace flank up to 3 meters could be observed; however, this flank disappears towards the east.

A field reconnaissance, from the village of As Sulayil to 'Irq Ar Rammak and from there to Wadi Al Hisy 40 kilometers southwards across the accumulation plain, showed a unique homogeneous plain except for the flat channel of Wadi Ad Dawasir, with wide areas covered by a desert pavement. The pebbles and gravels of this pavement consists of quartz and granitic and basic rocks, as well as various sedimentary rocks. Coming into the vicinity of the escarpment of the Jurassic strata complex, the large number of small wadis causes small erosion channels and scree fans. There the lag gravel shows an ever-rising percentage of limestone pebbles with diameters up to 25 centimeters. Over large areas limestone pebbles are so dominant that hardly any quartz or crystalline gravels can be found. In such places also the color of the land surface changes to a lighter shade.

A small pit showed a typical profile of desert pavement areas: at the surface an enrichment of pebbles and gravels through deflation of fine grained materials becomes discernible. This pavement covers a small sandy layer (about 5 centimeters thick) with a low percentage of coarse grains. The very inhomogeneous gravels are next, consisting of quartz and crystalline rocks as well as limestone gravels up to diameters of 25 centimeters. The upper horizon of this gravel layer (20 centimeters) comprises cement material which is terra-rossa-like, then turns into a higher percentage of calcium carbonate with increasing depth; these partly form crusts around the gravels. At present, cementation is of no significance since the calcium carbonate tends to disintegrate into powder.

Like the breaching of Wadi Ad Dawasir through the Tuwayq Mountains, nothing final can be said yet concerning the thickness and strata of accumulation because of the lack of both wells and drillings. With regard to the age of the sheet gravel surrounding Ar Rub' Al Khali, R. A. Bramkamp et al. (1956), D. A. Holm (1960), R. W. Powers et al. (1966), and K. W. Glennie (1970) think that the accumulation is related to pluvial periods of the Quaternary. Some authors (e.g., Z. R. BEYDOUN, 1966), however, assume a Pliocene age.

The field studies carried out in the areas of Wadi As Sah'ba (2.1.3.4.) and Wadi Ranyah (2.5.2.2.) substantiate the theory that coarse gravels were transported during a very humid period in the Upper Pliocene/Early Pleistocene. There is no doubt that local erosion and accumulation also occurred in younger humid phases during the arid periods of the Pleistocene. An investigation of young freshwater accumulations in the southwestern part of Ar Rub' al Khali would be of high value (2.6.). Places where Neolithic tools have been discovered in this part of Ar Rub' Al Khali (H. A. MCCLURE, 1974, 1976) suggest for this part of the Arabian desert a more humid climate for the post-glacial period than now exists.

### 2.5.4. Hydrochemical Investigations and Measurements of the Content of Isotopes of Wells in Wadi Ad Dawasir

(C. JOB, H. MOSER, E. PAK, W. RAUERT, W. STICHLER)

Wadi Ad Dawasir (Fig. 80) crosses the Tuwayq Escarpment at lat 20° 25′ N and disappears in the sands of Ar Rub' Al Khali. The Permian sandstones of the Wajid Formation outcropping in the west lie on top of the basement. A vast flat basin extends eastward to reach the Jurassic limestones of the Tuwayq Mountains. In this place, Quaternary clastic sediments cover Permo-Triassic limestones and fine-grained sandstones of the Sudair Formation. Around the margins of the wadi channel are Quaternary gravels, and towards mid-wadi, fine silty material covers the Dhruma Formation, mainly sandstones, or Jurassic strata consisting of limestones and dolomites. Towards the east, near As Sulayil, the wadi ends in a huge scree fan which lies on the Lower Cretaceous Biyadh Sandstone and Miocene sandstones and marls.

The majority of smaller oases are concentrated around the villages of Al Farah and Al Khamasin and the wadi channel. In most oases, hand-dug wells supply these oases with water. The wells near Al Farah (Nos. 157–161) reach depths of 20 to 30 meters in the Quaternary accumulation, thus reaching the surface of the Wajid Sandstone. Water occurs at 12 meters below land surface, and its temperature is between 28.9 °C and 30.8 °C. All these wells are strongly mineralized (28.6–84.6 me/l, Table 46). Sample No. 162 was taken from a drilled well that is reported to have a depth of 100 meters and to be cased down to 30 meters. This water has a temperature of 31.4 °C and a noticeably lower degree of mineralization (20.5 me/l) than those from hand-dug wells. In hand-dug well No. 165 and the wells in the wadi channel (Nos. 163, 168, 169) water occurs some 8 meters below land surface. Water temperatures range between 27.2 °C and 31.0 °C, and the waters are strongly mineralized (41.7–104.5 me/l). All Al Farah waters are saturated with oxygen whereas other waters reach just about 45 percent of the saturation value. Chemically, these are mostly Na-Ca-Cl-$SO_4$ waters. Samples 163 and 169 contain a high percentage of Mg. The ion concentrations vary independently from the total concentrations, thus preventing any grouping of the waters. No. 163 and No. 168 contain tritium; the remaining waters display almost no tritium content at all. Sample No. 168 contains strong portions of recent rainwater according to the tritium and the stable isotope content ($\delta$ D = — 5.7 per mill; $\delta$ $^{18}$O = — 2.00 per mill). All other waters belong to a considerably older storage area. Their isotope values range widely ($\delta$ D between — 26.3 and — 44.8 per mill, $\delta$ $^{18}$O between — 3.61 and — 6.08 per mill, Table 47). The waters from hand-dug wells are most inhomogeneous in every

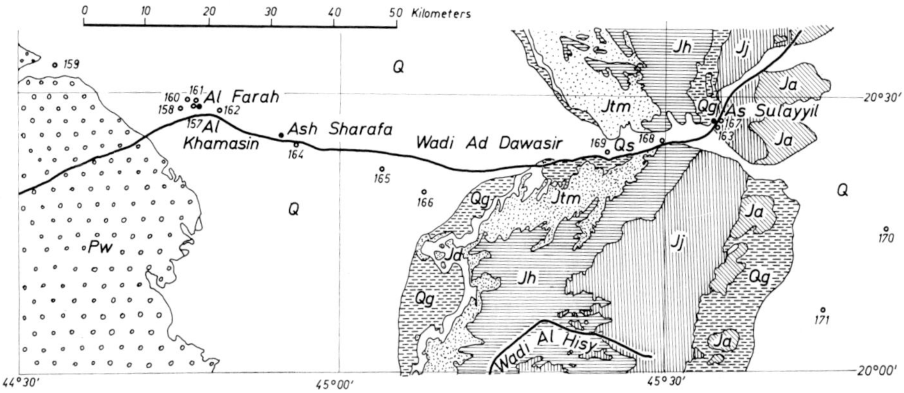


Fig. 80. Generalized geological sketch of the investigation area with markings of sampling spots in Wadi Ad Dawasir. After R. A. Bramkamp et al., 1956. Symbols: *Pw* Permian Wajid Sandstone, *J* Jurassic strata: *Jd* Dhruma Formation, *Jtm* Tuwayq Mountain Limestone, *Jh* Hanifah Formation, *Jj* Jubaila Limestone, *Ja* Arab Formation, Q Quaternary accumulation: *Qg* mainly gravel, *Qs* mainly silts

respect. Even neighbouring wells have different degrees of mineralization and ion distributions or differ considerably as to their contents of stable isotopes.

With regard to artesian wells, two groups can be differentiated chemically (Table 46). Water samples Nos. 164, 166 and 167 contain less gypsum than Nos. 170 and 171. The temperatures of the first group of wells increase from west towards east and these differences might correspond with the respective geothermal gradient of the Wajid Sandstone dipping eastward. All artesian wells have a low degree of mineralization (approx. 8–12 me/l), and contain, apart from No. 166, almost no oxygen. Artesian wells Nos. 170 and 171 lie east of the wadi end. In this location their temperatures are too low for the Wajid Sandstone. Consequently, they must have their origin in a higher horizon, a conclusion which is substantiated by their different chemical properties. Depending on the higher gypsum content, their total concentrations range between 14.6 and 16.6 me/l. As can be expected, all water samples collected from artesian wells are free from tritium. Their contents of stable isotopes range widely. The isotope data display no relation to the temperatures and concentrations of the waters. Any interpretation going further than that is therefore not practicable. The lowest isotope values of hand-dug wells

Table 46. *Temperature, oxygen content (mg/l) and chemical composition of the waters of Wadi Ad Dawasir.* Ion concentrations in µe/l, $H_3BO_3$ and $H_2SiO_3$ in mg/l. W stands for the type of well (cf., section 2.1.6.1.4.). The approximate location of sampling spots may be gathered from Fig. 80. SK and SA represent the sums of cations and anions. Sample collecting between February 21 and 27, 1975

| No. | W | °C | $O_2$ | $Na^+$ | $K^+$ | $Ca^{++}$ | $Mg^{++}$ | SK | $Cl^-$ | $Br^-$ | $SO_4^{--}$ | $HCO_3^-$ | $NO_3^-$ | $F^-$ | SA | $H_3BO_3$ | $H_2SiO_3$ |
|---|---|---|---|---|---|---|---|---|---|---|---|---|---|---|---|---|---|
| 164 | DA | 36.1 | 0.6 | 5,002 | 408 | 4,440 | 2,100 | 11,950 | 5,591 | 7 | 3,831 | 2,420 | 0 | 53 | 11,902 | 1.49 | 14.7 |
| 166 | DA | 46.0 | 2.2 | 4,567 | 470 | 3,697 | 2,020 | 10,754 | 4,770 | 1 | 3,724 | 2,129 | 0 | 7 | 10,631 | 0.91 | 17.8 |
| 167 | DA | 54.0 | 0.5 | 3,306 | 506 | 2,819 | 1,350 | 7,981 | 2,646 | 5 | 2,296 | 3,047 | 0 | 5 | 7,999 | 0.69 | 22.5 |
| 170 | DA | 42.3 | 0.1 | 5,219 | 357 | 6,620 | 2,420 | 14,616 | 5,454 | 6 | 6,731 | 2,360 | 19 | 19 | 14,589 | 1.03 | 15.0 |
| 171 | DA | 40.6 | 0.2 | 4,132 | 267 | 9,940 | 2,280 | 16,619 | 4,655 | 5 | 6,922 | 2,200 | 0 | 16 | 16,498 | 1.03 | 13.9 |
| 157 | HP | 29.3 | 8.5 | 20,878 | 119 | 19,200 | 2,240 | 42,437 | 25,822 | 17.5 | 12,731 | 2,150 | 1,610 | 16 | 42,347 | 1.77 | 20.9 |
| 158 | HP | 30.8 | 11.0 | 27,403 | 267 | 23,520 | 2,800 | 53,990 | 33,582 | 18.1 | 15,766 | 1,810 | 2,919 | 16 | 54,111 | 2.74 | 13.9 |
| 159 | H | 28.9 | — | 23,923 | 295 | 22,080 | 4,560 | 50,858 | 25,726 | 34.0 | 21,164 | 1,200 | 2,580 | 53 | 50,757 | 3.60 | 13.9 |
| 160 | HP | 29.4 | 7.2 | 13,040 | 101 | 13,100 | 2,340 | 28,581 | 17,346 | 13.9 | 8,345 | 1,580 | 1,274 | 24 | 28,583 | 1.43 | 16.7 |
| 161 | HP | 30.3 | 8.0 | 48,934 | 253 | 28,800 | 6,160 | 84,147 | 62,996 | 44.2 | 17,440 | 2,130 | 2,064 | 15 | 84,689 | 4.17 | 15.0 |
| 162 | DP | 31.4 | 8.5 | 7,829 | 163 | 10,300 | 2,360 | 20,652 | 12,406 | 13.5 | 5,683 | 1,640 | 927 | 12 | 20,682 | 0.8 | 12.8 |
| 163 | HP | 27.2 | 4.8 | 45,672 | 1,023 | 14,560 | 25,360 | 86,615 | 50,205 | 34.5 | 29,760 | 5,020 | 1,322 | 42 | 86,384 | 8.57 | 17.0 |
| 165 | HP | 31.0 | 3.4 | 68,500 | 1,074 | 25,360 | 9,440 | 104,374 | 64,168 | 71.6 | 35,400 | 2,620 | 2,177 | 97 | 104,534 | 14.00 | 30.0 |
| 168 | HP | 28.8 | 3.7 | 20,879 | 380 | 18,233 | 2,247 | 41,739 | 30,421 | 19.0 | 8,480 | 2,400 | 403 | 11 | 41,734 | 2.06 | 11.4 |
| 169 | HP | 29.5 | 3.6 | 23,923 | 408 | 31,760 | 21,760 | 77,851 | 55,546 | 54.0 | 19,960 | 1,350 | 1,032 | 11 | 77,953 | 2.34 | 18.6 |
| 151 | HP | 26.4 | — | 1,258 | 98 | 2,048 | 538 | 3,942 | 666 | 0.5 | 651 | 2,450 | 226 | 18 | 4,012 | 0.51 | 17.8 |
| 152 | H | 26.8 | 5.4 | 3,915 | 91 | 2,651 | 1,293 | 7,950 | 1,699 | 4.1 | 2,000 | 4,240 | 129 | 23 | 8,095 | 1.03 | 17.8 |
| 153 | H | 29.2 | 7.5 | 7,068 | 91 | 4,180 | 1,820 | 13,159 | 3,765 | 10.1 | 3,614 | 5,300 | 347 | 22 | 13,058 | 0.23 | 32.8 |
| 154 | HP | 28.2 | 8.5 | 4,197 | 104 | 3,414 | 1,044 | 8,759 | 2,113 | 6.5 | 1,807 | 4,460 | 399 | 23 | 8,809 | 0.63 | 26.1 |
| 155 | H | 31.2 | 9.0 | 2,932 | 100 | 1,928 | 755 | 5,715 | 1,036 | — | 1,175 | 3,320 | 111 | 22 | 5,664 | 0.80 | 24.2 |

(Nos. 158, 169) lie in the scattering range of artesian waters. According to the isotope contents, these waters are closely related to those of artesian wells [1].

Table 47. *Isotope contents of the waters of Wadi Ad Dawasir (samples taken between February 21 and 27, 1975).* The deuterium content (δ D) as well as the oxygen-18 content (δ $^{18}O$) is given in the relative deviation from the sea water standard SMOW. 1 T.U. corresponds to 3.2 pCi of tritium/l of water. $t_8$ is computed from $\delta D = 8 \cdot \delta\, ^{18}O + t_8$. As to the tritium contents the double deviation from the set standards is given. The respective mean errors are in δD ± 1‰ and in δ $^{18}O$ ± 0.2‰

| No. of sample | δ D (‰) | δ $^{18}O$ (‰) | $t_8$ | Tritium content (T.U.) |
|---|---|---|---|---|
| 164 | — 46.5 | — 6.52 | 5.66 | 2.8 ± 2.0 |
| 166 | — 42.9 | — 6.06 | 5.58 | 1.9 ± 2.5 |
| 167 | — 56.5 | — 8.02 | 7.66 | 1.6 ± 2.0 |
| 170 | — 49.0 | — 6.98 | 6.84 | 1.5 ± 2.0 |
| 171 | — 47.9 | — 6.73 | 5.94 | 0.7 ± 1.9 |
| 157 | — 38.5 | — 5.60 | 6.30 | 0.3 ± 1 |
| 158 | — 44.8 | — 6.08 | 3.84 | 0.7 ± 1.9 |
| 159 | — 44.3 | — 5.68 | 1.14 | 0.9 ± 2.0 |
| 160 | — 41.2 | — 5.64 | 3.92 | 0 ± 2.0 |
| 161 | — 35.7 | — 4.93 | 3.74 | 2.3 ± 2.0 |
| 162 | — 26.3 | — 4.39 | 8.82 | 0.5 ± 2.0 |
| 163 | — 33.7 | — 4.89 | 5.42 | 11.0 ± 2.2 |
| 165 | — 26.3 | — 3.61 | 2.58 | 1.5 ± 1.9 |
| 168 | — 5.7 | — 2.00 | 10.30 | 45 ± 3 |
| 169 | — 31.4 | — 4.71 | 6.28 | 2.8 ± 2.0 |
| 151 | — 6.8 | — 2.48 | 13.04 | 28 ± 2 |
| 152 | — 5.7 | — 1.97 | 10.06 | 35 ± 2 |
| 153 | — 7.7 | — 2.03 | 8.54 | 44 ± 2.3 |
| 154 | — 4.8 | — 1.87 | 10.16 | 30 ± 2 |
| 155 | — 6.2 | — 1.76 | 7.88 | 30 ± 2 |

Most hand-dug wells show less negative isotope values than artesian wells. In the $\delta D/\delta\, ^{18}O$ diagram (Fig. 81) the water samples 161, 163 and 165 lie close to the regression line of artesian waters, and those remaining (except 159 and 162) are not too far away. At first sight, one might, therefore, be tempted to explain the phenomenon of isotope enrichment as resulting from an evaporation process, beginning in artesian wells and continueing in hand-dug wells to an ever-increasing extent. Another argument raised at this point seems to be the positive relationship between the degree of mineralization and the enrichment of deuterium as well as $^{18}O$ ($r = 0.71 > 0.69$ for $n = 8$ and $\alpha = 0.025$). However, the $t_s$-values of waters from hand-dug

[1] These waters differ greatly both chemically and as to their δ $^{34}S$ content. According to measurements made by E. Pak, the water from the artesian well No. 164 displays a δ $^{34}S$ value of 9 ± 0.2 per mill. However, the hand-dug Al Farah wells Nos. 157, 158 and 160 have δ $^{34}S$ values between 13.1 and 13.4 ± 0.2 per mill in correspondence with a rock sample of gypsum collected from the outcropping sandstone (13.5 ± 0.2 per mill).

wells provide no clue whatsoever as to a potential connection with the degree of mineralization ($r_s = 0.26 < 0.59$ for $n = 8$ and $\alpha = 0.05$) as one would expect if evaporation took part. Both processes vary entirely independently. If, however, one makes a comparison of the total group of hand-dug wells to the artesian group, the first group will inevitably display a considerably higher degree of mineralization and isotopic enrichment. Moreover, it shows a lower $t_8$-value as a sign of stronger evaporation.

Two questions are of decisive importance concerning these waters: First, what is the reason for enrichment of stable isotopes in hand-dug wells as against artesian wells? And second, how can the correlation between mineralization and isotope content be explained, if there is no evaporation? It has been observed that the drilled well No. 162 (the water of which is not connected with the upper layers of the wadi floor due to a casing reaching down 30 meters) has a relatively low degree of mineralization yet it is strongly enriched isotopically. This enrichment cannot have occurred in the wake of evaporation because of its high $t_8$-value. It was indicated in section 2.3.2.2. that isotope shifting together with changes in the extent of mineralization are explainable as due not only to evaporation but also to the mixing of waters.

In Wadi Ad Dawasir such a potentiality arises through a shallow groundwater flow coming from the vast recharge area in the crystalline rock of the shield, following a channel filled with Quaternary sediments (Section 2.5.3.1.). The Quaternary filling of this channel consists, in its lower layers, mainly of coarse sands and gravels from the shield. It is covered, in its upper layers, by pebbles, fine sand and sabkhah sediments, together with evaporites up to a percentage of 62 percent. Waters taken from valley fillings of the shield in Wadi Ranyah situated some 200 kilometers from Al Farah were poorly mineralized (Table 46, 4–13 me/l) and showed a medium deuterium content of $\delta D = -6$ per mill and a medium $^{18}O$ content of $\delta\, ^{18}O = -2$ per mill. Assuming that these isotope values were likewise valid for the groundwater of Wadi Ad Dawasir originating in the shield, and assuming that the artesian well No. 164 is representative of the old sandstone water in the area of Al Farah, the mixtures of these two components should then lie on the broken line drawn in Fig. 81. This is indeed true of the isotope values of well No. 162, which is provided with a casing down to lower parts of the wadi filling. In contrast, except No. 157, all hand-dug wells (probably because of stronger evaporation), are grouped around a slightly flatter equation line drawn in Fig. 81 on the right ($\delta D = 7.95 \cdot \delta\, ^{18}O + 3.89$ per mill; $r_s = 0.96 > 0.87$ for $n = 8$ and $\alpha = 0.001$). This equation line shows that the shifting of isotopes may occur parallel to $\delta D = 8 \cdot \delta\, ^{18}O$, owing to a combination of mixing and evaporation, although it generally has been assumed that an inclination of $m = 8$ is a sure sign that no kinetically influenced evaporation has taken place.

These deliberations lead to the conclusion that the enrichment of isotopes as well as the mineralization in hand-dug wells may be explained by an admixture of recent shallow groundwater to old sandstone water, provided that the recent water in lower parts of the wadi filling displays a poorer degree of mineralization than that in upper parts. However, two objections have to be raised: First, the waters of wells Nos. 158 and 159 are more strongly mineralized than those of artesian wells, although their low isotope content suggests that recent groundwater has not penetrated. Presumably, these waters become mineralized just by contacting the upper layers of the wadi filling. Second, the waters of the hand-dug wells of Wadi Ad Dawasir are free from tritium except for Nos. 163 and 168, whereas the shallow groundwater of Wadi Ranyah does contain tritium. As the recharge area of Wadi Ad Dawasir lies farther away than that of Wadi Ranyah, it may be assumed

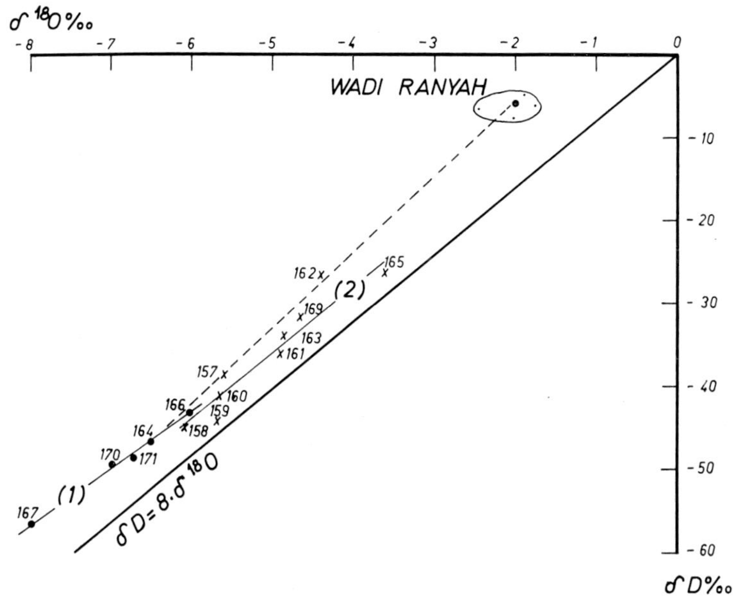


Fig. 81. δ D–δ $^{18}$O relation of the waters of Wadi Ad Dawasir and Wadi Ranyah together with the equation lines for artesian waters (*1*) and waters from hand dug wells (*2*). The broken line stands for mixtures of artesian waters with the groundwater originating in the Shield

that all Dawasir waters have become free from tritium due to their length of travel and, moreover, that they have their origin in the time before nuclear tests began. Since there are no floods running into Al Farah basin, any fast transport of major rainfall quantities from the shield can be excluded. The tritium enrichment due to rather poor precipitation infiltrating into the wadi subsoil does not generally suffice to give rise to a measurable content of tritium. The same applies to other areas such as Wadi Nisah and Riyadh Al Khabra. Only a relatively swift transport of recent rainfalls from a larger recharge area would be sufficient. This is true of the hand-dug well No. 163 in the vicinity of As Sulayil in which old, tritium-free sandstone water mixes with the recent runoff from the Jurassic limestones. Obviously, well No. 168 receives only this short-term runoff originating in the limestones. It can be proved by means of the isotope values of these two waters whether or not the mixing hypothesis is correct. In such a case, the δ D values should result in a mixing ratio that is proportionally about the same as the mixing ratio computable from the tritium values. Starting from the assumption that one mixing component is tritium-free artesian water with a medium deuterium content of $\delta D = -48.6 \pm 5$ per mill, and that the other corresponds to the shallow groundwater of well No. 168 containing 45 T.U. tritium and $-5.7\ \delta D$ per mill, a recent portion of $24 \pm 6$ percent would ensue from the tritium content of well No. 163, and a concurrent portion of $35 \pm 8$ percent would arise from its deuterium content.

Finally, it may be concluded that the mineralization of wells in Wadi Ad Dawasir most likely occurs on account of a mixing of recent, more or less salty and shallow groundwaters with qualitatively superior old waters originating in the Wajid Sandstone. Additionally, the direct contact of old waters with upper layers of Quaternary clastic sediments seems to have eventually caused mineralization, which furnishes another proof of how qualitatively superior waters may be spoiled by inadequate casings. The fact that even today old water occurs near the outcrop leads to the assumption that the Wajid Sandstone must be a vast storage area of potable water.

## 2.6. Ar Rub' Al Khali

(H. A. McClure)

### 2.6.1. Introduction

Fossil lake beds of the Rub' Al Khali serve as a framework on which to hang a time-stratigraphic column for the late Quaternary (H. A. McClure, 1976), and by which characteristics of the Quaternary in general can be inferred. Additional evidence for earlier phases of the Quaternary must be derived from correlation with similar sediments in surrounding areas, from the Tertiary tectonic history of the Arabian plate, and from other more indirect lines of evidence.

Descriptive details of the Quaternary of the Rub' Al Khali remain to be worked out. The following discussion is meant to serve as a general outline of Quaternary events in the Rub' Al Khali into which such details can later be fit.

### 2.6.2. Stratigraphy and Sedimentation

At the end of the Eocene, the last remnant of the Tethys Sea retreated from the Arabian Peninsula, possibly as a prelude to eastward movement of the Arabian plate. Abutment of the Arabian plate against the Asian plate eventually resulted in orogenic movements in Iran, climaxing in upthrusting of the Zagros Mountains about three million years ago (M. S. Garson, M. Krs, 1976). Although there is no obvious sign of orogeny as such in Arabia at this time, gentle tilting of the peninsula to the northeast is presumed to have taken place. Formations deposited in Arabia as a result of this activity were the Hadrukh, the Dam, and the Hofuf, of Early Miocene, Middle-Late Miocene, and Pliocene ages, respectively. Oligocene beds are unknown in central Arabia, but occur on the fringes of the peninsula. Present inclination of Hofuf Formation beds is on the order of one to two meters per kilometer. Sediments of at least partial equivalency to the Hofuf that were shed from foreland slopes and plains in surrounding areas constitute the Bakhtyari Formation of Iran and the Dibdibba and Bakhtyari Formations of southeast Iraq and Kuwait.

As a result of this activity and concurrent perhaps with Pliocene pluvial periods, the three great through-draining wadis of the Arabian Peninsula were incised. Wadis Ad Dawasir and As Sah'ba drained the west central shield and sedimentary area and fed the Rub' Al Khali basin, and Al Batin—Ar Rimah drained the north central shield and sedimentary area and debouched at the head of the Arabian Gulf. Fed by western watersheds, this drainage system eventually breached the Tuwayq sedimentary scarp and, re-inforced by drainage from the eastward dipping backslopes of the scarp, flooded the eastern portion of the peninsula with vast amounts of Neogene (Hadrukh, Dam, and Hofuf) sediments, derived from igneous, metamorphic, limestone, and clastic rocks. These sediments, grading from very coarse- to very fine-grained as they were spread out to the east, reach maximum thickness of 600 meters in the subsurface of the central Rub' Al Khali. The igneous and metamorphic fractions (most common, but not limited, to the Ad Dawasir system) originate from the western Shield; the limestone and quartz cobble and pebble fractions (most common to the As Sah'ba system) derive from backslope weathering of the Jurassic and Cretaceous limestone and sandstone scarps. Paleozoic and Cretaceous clastics, mainly sandstones, furnished the finer fractions of sand and silt which were deposited further out in the basin areas to the east. Simultaneously, limestone and clastic debris was shed northward into the basin from backslopes of the fringing Hadhramawt Mountains to the south. Significant amounts of material also were transported from the Oman Mountains across piedmont plains into the Rub' Al Khali from the east. This influx of alluvial sediment continued into the basin until Pleistocene aridity set in. Distribution of this alluvium in the Rub' Al Khali in its present surface form is totally unrelated to the old structural basin configuration. Topographic relief at the end of the Neogene indicates that the Rub' Al Khali ancestral structural basin was essentially levelled by the beginning of Quaternary time.

There is some indication that final breaching of the Tuwayq scarp and consequent influx of basement debris may not have occurred until the terminal phase of the Pliocene. Occurrence of basement-derived material is apparently limited to the upper part of the Hofuf Formation, and drill holes in the Rub' Al Khali show no significant amount of such material below the surface. The basement derived gravel that surfaces much of the Rub' Al Khali (desert reg) is, in fact, largely superficial (commonly one pebble layer thick). Exceptions are in the immediate areas of fan heads or along gravel trains and ridges, from one to two meters thick, that presumably represent residuals of former stream channels. A further indication of late Pliocene breaching of the scarp may be derived from the assumption that accelerated tilting of the Arabian Peninsula would have been consequent with main upthrust of the Zagros Mountains of Iran, as noted above, an event that did not take place until about three million years ago.

After cutting through the Tuwayq and subsidiary scarps toward the end of the Pliocene, base level was quickly reached and the through-draining wadis discharged their loads in extensive coalescing alluvial fans and plains. These deposits, today to a large extent blanketed by later eolian sands, are as much as 250 kilometers from fan head to outer perimeter in the case of Wadi As Sah'ba, and 325 kilometers in the case of Al Batin—Ar Rimah. The finer sand and silt fractions from this material that were spread out over the rest of the now virtually level Rub' Al Khali basin provided the source of sediment for later re-working by Pleistocene and Holocene arid wind regimes and sheet-flooding of alternating more moist periods of the Pleistocene. They provided the substratum on which deflation hollows and depressions were formed and filled with lake sediments of the more moist periods of both the Pleistocene and Holocene.

Quaternary sediments of the Rub' Al Khali are therefore basically re-worked Pliocene alluvial sediments, which in turn might be considered the unconsolidated equivalent of the approximate upper part of the Hofuf Formation. Unconsolidated surface Quaternary sediments of the Rub' Al Khali grade down into subsurface consolidated or semi-consolidated sediments of the "Neogene undifferentiated." Thus, no contact between Quaternary sediments and lower units can be readily and easily picked. The level at which consolidation begins varies considerably even locally, probably depending upon erratic fluctuations of past water tables. In some places, consolidated sediments (usually sandy limestones) are encountered a meter or so below the surface; in other places thirty meters or more below. Cementation, where present, is usually due to the presence of carbonates, and in some cases, considerable thicknesses (up to a hundred consecutive meters) of well indurated, hard drilling, sandy limestone are logged. Beds of soft, sandy claystone and sandy marl are fairly common also. Carbonate for cementation would have been derived from the fringing Jurassic and Cretaceous limestone formations to the west and from the Hadhramawt Tertiary highlands to the south. Secondary re-working of earlier (Pliocene-Pleistocene) lake bed marl and fresh water limestone deposits may likewise have contributed carbonate for cementation.

Quaternary thicknesses of the Rub' Al Khali cannot be consistently and reliably determined. However, to the extent that they can be separated from earlier sediments of the Neogene, they are comparatively thin. Exceptions are in the western border areas on weathered backslopes of the Jurassic and Cretaceous sedimentary scarps and in the fringing south where sediments were transported off the Hadhramawt arch. Minor Quaternary and possibly sub-recent local wadi development has taken place in these areas where limestone and sandstone debris of cobble and pebble size presently clog the various secondary wadis. Quaternary eolian sand accumulations that blanket most

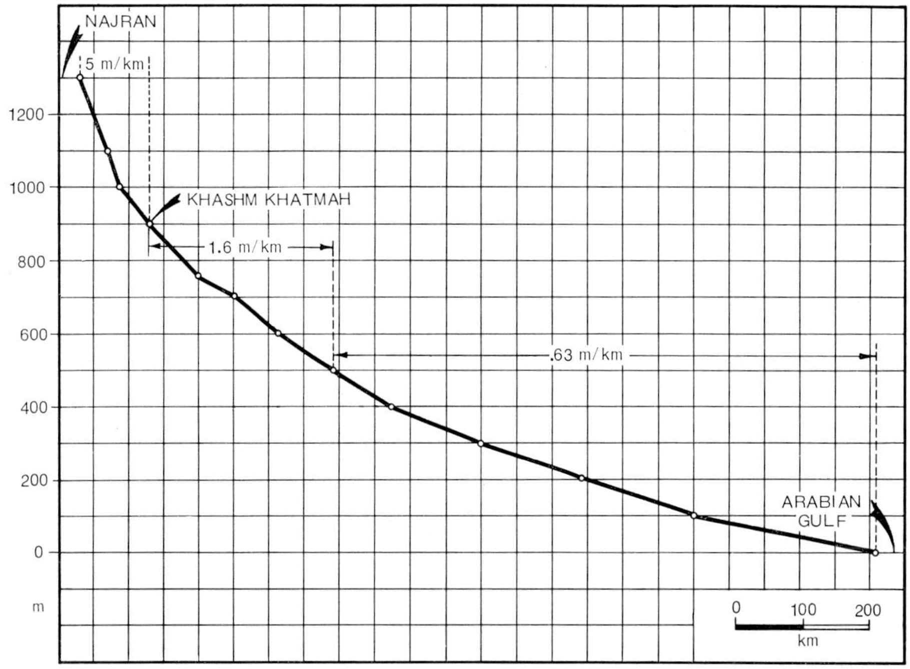


Fig. 82. Topographic profile of the Rub' Al Khali alluvial plain. Section approximately SW–NE from Najran to Arabian Gulf. Based on spot elevations in meters

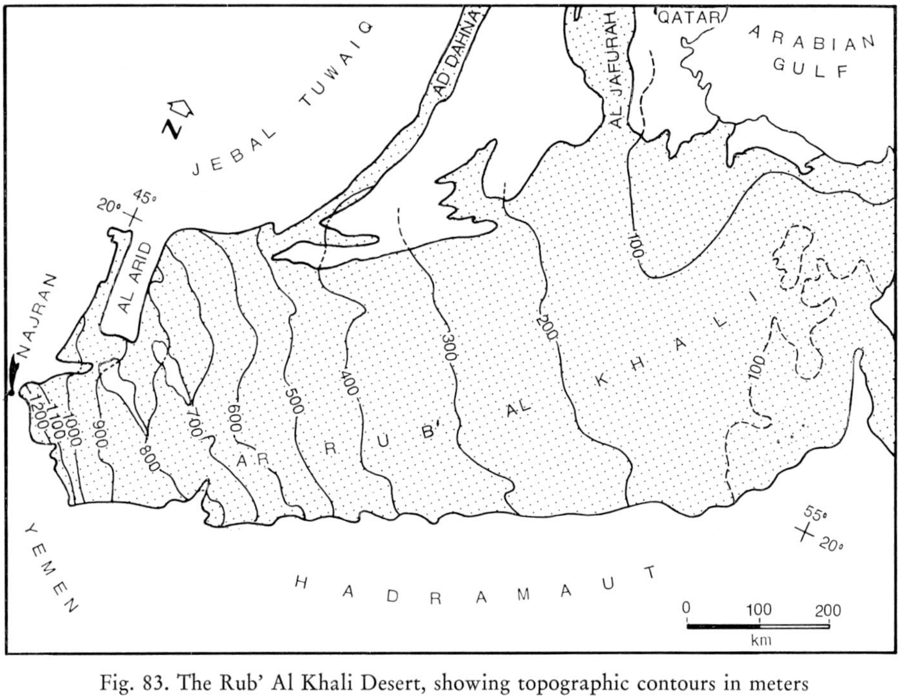


Fig. 83. The Rub' Al Khali Desert, showing topographic contours in meters

of the Rub' Al Khali range in thicknesses from a few meters in areas of sheet coverage to as much as 250 meters in areas of high dune accumulation, the maximum thickness of coverage being in the "sand mountains" of the northeast Rub' Al Khali.

### *2.6.2.1. Geomorphic processes*

Surface contours and radial profile (Figs. 82 and 83) as well as lithology and sedimentary characteristics indicate that the Rub' Al Khali still retains the basic features of an old plain of alluvium inherited from the Pliocene. Local Quaternary and sub-recent geomorphic processes have modified it slightly and marginally.

A principal geomorphic process of the Quaternary in the Rub' Al Khali has been wind erosion (deflation). This has resulted in finer fractions of the old alluvium being re-deposited into various kinds of eolian sand accumulations, such as those discussed by D. A. Holm (1960). Concentration of coarser alluvium gravels to form reg plains has also been a result of such deflation.

Aggradation since Pliocene time has been the dominant regional pluvial process, but degradation has been in operation on a secondary and more local scale. Aggradation can be demonstrated in the de-activation and in-filling of the three main through-draining wadis; degradation is still of some importance, as on the backslopes of fringing highland scarps where flash flooding even in the hyper-arid climate of today has some effect. Paleolithic flint tools are commonly associated with these backslope deposits, which are probably largely the result of Pleistocene pluvial periods. Limited sheet-flooding, as might be expected on the gently sloping topography of an old plain of alluvium, may be considered to have been more prevalent also during Pleistocene pluvials, and may have had a role in further eastward distribution of reg gravels.

The effects of sheet-flooding as observed today are much more pronounced near fringing fan head areas and piedmont slopes and where silt and gravel fractions are present on the surface in sufficient quantity to allow rainfall to be retained long enough to form floods. This does not happen in the areas of eolian sand accumulation where any present day precipitation is usually very quickly absorbed. An exception, however, that might be termed "sand-slope flooding", occurs on slopes of ordinary sand dunes during periods of rare, very high intensity rainfall. This kind of run-off results in concentration (settling out) of the finer gray silt fraction of dune material in puddles and pools at the bases of slopes and in hollows, much of "eolian" dune sand being, in fact, fine sand and silt of alluvial origin. This process is of very limited occurrence and of little importance in the hyper-arid climate of today, where such "pans" quickly desiccate, disintegrate, and blow away. The process, however, appears to have been of importance in earlier times of higher precipitation, such as in the Holocene sub-pluvial period, and may account for the formation of impervious substrata on which perched, playa, or "mud" lakes, characteristic of the Holocene in the Rub' Al Khali, developed.

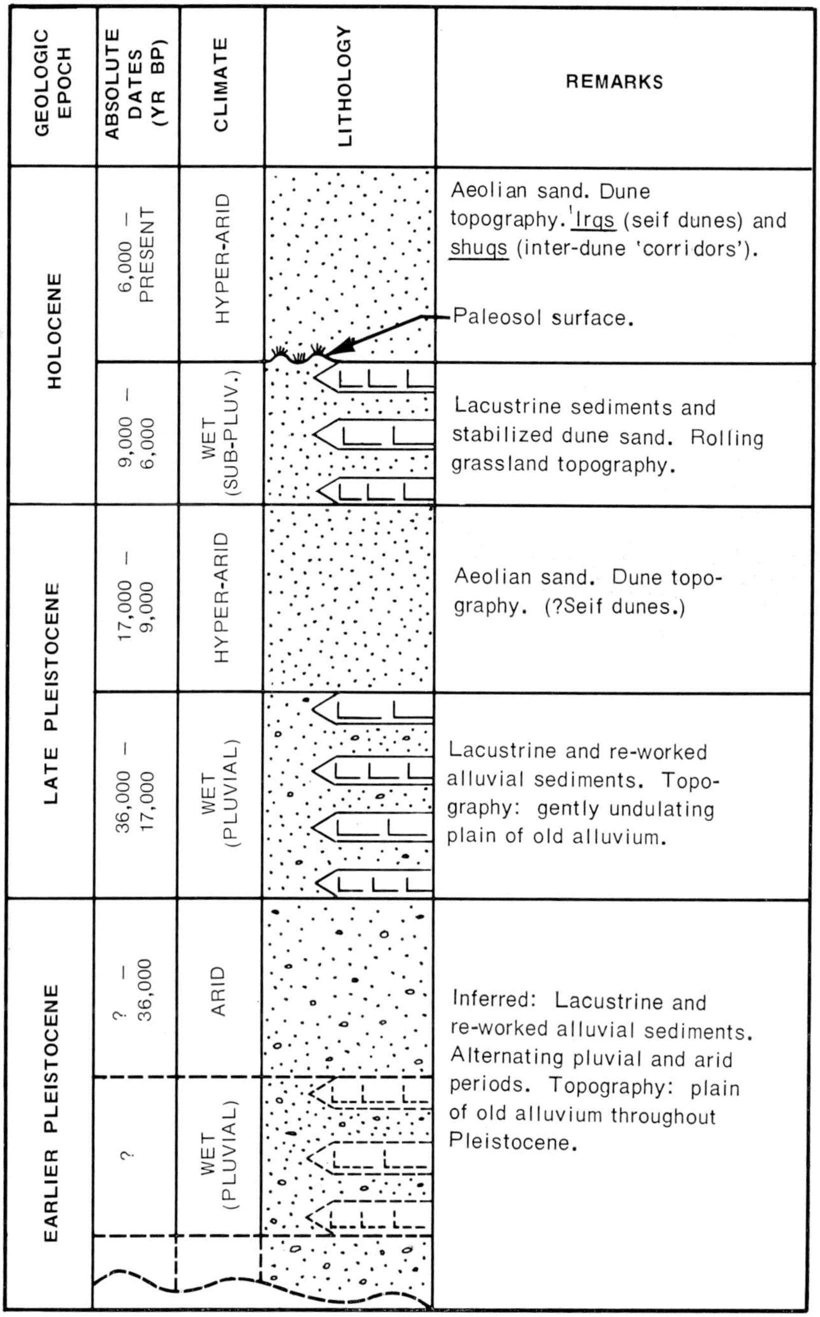


Fig. 84. Composite time-stratigraphic column of Quaternary sediments of the southwest Rub' Al Khali. (Unit thicknesses not to scale)

The present distribution of eolian sand approximately duplicates the distribution of the old Pliocene alluvium, suggesting that eolian re-working of these old sediments into present day sand sheets, dunes, ridges, and massifs of various kinds has been largely *in situ.*

### 2.6.2.2. *Late Quaternary history*

During wet phases of the Pleistocene, old Pliocene alluvium was partially cemented locally in deflation hollows and low areas and centripetal run-off

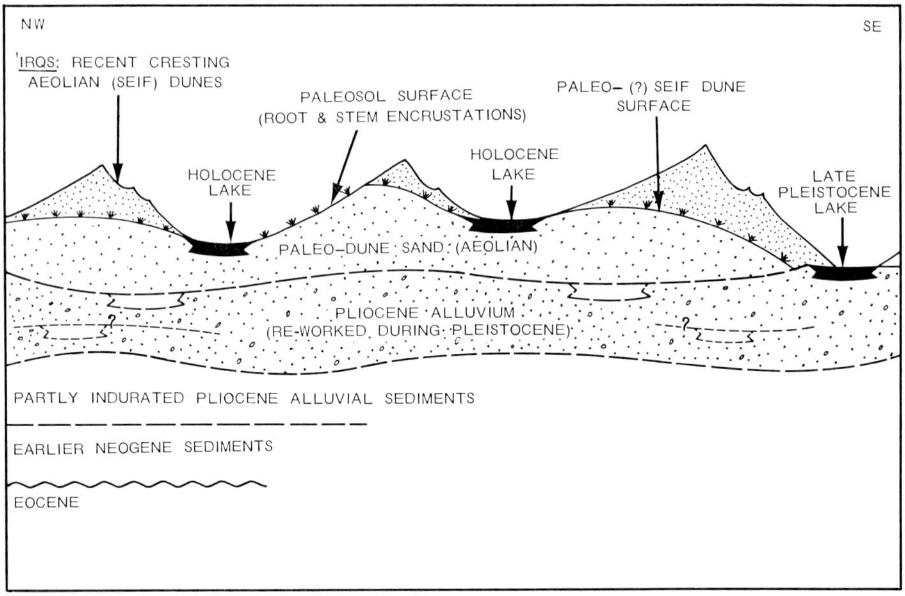


Fig. 85. Stratigraphic cross-section (schematic) of Quaternary sediments of the southwest Rub' Al Khali. (Not to scale)

filled these shallow depressions with lakes. The latest of this series of Pleistocene lakes is still preserved and has radiocarbon dates ranging from about 36,000 years B.P. to about 17,000 years B.P. (H. A. McClure, 1976). Evidence of earlier pluvial lakes has been removed by erosion on the surface, though some may have been preserved in the subsurface where gastropods, ostracodes, and *Chara* oogonia are sometimes logged in drill hole cuttings of the "Neogene" in deeper parts of the Rub' Al Khali basin.

Aridity of the present day intensity appears to have begun about 17,000 years B.P. (Fig. 84). Extensive scattered gypsum deposits of about this time attest to a general evaporation phase. The onset of desiccation may have been of an abrupt nature.

Deposition of red eolian sands apparently began after this period of hyperaridity set in. Lake beds of the 36,000 to 17,000 years B.P. period invariably rest on white or gray alluvial sands. The clear stratigraphic and paleoenviron-

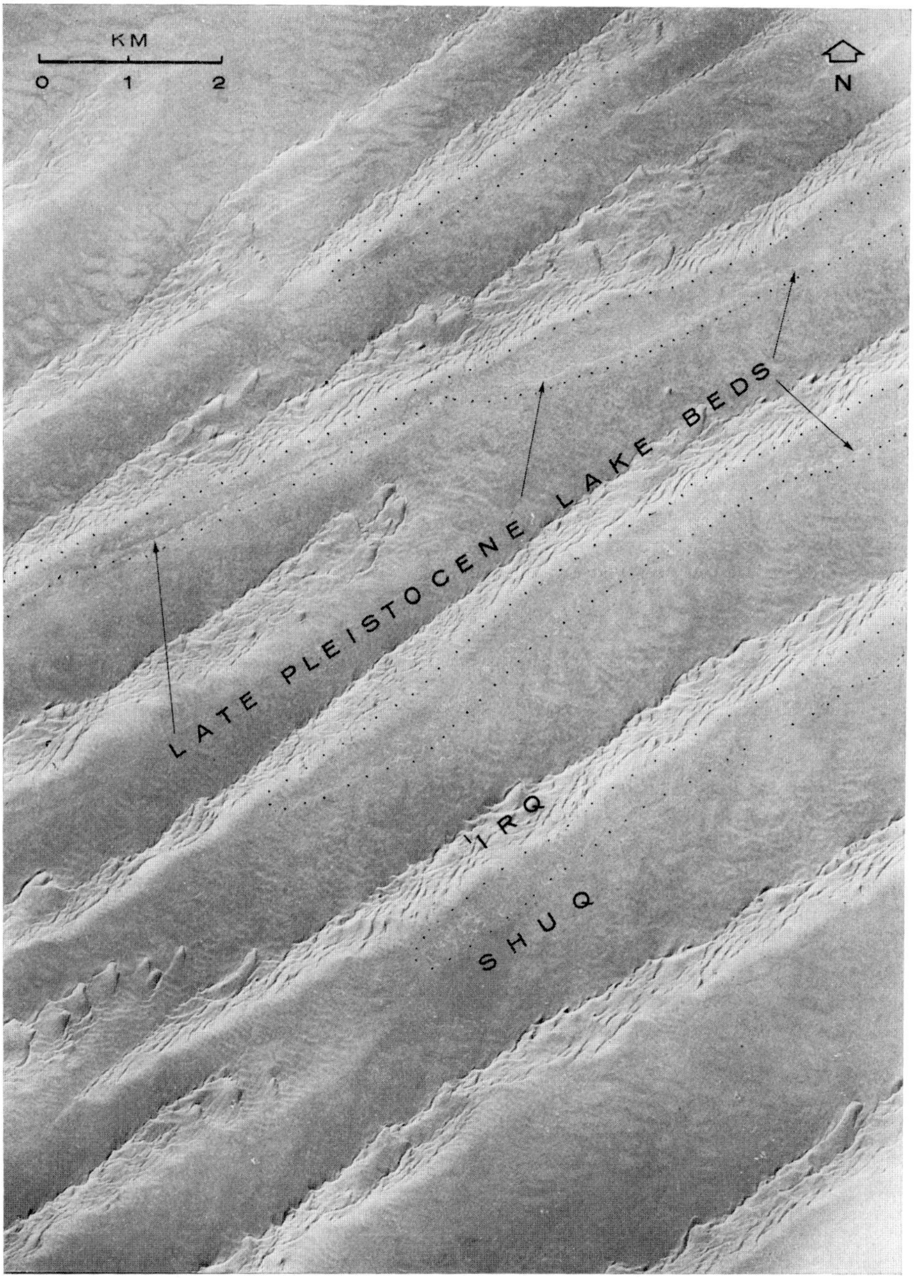


Fig. 86. Aerial photograph of part of the southwest Rub' Al Khali, showing late Pleistocene lake beds in the lees of seif dunes (*'irgs*). *Shugs* are the interdune "corridors", usually floored by reg gravels

mental context in which these red eolian sands occur may furnish a clue to the origin of present red desert sands in other areas of the world and to the origin of red beds in general. Hyper-aridity apparently plays a definite role, but it is notable that, even in the hyper-arid climate of today, some moisture is generally always present in sand dunes only a few feet below the surface.

A sub-pluvial phase with radiocarbon dates from about 9,000 to 6,000 years B.P., sandwiched between periods of red eolian sand deposition, then occurred. Weak pedogenesis, in the form of poorly indurated, calcareous or gypsiferous cemented red sand, took place on what appears to have been an old dune surface, possibly in the southwest Rub' Al Khali a seif (longitudinal) dune system similar to that of today. Some of the present more stable *zibar* (whaleback) sand surfaces may reflect this older eolian sand system also. Lineations of the present and presumed older dune systems appear markedly similar, and root and stem encrustations associated with the paleosol represent an old vegetation surface that has Holocene radiocarbon dates. If an *'irq* [1] topography with directional lineation similar to that of today was in existence prior to the Holocene sub-pluvial phase, a hyper-arid climate with a wind system similar to that of today might be assumed to have begun about 17,000 years B.P.

### *2.6.2.3. Late Pleistocene lakes*

Late Pleistocene lakes of the Rub' Al Khali (Figs. 85 and 86), though more substantial than those of the Holocene, which were definitely of playa type, were nevertheless warm water lakes of a shallow nature and relative short duration. Photosynthesis and $CaCO_3$ precipitation (productivity) were high in these lakes, and algal growth was common. Lithology is usually soft, gray, very calcareous marl, but occasionally sand facies occur. High lake levels seem not to have been sustained for very long at any one time. Thickness of individual beds seldom exceeds a meter, and at any one site more than two separate benches (presumably representing stand-stills) are rarely present. In geomorphic context, the beds most frequently occur as low terraces on the lee sides of dunes, where more recent deflation has been locally deepest. They are easily eroded by present winds, and local slumping due to the presence of an unconsolidated (alluvium) sand substratum is a common feature. Duricrust on the order of 5 to 10 centimeters thickness often caps the beds.

Ostracodes were common in these lakes (H. A. McClure, F. M. Swain, 1974) and fresh water molluscs such as *Melanoides tuberculata, Planorbis, Lymnaea, Unio,* and *Corbicula* frequent. Occasionally beds contain brackish water, hard-shelled foraminifera, indicating that alkali conditions set in as periodic desiccation took place in the terminal phases of some of the lakes. *Unio* and *Corbicula* probably represent the strictly fresh water end of a salinity continuum and the foraminifera (normally marine) the high salinity, desiccation end of the continuum. *Unio* is most often associated with sand

[1] Seif or longitudinal dunes in the Arabian Desert are locally called *'irqs* (Arabic singular *'irq*, plural *'uruq*). See D. A. Holm, 1960.

facies of the lakes. Ostracodes, *Chara,* and gastropods are common in the marl facies, presumably representing the middle phase.

Significantly, no fish remains occur in the sediments. The mammal fauna of the time indicates a savannah-grassland, arid to semi-arid environment, with bovid populations (*Bos primigenius* and *Bubalus*) at the drier extreme and *Hippopotamus* at the other comparatively more moist extreme (D. A. HOLM, 1960; H. A. McCLURE, 1974).

Fig. 87. Close-up photograph of a Holocene lake bed in the Mundafan basin area. Lithology is soft gray marl, containing gastropods, ostracodes, diatoms, and bovid bones

Spore-pollen content of the lake beds is meager, consisting largely of a few algal spores and grass pollen, a condition possibly reflecting basic Pleistocene aridity.

### *2.6.2.4. Holocene lakes*

Holocene lakes were largely playa type, though in at least one locality, the Mundafan basin on the southwest edge of the Rub' Al Khali (H. A. McCLURE, 1976), duration of standing water may have been longer than usual. Lithology is occasionally marl, like most of the late Pleistocene beds, but is most commonly gray "mud" (fine sand and silt), compact and slightly to very calcareous. This material is derived from settling out of the gray, fine sand and silt fraction of eolian sands during the puddling phenomenon ("sand-slope flooding") of dune run-off described above. Thickness of beds is usually on the order of less than a meter per bed, and more than several separate beds are rarely present at any one site. The beds often occur perched or "hanging" at different local elevations in sand hollows, some very high up, in the present seif dune system of the southwest Rub' Al Khali. They also commonly occupy the lee sides of dunes, possibly representing deposition in paleo-*shugs* (inter-dune "corridors"). Others appear to have been deposited in the hollows of rolling *zibar* dunes. The substratum on which these lake beds rest is invariably red eolian sand. The only known site where these beds occur with Pleistocene beds is in the Mundafan basin area. Beds are occasionally ashy or charcoaly and flint tools of Neolithic type are commonly associated, indicating contemporary presence of human habitation in the vicinity. Duricrust and such features as desiccation cracks and ripple marks occasionally occur.

Many of the beds are totally unfossiliferous, but others contain a molluscan and ostracode fauna similar to that of the Pleistocene, and a few beds, such as several of marl facies in the Mundafan, contain diatoms, sponge spicules, phytoliths, and algae. Spore-pollen content, as in the case of the Pleistocene lake beds, is extremely meager. Vertebrate bones, mostly bovid, are frequent.

Fig. 87 is a close-up view of one of the Holocene lake beds in the Mundafan basin area.

### 2.6.3. Paleoclimate

A general clue to the climate of the Quaternary as a whole is that little apparently happened to modify to a significant extent the old plain of alluvium inherited from the Pliocene. Since red eolian sand, presumably a product of hyper-aridity, does not apparently occur until after 17,000 years B.P., hyper-arid conditions are not indicated for any previous part of the Pleistocene. On the other hand, intense precipitation, had it occurred, would surely have reactivated or rejuvenated the old Pliocene through-draining wadi system; that apparently did not happen.

If the local late Pleistocene climate regime as deduced from lake bed evidence is a clue to earlier climates as well, the Pleistocene of the Rub' Al Khali as a whole can be considered a period of aridity to semi-aridity. Pluvials occurred at the same time as periods of maximum ice coverage in the rest of the world. Dry periods, but *not* hyper-aridity, probably alternated, occurring during interglacial sessions. Hyper-aridity then set in about 17,000 years B.P., concurrent perhaps with the approximate end of maximum ice coverage in the rest of the world, when the basic global circulation regime as presently known was possibly established. However, a northerly displacement of the southwest monsoon in the Holocene resulted in a sub-pluvial phase in the Rub' Al Khali from about 9,000 to about 6,000 years B.P. Retreat of this wind regime at the end of this period to its approximate present location, where it sweeps only across the southern tip of the Arabian Peninsula, then resulted in a return to hyper-aridity which persists today.

A further valuable clue to climate, as well as to age of Quaternary sediments of the Rub' Al Khali, might be inferred from the occurrence of Holocene lake beds in the middle of Wadi Ad Dawasir. These sites have flint tools of Neolithic type associated and radiocarbon dates of about 6,500 and 9,000 years B.P. Indication is thus that the main wadi had ceased to flow and had long been aggraded by this time. Present wadi banks are only one to two meters in elevation above the center. Assuming precipitation and runoff sufficient to activate tributary discharge and to aggrade the main wadi, it might be assumed that the lake period of the late Pleistocene was one of wadi aggradation and lack of through-flow. Further assuming that this pluvial was typical and that previous Pleistocene pluvials were not of a significantly greater magnitude in precipitation, with respect to the major wadi system, aggradation might be assumed to have been the dominant process throughout the Pleistocene. A period of rainfall considerably greater in magnitude than any of the Pleistocene would seem necessary to have carved out the three main through-draining wadi regimes of the Arabian Peninsula, though tilting of base level due to tectonic movement may have contributed some effect. Such precipitation probably occurred in Pliocene time, perhaps during the terminal phase of Hofuf Formation deposition.

Another indirect clue to climate and also age of the major wadi system is that flint tools, especially of hand-axe type, are noticeably lacking in association with the three main through-draining wadis, though they do occur in other geomorphic contexts in Arabia. It seems probable that the main wadi system had been carved out, then ceased to flow, and was in the process of infill and aggradation by secondary tributary drainage by the time man came onto the scene. Again Pliocene seems a reasonable age assignment for major wadi flow. Pliocene mastodon remains in Hofuf equivalent sediments have been reported by K. W. GLENNIE (1968) at Jabal Barakah on the coast of Abu Dhabi, and mastodon and rhinoceros remains have been radiometrically dated to about 10 million years in Iran (Tehran Journal, Feb. 12, 1977). Mastodon, rhinoceros, pig, and crocodile remains from the Dam Formation of central Arabia are certainly indicative of a lush (perhaps tropical savannah) climate at that time. As the Neogene forms a continuous sequence of lithologically and genetically similar sediments with no indication of time hiatus or significant stratigraphic break, a climatic continuum, as well as a faunal continuum, may be assumed. Pliocene rainfall, possibly enhanced by base level tilting, was probably more than adequate to have supplied the lush environment the above mentioned vertebrate fauna required and at the same time to have carved out the three great through-draining wadi systems of Arabia.

# 3. Summary and General Conclusions

## 3.1. Quaternary Sediments

(H. Hötzl, F. Krämer, V. Maurin)

### 3.1.1. Scope of Investigations

The focal points of the research work dedicated to the geology of the Quaternary in the area of the east Arabian Shelf Plateau were to investigate sedimentation and to determine conditions of erosion and deposition during the most recent geological past. Essential information on the geological processes and climatic conditions prevailing at that time was expected to be derived mainly from these investigations.

As is now known, the composition of sediments, the mineral content, form, size and sorting of clastic components, the kind of accumulation, and the characteristic texture are diagnostic factors for both accumulation area and accumulation conditions. Particularly in terrestrial deposits, the causal connection between their genesis and form, and between the causes of their formation, the kind of rock weathering, the transportation of weathered material by water and wind, and their accumulation becomes very clear. Investigations of the Quaternary strata complex of Arabia should therefore, provide not only information about the geological morphogenesis but also indications as to climatic changes in that area during the most recent geological past.

The objectives of our sedimentological and geological investigations were not only the making of a survey and mapping of various Quaternary layers, but also to gain insight into the processes during the whole Quaternary period by means of selected profiles, their chronological classification and correlation. As a starting point, we used the geological maps of the Arabian peninsula at the scale of 1 : 500,000 ("Miscellaneous Geologic Investigations, maps I-201 A to I-220 A") issued by the U.S. Geological Survey and the Kingdom of Saudi Arabia, Ministry of Finance and National Economy. These maps present the Quaternary sediments according to lithologic and genetic points of view—as far as this is possible with such a scale; however, a chronological subdivision of Quaternary rocks has not yet been made.

Our investigations centered on young marine coastal areas, sediments of lakes without outlets, young wadi accumulations, and soil and sinter formations in caves. The large wadi systems, too, attracted our special attention. An alternation of erosion and deposition phases shows their most immediate effects in such channels of surface runoff; consequently, indications for a chronological classification of activities during the Quaternary were expected to be found in these places. During the field survey, emphasis was laid upon sedimentological investigations. As far as possible, it was attempted to explain the genesis of sediments and their required conditions of formation. Samples were collected in order to determine the mineral components as well as the grain size distribution of clastic sediments and of various soil formations. Laboratory investigations focussed on analyses of grain size frequency, the determination of the carbonate, gypsum and salt content as well as mineralogic, granulometric, and morphometric examinations under the microscope. In a few selected samples, especially in clay minerals, roentgenographic investigations were conducted to determine the association of minerals [1].

The determination of the grain size frequency was mostly done by a combined sieve and wash analysis, using the ATTERBERG grain size specification. Samples with grain sizes of more than 0.063 millimeters were sieved dry; grains of less than 0.063 millimeters were sieved wet and then wash analyzed by the Aerometer method. A partial pretreatment of samples was carried out in order to remove salts soluble in water and to prepare carbonate-cemented clastic sediments, if necessary. Precise details as to the analysis of the grain size frequency and investigations of sediment material may be taken from adequate manuals (e.g., G. MÜLLER, 1964).

The determination of the carbonate content was carried out by acid treatment; a low concentration of hydrochloric acid was used to forestall any harm to the clay material. The microscopic examinations served to determine the minerals belonging to the grain mixture, the grain roundness and the forms of the impact scars. From the impact scars, in particular their matting and the existence of cut structures, additional information on the fluviatile and eolian transport could be expected. Thin sections were made from cemented sediments, mostly from formations of calcareous crust. Microscopic analysis was centered on mineralogic, granulometric, and morphometric examinations.

## 3.1.2. Sedimentation in Wadis

### *3.1.2.1. General conditions of accumulation*

The sedimentation process in a wadi is the result of special morphodynamics in arid areas for which physical weathering processes, the geological activity of the wind, and sporadic runoffs from intensive episodic rainfalls are of primary importance. The suddenly rising wadi floods set huge masses of sediments into motion, entailing mushy loads of detritus and mud. As the surface water runoff infiltrates into the dry subsurface the floods rapidly lose their transport power and the suspended detritus is deposited either in the wadi channel itself or in the anterior shallow pan.

The deflation of fine material as well as the great bulk of eolian sand accumulations finally shape the highly varied complex of alluvial and eolian sediments in wadi fills. General summaries of these processes together with descriptions of the typical forms of sediments have been made by G. SOLLE (1966), K. W. GLENNIE (1970), and H. E. REINECK, I. B. SINGH (1973).

The following discussion focusses mainly on the big wadis in the central part of the Arabian peninsula. Investigations were conducted into the young sediments and their conditions of sedimentation in the upper sections of Wadi Ar Rimah, Wadi Hanifah, Wadi Al Luhy, Wadi Birk, and Wadi Ad Dawasir, and in the lower parts of Wadi Bishah and Wadi Ranyah. With regard to the recent erosion and accumulation activities in wadis, the large valley formations that originated from vast river systems during humid periods hold a

[1] We owe particular gratitude to Professor Dr. SMYKATZ-KLOSS of the Institute of Mineralogy of the University of Karlsruhe for conducting the roentgeno-fluorescence analyses.

special significance. Their original function as west-to-east river systems, by which large basins and depressions were filled with material, has completely changed. Under recent arid conditions, accumulation areas are confined to the wadi channels themselves, since only a very small quantity of episodic precipitation falls. The channels have become the actual area of sedimentation of erosional material transported by sporadic runoffs.

In the basins situated immediately in front of the big wadis as well as the intercalated depression (reached by runoffs from the wadis in very rare cases only), deflation and corrasion or else accumulations of drifting sand are dominant. This becomes even more apparent in wide hollows between the sporadic cuesta escarpments recently followed by vast stretches of sand deserts such as Nefud As Sirr, Nefud Ath Thuwayrat and Ad Dahna. They partly block the course of episodic runoffs coming from the wadis and cause the formation of lakes without outlets in the wadi channels. The continuous surface runoff may additionally be interrupted by dune areas in these wadis or by lateral alluvial fans. In a few cases, flat valley watersheds (e.g., in Wadi Birk) have developed in wadi channels.

Today the big wadi channels are completely dominated by accumulation processes, whereby the constant interaction of eolian and episodically fluviatile activities have led to close co-existence of entirely different sediments. Wells and well-profiles in the investigated wadis reveal that the above-mentioned conditions of accumulation must have prevailed during almost the whole period of young sedimentation in these channels. Other interim phases with better runoff conditions are expressed by intercalated pebble and gravel layers in the profiles. In back water areas smaller fresh water lakes (e.g., in Wadi Al Luhy) were sometimes formed.

In contrast to the broad valleys, the shorter wadis with their limited recharge areas in their upper parts furnish examples of very active headward erosion working even today. It is only in their lower parts and in the alluvial fans that the typical sedimentation of fanglomerates can be found. The active headward erosion can even be observed at rather low altitudes of just a few meters. However, the erosional extent is dependent on the local relief. Examples of this phenomenon can be cited from the entire central and eastern Arabian area. The narrow "gullies" at the eastern border of As Summan plateau should be mentioned, as well as the deep-cut wadis in the area of the Tuwayq Mountains and the short crevice-like channels in the recharge area of the lower Wadi Bishah, mainly consisting of volcanic or metamorphic rocks.

### *3.1.2.2. Description of sediments*

#### *Fanglomeratic detritus*

Highly characteristic of the sporadic runoffs in arid regions are accumulations comprising mostly unsorted materials of angular or subrounded shape. Depending on the original rock and on runoff conditions, they are more or less strongly intercalated with clayey-silty material, which together with sinter precipitations may rapidly contribute to a certain degree of cementation of these sediments. The thickness of the various layers is generally 0.3 to 2 meters; frequently, repeated detritus layers lie on top of one another in smallish wadis. Both grain size parameters and grain size distributions (samples 17, 18, 19, 28) illustrated in Table 48 and Fig. 88 show their poor sorting and skewness. The samples investigated comprised relatively fine-grained detrital sediments; the median values may reach a size of a few centimeters in accumulations near denudation areas.

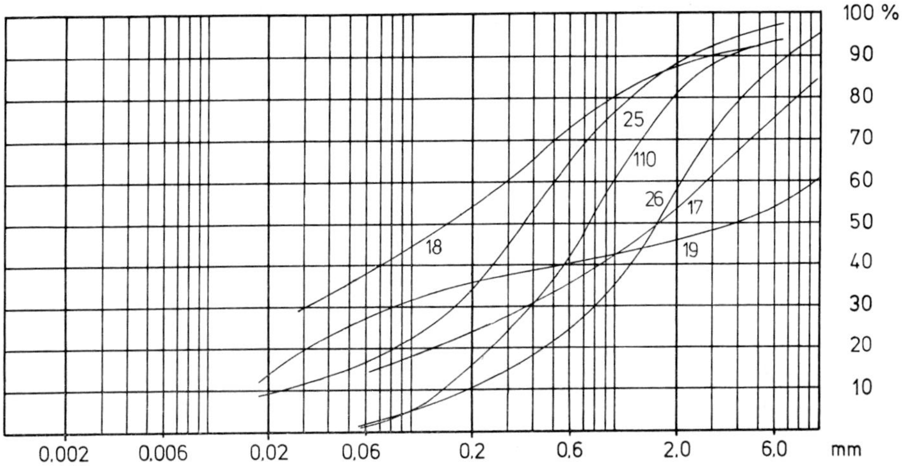


Fig. 88. Cumulative grain size distributions of wadi fan material and wadi gravel. (*17* terrace Wadi Sha'bah, tributary of Wadi Ar Rimah; *18, 19* terrace at Aqlat As Suqur, Wadi Ar Rimah; *25* alluvial fan, surroundings of Khafs Banban; *26* terrace 20 kilometers W of Ar Rass, Wadi Ar Rimah; *110* alluvial fan, Wadi Ishghab tributary of Wadi Ad Dawasir

Corresponding to the flat valley cross-sections and to minor relief differences, the amount of these fanglomeratic detritus sediments rapidly decreases in the accumulations of the big wadis. They form marginal talus fans interlayered with fine-clastic eolian and fluviatile sediments towards midwadi. However, the coarse-clastic sediments become increasingly dominant in small and narrow tributaries (Fig. 89). If these tributaries follow a longer course, the detrital sediments may be subject to multiple redeposition, thus undergoing various processes of sorting and rounding. In the lower part of Sha'ib [1] 'Uthaylan, cut into the cuesta backslope of the Jurassic strata complex, are found gravels as well as detritus which are poured into Wadi Birk in the form of a flat alluvial fan even today. In Wadi Al Hisy, another channel discharging from the Jurassic cuesta escarpment 30 kilometers south of Wadi Ad Dawasir, alluvial material up to a size of 20 centimeters, beside sands and gravels, is transported to the nearby accumulation plain by episodic runoffs.

*Drifting sands*

Eolian accumulations contributed a considerable share to filling up big wadi channels. Horizontally laminated sand layers and crossbedded sand-

[1] Sha'ib is the Arabic word for a small wadi.

banks with a pronounced foreset bedding as well as massive dune complexes reaching up to several meters can be observed. The primary structures mostly have been destroyed by plant roots or else by minor fluviatile redeposition. Frequently, the sands of horizontal laminations have a thickness of a few centimeters; sporadically drifting gravel grains have become embedded in these sands. Likewise, such gravel components occur in sandbanks which are subsequently enriched in the upper part of the foreset bedding. Their accumu-

Fig. 89. Coarse detritus of an alluvial fan in a tributary of the lower part of Wadi Bishah. (Photo: H. Hötzl, 1975)

lation was effected in the form of flat eolian sand covers, flat dunes and drifting sand hills, depending on vegetation (Fig. 90).

While these flat sediments primarily originate from local deflation in the wadi area itself, the vast dune complexes first of all arise from being blown in from outside as so-called "sand drifts" (cf., H. E. Reineck, I. B. Singh, 1973). The position and arrangement of the vast dunes in some favoured wind directions indicate the sand drift process, whereby the sand accumulates on the immediate lee side of the wadi channel and/or on the windward side of the opposite bank. Impressive examples are to be found in the lower part of Wadi Bishah, in Wadi Birk and Wadi Al Hawtah, where not only the described deposition can be seen together with their corresponding dune fields, but also the portion of drifting sands in the young wadi filling.

Of a quite different nature are the vast eolian sand accumulations in Wadi Ad Dawasir west of the Tuwayq Mountains. Narrow, elongated sand dunes have formed in the remarkably wide accumulation of this wadi, the east-west trending axis of which lies parallel to the prevailing wind directions. These wadi accumulations are still active, and masses of eolian material are transported along the longitudinal dunes.

Both texture and structure of eolian sands are more or less homogeneous. The common grain size distribution becomes recognizable through the good sorting, as shown in samples No. 31 and 111, Table 48, Fig. 91. The median values range between 0.2 and 0.5 millimeter. Moreover, silty to fine-

Fig. 90. Recent dunes in Wadi Birk 10 kilometers downstream from the village of Al Hawtah; the dunes are partly consolidated by vegetation, partly eroded by sporadic floods. (Photo: V. MAURIN, 1975)

sandy, loess-like dust deposits, which will be discussed in the section on wadi sabkhah sediments later on, were also found. Local conditions have left a strong impact on the mineral composition of drifting sands in wadis; in this area calcite amounts to approximately 40 percent; calcite may be almost entirely lacking in dunes of the actual sand desert.

*Fluviatile sands and gravels*

Fluviatile sand accumulations are now found in the narrow sections of the big wadis where they have settled in the low water wadi floors of pronounced runoff channels. The eastern part of the breaching of Wadi Ad Dawasir through the Tuwayq Mountains, Wadi Birk in the area of the oasis of Al Hawtah, and Wadi Hanifah near Al Ha'ir south of Ar Riyadh should be cited as characteristic examples. Typically these sandy stretches occur immediately downstream of deep-cut, low water channels and local terraces as well as around the mouths of tributaries that obviously carry larger runoff quantities after sporadic rainfalls. In these cases, material shifted from the wadi itself is primarily involved. The wadis are eroded sideward and downward due to floods; the eroded sediment is transported farther on and in most cases redeposited after short distances.

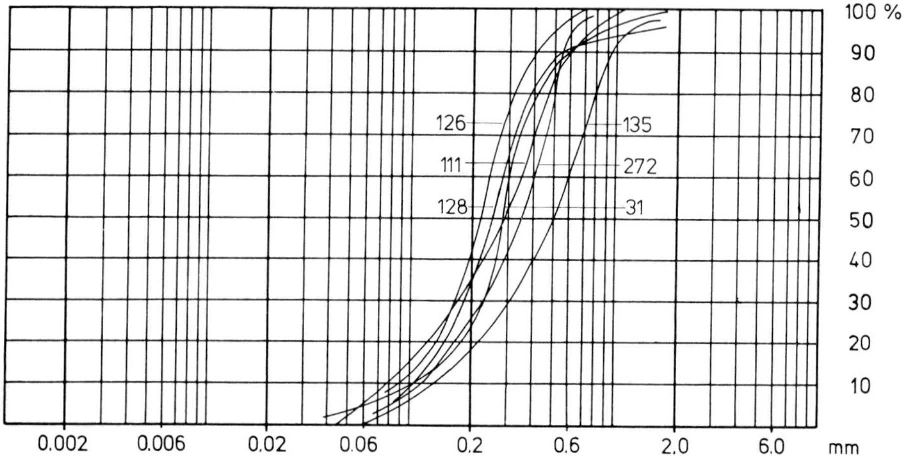


Fig. 91. Cumulative grain size distributions of eolian and fluviatile wadi sands. (*31* eolian sand accumulation at Al Ha'ir, Wadi Hanifah; *111* dune sand 25 kilometers W of Al Khamasin, Wadi Ad Dawasir; *126* fluviatile transported dune sand, recent, Wadi Birk; *128* fluviatile sand accumulation, Wadi Birk; *135* dune sand, sandpit west of Al Hofuf; *272* dune sand, Ad Dahna)

These sands are well sorted. The median sizes generally range between 0.1 and 0.6 millimeter. The strong agreement of these characteristic values with those of eolian sands, their similar degree of rounding and the considerable amount of frosted grains document the high portion of redeposited drifting sand. Such transport products are clearly visible in samples 126 and 128 (Table 48). The erosion of fine portions in the course of fluviatile transport becomes noticeable in the lower coefficient of uniformity as against eolian sands (2.2. as against 4 to 5).

In the profile sequence fluviatile and eolian sediments are arranged in a close rhythmic bedding; their structures and textures, however, are different. Remarkable changes in grain size can be observed in the various layers of fluviatile sands, whereby normal grading often occurs. In a few cases, small gravel layers may be interbedded. Numerous refilled pot-holes and channels are detectable at the base. The internal structures of the various banks display either even or arched cross-bedding, foreset beds partly with grain size grading, festoon-like climbing ripples and convolute structures. Horizontally laminated fine sands are relatively frequent and hardly to be distinguished from the homologous eolian sediments unless they are accompanied by coarser gravel layers (Fig. 92).

Fig. 92. Horizontally laminated fine- to medium-grained fluviatile sand covered by cross-bedded coarse sand with gravel lenses. Lower strata of the terrace near the village of Al Hawtah, Wadi Birk. (Photo: H. Hötzl, 1975)

Fig. 93. Clayey silty dikaka sediments characterized by a dense network of root tubes filled with gypsum. Terrace in the upper part of Wadi Hawtah. (Photo: H. Hötzl, 1975)

According to E. D. McKee et al. (1967), such textures are typical of accumulations resulting from suddenly rising, rapidly flowing floods in arid and semi-arid regions, where only fine-grained erosional detritus is available in their recharge areas. The horizontally laminated sediments are deposited in the main stream area where the highest flow velocities are reached. Cross-bedding forms marginally, and ripple marks develop in more protected areas.

Both gravels and pebbles are almost entirely absent in recent sedimentary processes operating in the big wadis. When they do occur, they are dependent on the corresponding depositions from wadi tributaries such as the

Fig. 94 *a*. Shallow eolian longitudinal ripples on the wadi floor near the village of Muqabil, Wadi Ranyah. (Photo: V. Maurin, 1975)
Fig. 94 *b*. Cuspidate ripple marks developed by an episodic flood March 1974 near the village of Al Ha'ir, Wadi Hanifah. (Photo: V. Maurin, 1974)

mouth of Sha'ib 'Uthaylan into Wadi Birk or else local transport of smaller gravel banks. Coarse-clastic accumulations occuring at the base of wadi fills and in a few sporadic interbedded strata reaching a thickness of several meters provide indications as to intermittently changing conditions of sedimentation.

*Wadi sabkhah sediments*

The term "wadi sabkhah sediments" is introduced as a comprehensive term for fine-sandy, silty, clayey, and evaporitic sedimentological sequences that are accumulated in pan-like wadi sections owing to the interactions of eolian and episodically fluviatile processes. The following sediments are covered by this term:

1. Salt and gypsum layers and crusts of small thickness; they originate from evaporation of runoffs collected in various wadi sections or from water rising from the subsoil via capillarity or via the so-called "evaporation pumping process" (H. J. Hsü, J. Schneider, 1973).

2. Salt and/or gypsum clays, in which the evaporitic components are lenticularly enriched or may be embedded as regularly distributed veins and strings or even be present in the form of fine crystals. They come into existence on account of muddy deposits saturated with water of inactive runoffs, whereby gradual evaporation causes components of an evaporitic nature to crystallize.

Episodic rainfalls also contribute their share to the above mentioned process. Usually, a high proportion of silts as well as sandy components that were brought in by the wind and bound by adhesion occur in them.

3. Fine-sandy, clayey, in some instances mainly marly, silts of a different genesis. Beside loess-like sediments according to T. J. SMALLEY, C. VITA-FINZI (1968) and normal eolian dust accumulations, deposits from drifted suspensions of episodic runoffs in calm and slowly flowing waters have to be mentioned in particular; in a few cases, limnic deposits occur also.

4. Fine sands that have experienced manifold fluviatile and eolian transport processes.

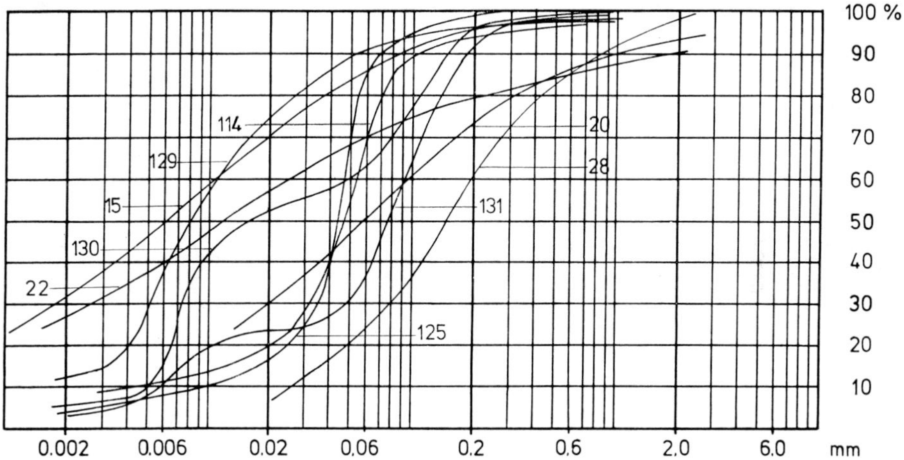


Fig. 95. Cumulative grain size distributions of wadi sabkhah sediments. (*15* recent saltclay, Wadi Sha'bah; *20* stillwater silt, "snail-horizon", terrace profile Wadi Al Luhy; *22* grey stillwater silt, terrace profile Wadi Al Luhy; *28* silty fine sand, terrace profile Wadi Hawtah; *114* silt with high percentage of salt and gypsum, endorheic basin, Wadi Ad Dawasir; *125* silt and fine sand, recent alluvial accumulation, Wadi Birk; *129* silt with gypsum, terrace profile, endorheic basin sediment, Wadi Birk; *130* sandy silt, fluviatile, with ripple marks, terrace profile, Wadi Birk; *131* silt and fine sand, terrace profile, Wadi Birk)

5. Dikakah sediments (according to K. W. GLENNIE, B. D. EVAMY, 1968); these sediments are characterized by a dense, network-like arrangement of fossil roots the existence of which may be ascribed to fields rankly spread with herbs and bushes growing on the silty floor of a wadi that retains humidity for some time. The root tubes are sometimes stuffed with fine sand; most frequently, they are filled with fine and coarse crystalline gypsum (Fig. 93).

The internal structure of these sediments varies in accordance to their various geneses. Generally, an indistinct, shallow-waved stratification or lamination prevails which may turn into a entirely different, structureless body. Of particular interest are adhesion ripples, salt and clay polygons, sand dikes, mud cracks with sand filling, thin sporadic sediment cycles, clay films, broken clay films with flat and curled clay flakes, and shallow ripple marks (Fig. 94 *a*, *b*).

Table 48. *Grain size and grain morphological parameters of Quaternary sediments from central and eastern Saudi Arabia* (degree of roundness after G. Müller, 1964: 1 = very angular, 2 = angular, 3 = subangular, 4 = subrounded, 5 = rounded, 6 = well rounded)

| No. of sample | Description of samples | Median (Md) in mm | Sorting $So = \sqrt{Q_3/Q_1}$ | Skewness $Sk = Q_3 \cdot Q_1/Md^2$ | Coefficient of uniformity U = P 60/P 10 | Ratio of frosted and polished grains | Roundness of frosted grains | Roundness of polished grains |
|---|---|---|---|---|---|---|---|---|
| | *Fanglomerates:* | | | | | | | |
| 19 | Fanglomeratic gravels of the accumulation terrace near Aqlat As Suqur, Wadi Ar Rimah | 5 | 21.0 | 0.04 | 540 | 2 | 3 | 3 |
| 18 | Terra rossa like weathering products of fanglomeratic gravels (location 20–30 cm below the desert pavement), terrace near Aqlat As Suqur, Wadi Ar Rimah | 0.18 | 5.4 | 0.37 | 58 | 0.5 | 5 | 3 |
| 17 | Terra rossa like weathering products of fanglomeratic gravels (location 20–30 cm below the desert pavement), terrace Wadi Sha'bah | 1.6 | 5.2 | 0.43 | 87 | 0.5 | 2 | 2—3 |
| 26 | Fanglomeratic coarse sand mainly granite detritus, 20 km west of Ar Rass, Wadi Ar Rimah | 1.6 | 2.7 | 1.05 | 10 | 1 | 5 | 2 |
| | *Eolian sand:* | | | | | | | |
| 31 | Eolian sand near Al Ha'ir, Wadi Hanifah | 0.35 | 1.5 | 0.73 | 4 | 1 | 4—5 | 3—4 |
| 111 | Dune sand, 25 km west of Al Khamasin, Wadi Ad Dawasir | 0.29 | 1.8 | 0.8 | 4.4 | | | |
| 135 | Dune sand, "sand pit" west of Al Hofuf | 0.45 | 1.7 | 0.83 | 4.8 | | | |
| 272 | Dune sand, Ad Dahna, about 100 km NE of Ar Riyadh | 0.32 | 1.3 | 0.67 | 4.5 | | | |
| | *Fluviatile sand:* | | | | | | | |
| 126 | Fluviatile transported dune sand in the present runoff channel, Wadi Birk, 15 km north of Al Hawtah | 0.22 | 1.4 | 0.93 | 2.3 | | | |
| 128 | Sand layer, 926—941 cm below terrace surface, Al Hawtah, Wadi Birk | 0.25 | 1.6 | 1.0 | 2.15 | | | |

| | | | | | | | | |
|---|---|---|---|---|---|---|---|---|
| | *Wadi sabkhah sediments:* | | | | | | | |
| 15 | Salty clay, present sedimentation in a lake without outlet, Wadi Sha'bah | 0.006 | 3.1 | 2.5 | 10 | | | |
| 20 | "Snail clay", about 2 m below terrace surface, Wadi Al Luhy | 0.055 | 3.4 | 1.0 | 19 | 1—2 | 4—5 | 2—3 |
| 22 | Clayey silts of dark grey colour, terrace, Wadi Al Luhy | 0.01 | 2.2 | 2.0 | 50 | 1—2 | 4 | 3 |
| 28 | Silty fine sand, terrace, Wadi Al Hawtah | 0.15 | 2.6 | 1.3 | 5.0 | 1 | 4—5 | 4—5 |
| 114 | Silt with high percentage of salt and gypsum, lake without outlet 5 km west of Al Khamasin oasis, Wadi Ad Dawasir | 0.044 | 1.4 | 0.74 | 9.1 | | | |
| 125 | Silt and fine sand, present runoff channel, beginning lake without outlet, 20 km north of Al Hawtah, Wadi Birk | 0.048 | 1.3 | 0.74 | 3.4 | | | |
| 129 | Medium grained silt with gypsum, 625–638 cm below terrace surface, Al Hawtah, Wadi Birk | 0.009 | 2.44 | 0.4 | 14.0 | | | |
| 130 | Sandy silt, 570–603 cm below terrace surface, traces of wind ripples, Al Hawtah, Wadi Birk | 0.018 | 4.08 | 2.9 | 11.0 | | | |
| 131 | Silt and fine sand, 250 cm below terrace surface, Al Hawtah, Wadi Birk | 0.07 | 1.8 | 1.2 | 14.0 | | | |
| | *Various sediments:* | | | | | | | |
| 25 | Alluvial fan near Khafs Banban | 0.45 | 3.0 | 0.44 | 25 | 0.5 | 2 | 5 |
| 106 | Soil weathering presumably not transported, 15 km west of Rawdah, Wadi Ranyah | 0.005 | 6.5 | 15.0 | 20 | | | |
| 110 | Alluvial fan, Wadi Ishghab, NW of Wadi Ad Dawasir | 0.8 | 2.4 | 0.7 | 7.14 | | | |
| 138 | Calcareous mud, pit NW of Al Hofuf | 0.045 | 2.6 | 0.7 | 23.3 | | | |
| 214 | Fossil beach ridge in front of the 10 m terrace, Gulf coast north of Salwah | 0.65 | 2.2 | 0.38 | 7.5 | | | |

Table 49. *Mineral composition of present and fossil wadi sediments*
(××× = main constituent, ×× = secondary constituent, × = subordinate constituent)

| No. of sample | Description of samples | Quartz | Feldspar | Calcite | Gypsum | Halite | Kaolin | Illite | Montmorillonite | Mixed-layer clay minerals | Residue insoluble in conc. HCl, weight % |
|---|---|---|---|---|---|---|---|---|---|---|---|
| 15 | Salty clay, present sedimentation into a lake without outlet, Wadi Sha'bah | ×× | × | × | ×× | ××× | ×× | × | | | 45 |
| 17 | Terra rossa-like weathering products of fanglomerate gravels (20–30 cm below the desert pavement), terrace, Wadi Sha'bah | ××× | ×× | × | ×× | × | × | × | | × | 78 |
| 18 | Terra rossa-like weathering products of fanglomeratic gravels (location 20–30 cm below the desert pavement), terrace near Aqlat As Suqur, Wadi Ar Rimah | ××× | ×× | ×× | × | × | × | × | × | | 63 |
| 19 a | Sand and gypsum, terrace near Aqlat As Suqur, Wadi Ar Rimah | ××× | ×× | × | ××× | | | × | × | | 41 |
| 20 | "Snail clay", about 2 m below terrace surface, Wadi Al Luhy | ××× | × | ××× | × | | | × | × | | 44 |
| 22 | Clayey silts of dark grey colour, terrace, Wadi Al Luhy | ××× | × | ××× | × | × | × | × | | × | 21 |
| 24 | Clay 8 m below terrace surface, Wadi Hanifah, 6 km north of Al Ha'ir | ××× | | ××× | | | | ×× | × | × | 52 |
| 25 | Alluvial fan near Khafs Banban | ××× | × | ×× | | | | | × | | 81 |
| 26 | Fanglomeratic coarse sand mainly granite detritus, 20 km west of Ar Rass, Wadi Ar Rimah | ××× | ××× | × | × | | | | | | 94 |
| 28 | Silty fine sand, terrace, Wadi Al Hawtah | ××× | × | ××× | × | | | | | | 60 |
| 31 | Eolian sand near Al Ha'ir, Wadi Hanifah | ××× | × | ××× | × | | | × | | | 65 |

The grain size distributions yielded rather flat curves comprising the entire range of clay, silt and fine sand (Fig. 95). The median values (Table 48) range between 0.006 and 0.15 millimeter. The mineral composition strongly depends on the local rock concerning the actual detrital quartz, feldspar, calcite, dolomite and mica components in particular. In Table 49, the mineral components of the various wadi sabkhah samples (15, 19 a, 20, 22, 24, 28) determined by means of the roentgenofluorescence analysis are given in a semi-quantitative order. The portion of components soluble in hydrochloric acid generally comprises 40 to 60 percent by weight.

The numerous and very thinly-laminated sequences of these distinctly different types of sediments reaching thicknesses of only centimeters or millimeters, as well as their tight lateral changes make any dimensional survey or any differentiation of the various sediment layers impossible. It is in view of a chronological parallelization that the integration of fine-clastic and/or evaporitic materials under one comprehensive term seems inevitable.

The selected term of "wadi sabkhah" should reflect the predominantly evaporitic character of wadi sediments accumulated under special circumstances. The substance of the notion is almost entirely identical with that of "inland sabkhah" coined by K. W. GLENNIE (1970). In contrast to this, layers of wadi sabkhahs may locally comprise limnic fresh water formations of small thickness; the absence of salt or gypsum has also been observed.

Besides coarse-clastic fanglomerates, eolian and fluvial sands, wadi sabkhah sediments thus represent a third independent, environmental component in the wadi which may be found all over the big wadi systems.

## 3.1.3. Sediments in the Area of Vast Accumulation Plains

Large parts of the Arabian Shelf are covered by young alluvial and eolian clastic sediments as well as by crust residuals. It was not possible to examine the large areas of these sediments thoroughly and in detail during the "Quaternary Project". Some surveys of profiles such as the delta accumulation of Wadi As Sah'ba or the duricrust in the region of As Summan Plateau were compared and they give partly new interpretation as to their development, but they could not demonstrate in a representative way the manifold formations and structures. The investigation in the wide dune areas had to be postponed completely. The most important of the clastic sediments and their areas of distribution will be described in the following section.

### *3.1.3.1. Alluvial fans*

The three extensive centrifugally-arranged wadi systems of the central Arabian area, i.e., Wadi Ar Rimah, Wadi As Sah'ba and Wadi Ad Dawasir, after breaching through the cuesta landscape of Mesozoic and partly Lower Tertiary strata, have built gravel fans of enormous size. These fans spread delta-like towards the east and represent massive accumulation plains with low and constant gradients of about 1 permil and lengths of more than 300 kilometers.

The most famous and, by reason of its extent, most impressive one—because it only has small dune areas—is Ad Dibdibah plain. It is the accumulation plain of Wadi Ar Rimah and is cut through at the present time by Wadi Al Batin, the flat continuation of Wadi Ar Rimah. It is directed towards the northeast and covers, with its total surface of about 60,000 square kilometers, parts of southern Iraq, northeastern Saudi Arabia and Kuweit.

The gravel fan of Wadi As Sah'ba is as large, but wide parts of it are occupied by the sand dunes of Al Jafurah. Its longitudinal extension is directed to the east and it reaches the Arabian Gulf north and south of the Qatar peninsula. Towards the south it merges into the accumulation plain of Ar Rub' Al Khali basin margin. The western part of this basin, with a spread of 400 by 500 kilometers, seems to have a uniform coherent gravel cover, which is overlain for the most part by young eolian sand accumulations. Only on the margins—especially in the north and southwest—and on a few narrow strips parallel to the main dunes in the interior of the basin does the gravel layer crop out. These fluviatile components were deposited for the most part by the Wadi Ad Dawasir system, which merges from the northwest into Ar Rub' Al Khali basin and by the drainage systems of the southwest, which descend from the southern 'Asir highlands, and the northward directed wadis of the Hadhramawt mountains.

There is another occurrence near the border of Iraq and Saudi Arabia, known as As Sahn plain (R. W. Powers et al., 1966) that can also be compared genetically. In spite of its isolated situation today, the gravel components allow the conclusion that there may originally have been a much larger gravel fan, too.

Additional gravel fans of smaller extent, but all the same still important, are found in the central Arabian cuesta landscape, as east of Wadi Maraghan, Wadi Al Jadwal (Al Aflaj), Wadi Atk, or in the region of As Sa'irah about 150 kilometers east of Hail. There is another vast gravel fan in the basin situated in front of the Tuwayq Escarpment, which is bedded in the less resistant Permo-Triassic sediments and extends almost 1,000 kilometers. The composition and development of the gravels of this deposit are, however, very different.

Concerning the climatically dependent exogene development of the landscape during the Quaternary, the above mentioned delta gravel fans in the east of the Arabian Peninsula undoubtedly have been of greatest importance. All these accumulations have been assigned to the Pleistocene (D. A. Holm, 1960; W. Fuchs, et al., 1968; K. W. Glennie, 1970) according to the former generally accepted conception of intense rain periods, the "pluvials", which should, in the semitropical belt, run parallel to the Pleistocene cold periods of the middle and polar latitudes. But they have been used on the other hand as a proof for the pluvial conditions in the Pleistocene, although there were no exact data for their time classification till now.

Our investigations (cf., sect. 2.1.3.) dated the age of these delta gravel fans with the Late Pliocene/Early Pleistocene. The conclusions drawn from their formation and development, concerning the climatic conditions during the time of development and the following Quaternary, show a result which differs from the conception that previously existed.

All these gravel deposits in the east have in common their delta form, their wide extent, their diverging gravel covers on the surface and their generally low thickness of the gravel layers (often less than 1 meter) as well as their composition and grain size distribution. Their topography, low outward gradient and gravel ridges originating from deflation and former flow channels (relief inversion) were thought by D. A. HOLM (1960) to indicate authentic deltas on the margin of a coastal plain of a sea which at that time reached further to the west. Personal observations in the area of Wadi As Sah'ba delta (sect. 2.1.3.4.) confirm this conclusion. The existence of young marine sediments at altitudes of 140 meters above the present sea level prove the Upper Pliocene marine transgression which can be correlated in time as well as by its altitude with the beginning of the delta. The further development of the gravel fan coincided mostly with the Late Pliocene/Early Pleistocene marine regression. This becomes apparent through the partly simultaneous vertical erosion of the main runoff channels into the newly accumulated upper parts of the fan. Together with the marine regression the gravel fan expanded more and more over the widening coastal plain. The river load was distributed in the form of braided streams over the entire gravel fan and thus enlarged the delta body evenly towards the sea.

The delta channels and the later flow channels of the runoff, which are distributed over the gravel fan, also explain the existence of the long slim gravel ridges and gravel accumulations, which are often found only on the top of such ridges. Similar observations have been made on the Wadi As Sah'ba gravel fan (cf., sect. 2.1.3.4.) and on Ad Dibdibah plain (W. FUCHS et al., 1968). These ridges overtop the surrounding deflation area as high as 20 meters because of the intense deflation. K. W. GLENNIE (1970) also referred to this formation of the gravel channels with impressive examples from the Oman region.

This is also an important explanation for the low thickness of the gravel cover, which apart from a few gravel channels of some meters depth, mostly measures less than 1 meter and partly consists of only a single detritus layer on the surface. Between the channels on the flat shore areas with poorly sorted but mainly fine clastic sediments the deflation of these may cause an enrichment of gravels. On the other hand the gravels scattered over the areas may also have been transported by floods from the gravel ridges. W. FAIRBRIDGE (1972) referred to the formation of thinner allochthonous gravel layers out of braided stream sediments by deflation.

The grain size distribution of these sediments shows a wide spectrum from clay and silt up to 40-centimeter cobbles (cf., D. J. DOEGLAS, 1962). Usually the pebbles are limited to the described channels, while fine clastic components lie in the wide areas between the channels. Especially in the sedimentation of the old channels, i.e., the present gravel ridges, a cross bedding can often be observed. D. A. HOLM (1960) already pointed to the delta growth.

The composition of the gravels is manifold. Beside the sediment components of the Paleozoic to Tertiary complex of strata, there prevail above all quartz pebbles as well as magmatic and metamorphic rock components from the area of the Shield. D. I. MILTON (1967) and W. FUCHS et al. (1968)

list in the area of Kuwait rhyolites, felsites, porphyrites, granophyres, aplitic granites, granites, quartzites and serpentinites. The basalt pebbles of As Sahn plain further northwest are mentioned by R. W. POWERS et al. (1966). They assumed the basalts of Al Harrah in the border area of Jordan and Saudi Arabia to be the origin of these components.

Gravels of similar composition can also be found in the gravel fan of Wadi As Sah'ba. Near Haradh at the narrow beginning of this fan there also exist different limestone and sandstone components and chert, as well as a few rounded residuals of silicified, probably Cretaceous, wood; but here, too, appear different quartz pebbles—partly derived from the transport and denudation of older sediments—as well as pebbles of rhyolites, felsites, porphyroides, basalts, granites, gneisses and amphibolites from the area of the crystalline Shield. While cobbles near Haradh are as big as 25 centimeters in diameter, the maximal grain size is reduced towards the fan margin as in the southeast of Al Hofuf as well as near Salwah, to less than 10 centimeters. In these areas with predominant pebble size particles the quartz components, which are more resistant to attrition, predominate.

The cementation of these gravels varies widely. The cement material can be calcium carbonate or gypsum. Thus, the older delta accumulations near Haradh, where the young channel was already embedded, partly consist of conglomerates or sandstone. Towards the edges of the fan, however, the gravels, especially the pebbles of the channels, show only young crusts of calcite with little cementation. In addition to this, the thin pebble layers near the Bay of Salwah show a strong secondary crystallization of gypsum especially in those parts underlain by gypsum strata.

The sediment structures and the origin of the gravels prove an enormous fluviatile material transport in the period of the gravel accumulations in the delta fans. This demands a significant depth of precipitation especially in the crystalline Shield area where most of the gravels originate.

A further indication of the climate is given by the development of the gravel fans with their braided channel patterns. According to the studies of D. J. DOEGLAS (1962), J. M. COLEMAN (1969), W. B. BULL (1964) and E. BLISSENBACH (1954), it follows that such sediments can only be built when there are waterflows with enormous sediment loads and extremely fluctuating discharges just as appear during seasonal precipitations. The flat gradient of the gravel fans and their great extension as well as the boulders which are as big as 40 centimeters in diameter, represent a corresponding discharge of the intermittently occurring water floods. Accordingly there can be assumed a semihumid or semiarid climate at least in the upper course and in the recharge area of these drainage systems during the period of the Late Pliocene/Early Pleistocene marine regression. This coincides with the time of accumulation of the delta fan.

### *3.1.3.2. Sand dunes*

About half of the Arabian Shelf, i.e., about one third of the entire Arabian Peninsula is covered by eolian sand accumulations (Fig. 96 and 97). Ar Rub' Al Khali which is the largest continuous dune area in the world, extends

Fig. 96. Dune complex in the central Rub' Al Khali. (Photo: J. G. Zötl, 1972)

Fig. 97. Dune area in the Jafurah east of Al Hasa. (Photo: J. G. Zötl, 1974)

over 600,000 square kilometers; the other big areas amount to a further 180,000 square kilometers. The location and the distribution of large dune regions are described in the section on the geomorphology of the Arabian Peninsula (1.2.). The most important dune forms of Ar Rub' Al Khali are described in chapter 2.6.

The geological action of the wind in the Quaternary was the most effective factor of the exogene powers for this period with respect to denudation and accumulation. Changes in the morphological features of the Arabian Peninsula during this time were decisively influenced by the eolian processes. Although all this is very important for understanding the geologic, geomorphologic and climatic conditions during the Quaternary our knowledge of the process of dune development with time, the origin of the material, and the possible variations of the prevailing wind directions, is very sparse.

First detailed descriptions of these dune areas can be found in the travel reports of H. S. PHILBY (1933), W. THESIGER (1948/1949) and others. D. A. HOLM (1953, 1960) dealt with the topography and distribution of the dunes, while G. F. BROWN (1960) and J. A. GREENWOOD, D. BLEACKLEY (1967) discussed the dependence of the distribution areas on the prevailing directions of the wind and the general atmospheric pressure. References to the configuration of the dune areas in the 1 : 2,000,000 geological map of the Kingdom of Saudi Arabia are given by R. W. POWERS et al. (1966). Up till now, no sedimentological studies have been carried out—apart from the work of K. W. GLENNIE (1970), K. W. GLENNIE, B. D. EVAMY (1968), and V. SKOCEK, A. A. SAADALLAH (1972) which only in parts deal with the Saudi Arabian area. Two analyses of grain size distribution of sands from Al Jafurah (sample No. 135) and from Ad Dahna (sample No. 272) conducted as a comparison to the eolian wadi accumulations, are shown in Fig. 91 and the respective indices in Table 48.

As mentioned in the introduction to section 3.1.3. the investigation of the sand dunes during the Quaternary Project was postponed. A detailed stratigraphic study of the dune formation, provided that even older dune structures are still visible in spite of the continuous eolian movements, would be not only of regional importance but it could also render interesting conclusions concerning the planetary wind systems in connection with the Pleistocene changes of climate.

### *3.1.3.3. Lacustrine deposits*

Quaternary fresh water deposits have been described from the areas of Iraq, Syria, and Jordan, as well as in the Jordan graben which joins the Arabian Peninsula in the north. R. G. S. HUDSON et al. (1957) and K. M NAQIB (1967) point to lacustrine fine clastic deposits west of the Shatt Al Arab. The vast Middle and Upper Pleistocene lake sediments in the basin of Damascus have been examined by K. KAISER et al. (1973) in a thorough study. The young fresh water limestones and the similar complex of strata in the depression of El Jafr were reported by R. HUCKRIEDE, G. WIESEMANN (1968). Lacustrine formations, e.g., Ubeidija, Hula and Lisan Formations are interbedded frequently in the Quaternary sedimentation of the Jordan graben, of which there are detailed stratigraphic and lithologic studies (e.g., L. PICARD, 1965; O. BAR-JOSEF, E. TSCHERNOV, 1972). Investigations on the limnofluviatile deposits in the area of the Sinai peninsula were carried out by D. NIR (1970).

References to the area of the Arabian Peninsula itself are scarce. Z. R. BEYDOUN (1966), Z. R. BEYDOUN, J. E. GREENWOOD (1968), H. WISSMANN et al. (1943), and W. WENZ (1943) describe lacustrine sediments consisting of chalky marl and limestones with *Characees* and gastropods as well as calcareous tufa in the area of Hadhramawt and Dhufar. They also refer to sandy clays and silts which have accumulated partly in terraces against the dunes on the south and southwest margin of Ar Rub' Al Khali (see chapter 2.6.2.). According to R. W. POWERS et al. (1966) teeth and bones of vertebrates of the Pleistocene (?) and Lower Holocene are enclosed. Numerous Paleolithic and Neolithic artifacts from this area have been described (F. E. ZEUNER, 1954; H. FIELD, 1958; G. CATON-THOMPSON, 1964; H. A. MCCLURE, 1971, 1976). Charcoal residuals in connection with these gave a carbon-14 age of 5,090 ± 200 years (according to R. W. POWERS et al., 1966).

The writers found Quaternary lacustrine fresh water sediments in Wadi Al Luhy about 20 kilometers south of Ar Riyadh and in the area of Al Hasa oasis near the Gulf coast. The findings in Wadi Al Luhy have already been referred to in connection with the wadi sediments (cf., sect. 3.1.2.2.). The clayey or even sandy sediments, which in some parts show certain amounts of coaly organic components, in other parts numerous gastropods especially of *Melanoides tuberculata* (MÜLLER), may have been deposited as a consequence of a damming up of Wadi Al Luhy because of the increased sedimentation in the nearby main Wadi Hanifah. Carbon-14 dating of the shells of gastropods gave an age of 8,400 ± 140 years (see 2.4.2.).

In the oasis area of Al Hasa there were series of mainly filled up small fresh water lakes which were drained in the course of the completion of the new drainage project. They received their tributaries from the springs of Al Hofuf area whose runoff towards the sea is blocked by the sand dunes of Al Jafurah. New exposures in the oasis area show that the lakes had a considerably wider expanse at times in the Early Holocene and the Pleistocene.

In course of the construction of the southern drainage channel about 15 kilometers east of Al Hofuf there were exposed at a depth of two meters clays and siltstones which included numerous gastropods of *Melanoides* and *Bulinus* genera. The carbon-14 age of these shells is 2,180 ± 210 years (cf., Table 7).

Calcareous mud appeared in the northwest part of Al Hofuf during the excavation of the pumping station B for the sewage channelling. These carbonate sediments, 2.5 meters thick, lie on green-grey to red spotted clays of the Dam Formation. They are covered by a thin layer of anthropogene rubbish. The cementation under this humid condition was very loose. The investigation of sample No. 138 gave the grain-size distribution shown in Fig. 98 and Table 48. Besides eolian quartz grains there were found impressions of plant-stems and shells of ostracods not yet fully identified.

Further findings of lacustrine sediments are in the northern part of the oasis area of Al Hasa. Directly east of the place Al Mrah, coaly substances of plants are included in grey limnic sediments on a terrace about three meters above the level of the actual northern drainage channel. The topography of this area allows the conclusion that this regional sector had been a small flat peninsula on the shore of a former lake. This is also confirmed by numerous Neolithic chert artifacts. The existence of artifacts, especially spear heads, which had been used for fishing, has been reported from other parts of Al Hasa oasis area, too (H. A. McClure, 1971; W. A. Goellner, 1972; cf. also H. Kappel, 1967).

The lacustrine sediments of the Quaternary in the areas of the Arabian Peninsula were deposited during the more humid climates of the Pleistocene and Holocene. The increased surface runoff as well as the greater discharge of the springwater, led to the creation of fresh water lakes in depressions or in runoff channels which were dammed up by dunes, especially on the margins of large dune areas. The quantity of precipitation did not, however, suffice to re-establish completely the pattern of the old river systems.

### *3.1.3.4. Duricrust*

Vast areas of the Arabian Shelf are coated by a cover of calcareous duricrust which characterizes the landscape. This duricrust is the result of certain climatic conditions mainly during the Upper Pliocene and Early Pleistocene (cf., sect. 2.1.4.4.). Its further development during the humid phases of the Pleistocene and Holocene may have been relatively slow, while the surface denudation has been quite considerable during the intermediate arid phases of the Quaternary as well as under the present circumstances.

The main distribution area of this calcareous duricrust is the region in the north of Wadi Ar Rimah, where it occurs most distinctly in the areas around the Great Nefud especially in the regions of Al Huj, Al Jawf-Sakakah, Al Widyan and Al Hathalil. The second vast area of distribution is the far-stretched As Summan Plateau which spreads in the form of a bow, concave to the east, from the border between Iraq and Saudi Arabia to Ar Rub' Al Khali in the south. In As Summan Plateau mainly the marl of the Upper Tertiary (Miocene/Pliocene), the calcareous sandstone and sandy limestones of the Upper Tertiary (Miocene/Pliocene) and limestone of the Lower Tertiary are covered by these crusts. Beside these Tertiary sediments, in the first mentioned area, Cretaceous limestones, and, in the area of Al Jawf and Al Huj, the Paleozoic sandstones are coated by these duricrusts, too.

R. P. Miller (1937) was the first to mention these extensive caliche occurrences. Owing to their wide distribution he called them Summan Formation. R. P. Miller emphasized that the slim ridges especially are overlaid by a massive duricrust. He explained these far-streched, partly dendritic forms as valley systems which have been built by relief inversion.

On the 1 : 500,000 geological maps of Saudi Arabian Kingdom prepared by the U.S. Geological Survey and ARAMCO, the spatial distribution of the duricrust in northern Saudi Arabia was also delineated (R. A. Bramkamp, L. F. Ramirez, 1959 a and 1963; R. A. Bramkamp, L. F. Ramirez et al., 1963 a and 1963 b). There are detailed morphological, petrographic, and chemical investigations on calcareous duricrusts from the Shedgum area at the eastern margin of As Summan Plateau near Al Hofuf by R. W. CHAPMAN (1971 and 1974, cf., sect. 2.1.4.). The following digest mainly refers to personal observations in the areas of As Summan Plateau (near Ma'aqala), of Haradh and Al Hofuf.

In spite of the varying thickness and various underlying rocks the duricrusts show the profile sequence already known on other continents (e.g., C. C. REEVES, 1970; N. P. JAMES, 1972; A. GOUDIE, 1973; W. D. BLÜMEL, 1976).

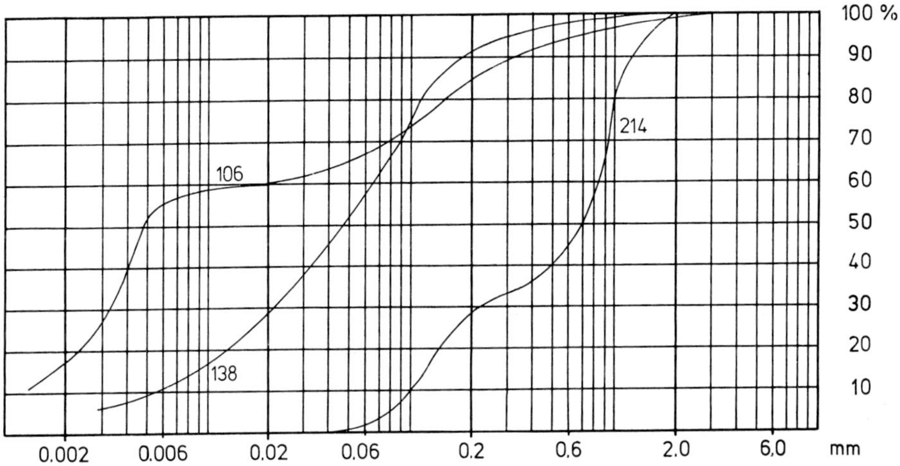


Fig. 98. Cumulative grain size distributions of various sediments. (*106* residual loam of basalt weathering, Wadi Ranyah; *138* lacustrine calcareous mud, Al Hofuf; *214* Pleistocene beach ridge, Gulf coast, north of Salwah)

Basically there must be distinguished a very hard surface and a relatively soft lower part. The last mentioned gradually merges into the underlying rocks. An even more differentiated division of the profile sequence according to A. GOUDIE'S (1973) "evolutionary classification of calcrete" is possible by examining the stage of development proceeding from the lower to the upper strata.

The process of crust formation starts in the unweathered original rocks with an increasing disappearance of the former sediment structure. However stratification of beds caused by different material can still be distinguished within the upper part of the crust. The change of sediment structure already mentioned is caused first of all by solution processes which lead to a loosening of grain cementation. In the calcareous sand and siltstones of As Summan Plateau this usually takes place in a very regular form; not so in the compact limestones, e.g., the Umm er Radhuma Formation, where there can be noticed a selective rock weathering according to the secondary permeability along the joints. The solution emerging from the joints and the capillary veins at first leave unweathered parts with irregular forms in the center so that the lower zone of the crust looks like a breccia. Sometimes it can become more distinct by different colors as the consequence of bleaching of the weathered zones by the solution of ferric oxides on the one hand while brown limonitic crusts are formed around the unweathered limestones on the other. This will be decreased later by infiltration of the solution. The last mentioned process can be observed to a smaller extent also in the sandstone where small concretions of ferric oxides appear in the lower zone of the crust (cf., also R. W. CHAPMAN, 1971). Intercalations of clayey silt beds sometimes keep their red tint of the ferric oxide to the upper zone of the crust because of their low permeability.

Still in the middle part of the lower crust begins a secondary supply of calcium carbonate, the renewed precipitation of which causes "clotted fabrics" and irregular "nodules". The new formation of calcite takes place in the form of micrite which gathers as a film preferentially around quartz, feldspar, and concretions of ferric oxide, but also around the still unweathered remnants of limestones. In the upper profile sequence the enrichment of carbonate increases so that the originally diffusely dispersed nodules are also more abundant. Beds enriched with nodules are caused by the original composition of the different layers. It proves in this case the strong dependence of the crust formation on the original rock material. Together with the increasing formation of nodules in the upper part of the lower zone, the noncalcareous components such as quartz, feldspar and ferric oxide concretions are substituted by calcium carbonate. Old remnants of limestone are changed completely into micrite or microsparite (Fig. 99 a, b, c). In the transition zone to the upper crust the nodules gradually fuse and at first produce an intermittent hard crust. The final $CaCO_3$-substitution results in the formation of a uniform indurated "calcrete hardpan".

In spite of its firmness the structure of the upper zone of the crust is mostly nodular, although in the uppermost parts it may also be indistinctly lamellar (Fig. 100). The form is obviously influenced by the nature of the underlying rock. On marl therefore a thin bedded formation prevails; on sandstone, a nodular one. On limestone the surface frequently is cavernous; this can be seen on the Lower Tertiary Umm er Radhuma limestones and also in the area of Jurassic limestones of the Central Arabian cuesta landscape.

---

Figs. 99 *a, b, c*. Calcareous duricrust: *a*) Thin section of a hard nodular upper crust, sample from As Sulb area, 15 kilometers SW of Ma'aqala. *b*) Detail of Fig. *a*; the nodules consist of recrystallized microsparite (grey) with patches of micrite (dark); the irregular quartz grains (light) are partially replaced by calcite, producing serrated grain boundaries. The nodules are bordered by an irregularly laminated brown micritic crust. In the area between the nodules the rounded and subangular quartz grains coated by a small rim of fibrous palisade calcite crystals are embedded in a drusy mosaic sparite (cf. Fig. *c*). Alterations of this original matrix are caused by micritization (dark crust partially developed around the quartz grains) starting on the border between fibrous calcite rims and the drusy mosaic sparite. *c*) Thin section of a non-nodular hard crust from the Shedgum area, NW of Al Hofuf. Quartz grains with a rim of fibrous calcite embedded in drusy mosaic sparite. Beginning of micritization and replacement of quartz by calcite. (Photos: H. Hötzl, 1975)

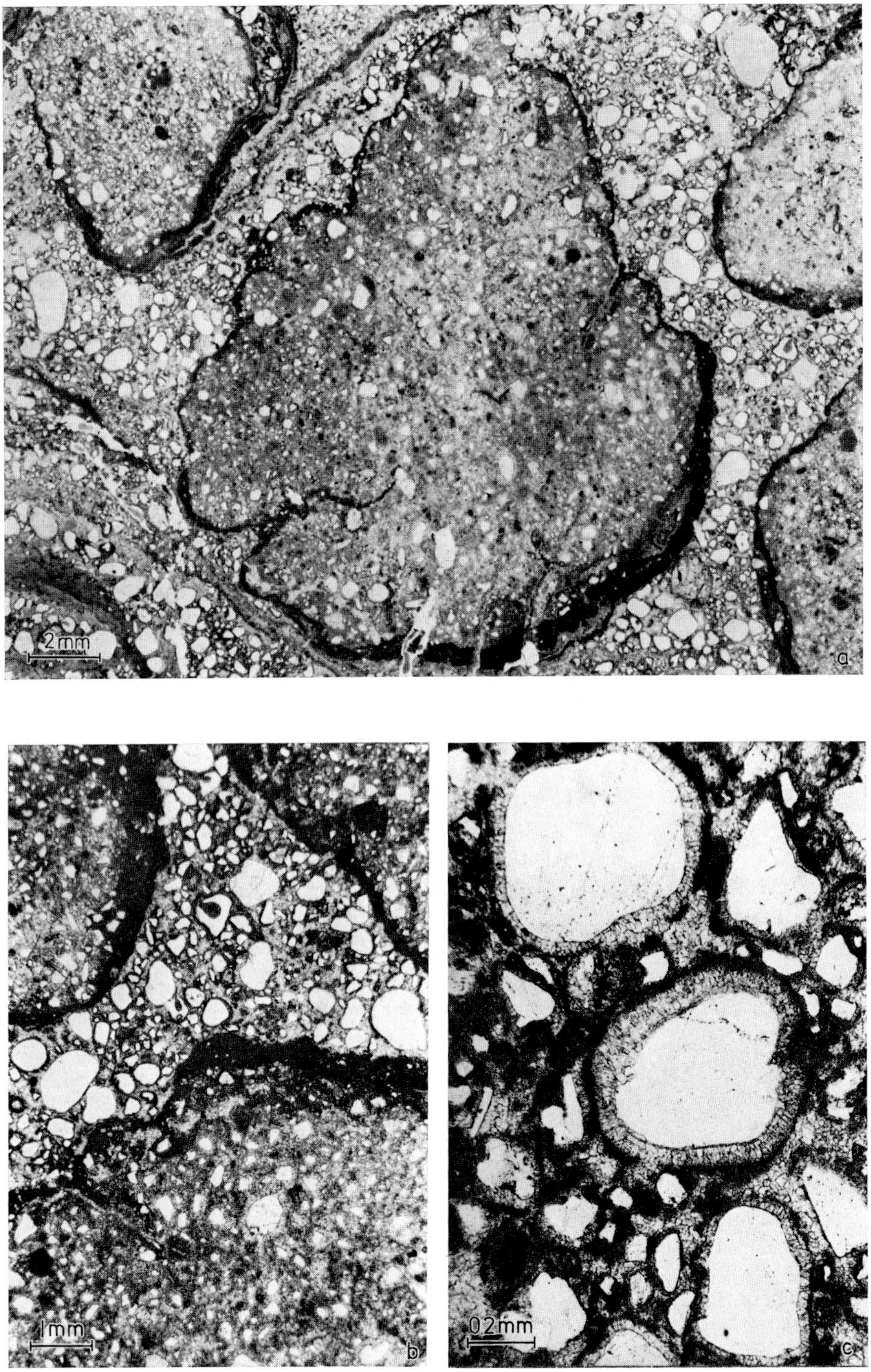


Fig. 99 *a–c*

As mentioned before, the basically different solubility along joints is manifested in this case. In sandstones it causes a homogeneous impregnation of the rock, while in limestones the solutions circulate in the joints and capillary tubes and therefore saturate in the first place the areas around the joints. Rocks surrounded by such zones often maintain noncrusted centers, which may be dissolved or exposed later under different climatic conditions. In consequence, in thinner crusts the usually distinct differentiation of a hard upper and a soft lower crust disappears.

Frequently there are loose pieces of rock scattered on the surface. This is caused by weathering typical of an arid climate as in present times. The influence of this weathering leads at first to the disintegration into debris directed parallel to the lamellar stratification and subsequently to a further destruction into single fragments. According to A. Goudie (1973) this is a symptom of senility, which does not express so much relatively old age but the termination of crust formation on account of climatic changes.

The total thickness of the duricrust, as well as the relation between the upper and the lower zones of the crust, depends on the underlying rocks, the slope gradient, the period of formation and other not yet determined factors. On As Summan Plateau the thickness generally measures between 2 and 5 meters but may amount to 7 meters or more as in the cave areas of Dahl Abu Marwah near Ma'aqala. The main part usually belongs to the soft lower crust which often shows excavations and shelter caves. The upper zone of the crust mostly measures 10 to 70 centimeters, but may reach 2 meters or more, as, e.g., on lacustrine limestones. On the Jurassic limestones of the Central Arabian cuesta landscape there are mostly thin hard crusts, the soft lower zone of the crust is only present in the cavernous lower layers.

According to numerous investigations on the various continents, incrustation processes may occur under different conditions and in different ways. The ideas given in the literature were summarized by A. Goudie (1973). Generally the development is connected with the weathering of the ground. The formation of the crusts can take place on or below the surface. The wide-spread crusts on the Arabian Shelf mentioned before probably developed for the most part directly at the surface. This is different from young calcareous crusts described in the section about Wadi Ranyah and Wadi Bishah for which there are carbon-14 datings. They are subsurface formations according to an authentic $C_{Ca}$-horizon.

The $CaCO_3$ necessary for the formation of the crusts is chiefly supplied by the local rocks and it is transported by ascendent waters. According to A. Goudie (1973) this is therefore a matter of genuine *in situ* formation and the crusts accordingly represent the residual products of weathering. Unsaturated solutions at first infiltrate to a certain depth, dissolve the carbonate there and then rise under the influence of capillarity and of "evaporation pumping" to the surface, where because of evaporation and maybe the action of microorganisms (cf., E. Krumbein, 1970) the $CaCO_3$ is again precipitated.

If the essential process of the duricrust formation on As Summan Plateau and probably also the occurrences in the east of the Great Nefud are characterized thus, perhaps other processes were (and are) also more or less active.

L. R. Gardner (1972) referred to the importance of the eolian process as a lime supplier for the duricrust, above all in regions near the desert. "Frosted quartz grains" were found only in the upper zone of the crust in the examined samples and even in places where they are generally never found in the local rocks. An explanation could be wind transported fragments which were cemented later by precipitation from the rising solutions. The eolian transport can be proved this way, but the percentage of lime supply, which possibly exists, cannot be estimated from it.

Fig. 100. Duricrust section of Al Hasa area. The lower part (light) is less consolidated through solution of calcite. Towards the surface an increasing number of irregular nodules can be seen. In the hard upper crust the nodules are cemented and evenly consolidated to a hardpan calcrete. (Photo: V. Maurin, 1973)

The listed processes of formation of calcareous crusts demand very special climatic conditions and therefore allow inferences on the time of development. On one hand it requires a special depth of rainfall, the infiltration of which into the ground causes a certain degree of weathering, and on the other hand it is necessary that the yearly mean average of evaporation exceeds the depth of rainfall; these conditions exist mainly in semiarid climates. A yearly rate of 200 to 600 millimeters of rainfall can be taken for an average (A. Schäfer, 1975; A. Goudie, 1973; F. Kabelac, 1962). N. P. James (1972), referring to Barbados, pointed out that this rate can be higher, provided there is a deficit of rainfall owing to a strong evaporation.

The existence of eolian "frosted quartz grains" in the upper zone of the crust strengthens the assumption of a semiarid climate with seasonal rainfalls.

R. W. CHAPMAN (1974) also came to the same conclusion after his investigations in the Shedgum area near Al Hofuf. Yet his deductions concerning the Quaternary climate and its fluctuations—pluvials with fluviatile erosion and interpluvial with duricrust formation—are doubtful, because he classified the duricrust as too young and also explained the marine erosion terraces of the Upper Pliocene as due to different Pleistocene fluviatile phases of denudation.

### 3.1.4. Marine Sediments and Sabkhahs in the Coastal Region of the Arabian Gulf

Still during Pleistocene the sea levels influenced by eustatic changes were in some parts considerably higher than the present one, in spite of the important Late Pliocene/Early Pleistocene regression following the Upper Pliocene transgression. Numerous marine abrasion terraces of present altitudes up to 70 meters above sea level, e.g., in the area of the Bay of Salwah (cf., sect. 2.1.2.) or even on the northern part of the Bay of Kuwait (W. FUCHS et al., 1968)—with partly unambiguous beach sediments—give a clear proof of it.

In the area of the flat coastal region between the Bay of Salwah and Shatt Al Arab extensive areas as much as 100 kilometers inland from the present shoreline, were therefore at times, depending on the altitude of the sea level, covered by the sea. This is the reason why Pleistocene shallow-sea deposits, e.g., lagoonal or littoral facies, can be found especially in these areas. Unfortunately, there are no special sedimentological investigations of this area, which is occupied today by wide sand dunes but also by vast sabkhah plains, remnants of the marine regression. Deposits of marine calcarenites are known from the regions all along the coast between Ras Tannurah and Ras Al Misha'ab, from Wadi Al Miyah, from the area around Al Qatif and Ad Dammam as well as from parts east of Al Hofuf and from the coast between Dhahran and Salwah (cf. also D. A. HOLM, 1960).

These calcarenites consist of calcareous detritus, öoids, pellets, shell fragments and quartz grains which are consolidated by a calcareous cement (Fig. 101). They are partly under- or overlain respectively by eolian sands and therefore indicate repeated Pleistocene sea level fluctuations. M. STEINECKE (according to D. A. HOLM, 1960) suggested the name "Bahr Formation" for these deposits after Jabal Bahr near Al Jubayl. The name has not been used commonly; that is why it was not included in the Lexique Stratigraphique International (vol. 3, fasc. 1061). Comparable sediments are also found in the region of the Trucial Coast, on the coast of Masqat and Dhufar as well as on a few islands of the Arabian Gulf (G. E. PILGRIM, 1908; G. M. LEES, 1928; E. A. SHINN, 1969; Z. R. BEYDOUN, J. E. GREENWOOD, 1968; P. KASSLER, 1973). There they are listed under the name of "miliolites" or "miliolitic limestone" and almost all of them are referred to the Pleistocene.

The middle Holocene beach sediments, which only appear on terraces two to three meters above the present sea level, must be differentiated from these Pleistocene calcarenites. Generally they are to be found only close to the present shore line. They originated here during the middle Holocene highest sea level. Their time classification is confirmed by numerous carbon-14 dates (cf. sect. 2.1.2. as well as e.g., G. Evans et al., 1969; H. Kapel, 1967). They are constructed by beach ridges, serpulid reefs, *Cerithies* pavement, calcareous sands mainly composed of shell fragments, and shell banks. The fauna associates with the present faunal forms. A detailed description of the sedimentary sequence in an outcrop along the Abu Dhabi coast was published by G. Evans et al. (1969); above dune sands there are grey calcareous quartz sands (built by marine erosion of the sand dunes and by their mingling with marine grain components), then follows muddy carbonate sand and on top of it there are algal limestones, which probably developed during the regression in the area between the tidal fluctuations.

Clear indications of a higher sea level reaching further inland are given by the vast coastal sabkhahs on the whole area of the Gulf coast. These are mainly old bays, which have gradually been filled up by eolian deposits as well as silty and clayey sediments, carbonates, gypsum, anhydrite, and salts. Sometimes supratidal algal crusts or layers of shell detritus can be found depending on the level of the groundwater table and on the capillary fringe. Today the sedimentation caused by precipitation of halite and/or gypsum and by adhesion of eolian quartz grains and deflation processes counterbalance each other.

Initial steps towards a sedimentological and geochemical investigation of the above mentioned sabkhahs on the Saudi Arabian coastal section were made by D. H. Johnson et al. (cf., sect. 2.1.5.). Results concerning the age of the sabkhah formations as well as the process of marine regression cannot be offered before drillings and investigations of the complex of strata have been undertaken. The numerous studies of sabkhah formations in the area of Abu Dhabi (e.g., G. Evans et al., 1964; G. P. Butler, 1969; P. Bush, 1973; K. J. Hsü, J. Schneider, 1973) make evident that these sabkhahs near the coast are built above Pleistocene eolian sands. The direct interfingering and transition to subrecent and recent deposits in the tidal flat area took place during the middle Holocene highest sea level stage.

The position of sabkhahs situated further inland, such as exist at different altitudes north of Al Jubayl, 80 kilometers inland, as far as Wadi Miyah, or between the Gulf coast and the Al Hofuf-Abqaiq area is not yet clear. D. A. Holm (1960) and K. W. Glennie (1970) believed that they are not authentic inland sabkhahs formed in deflation basins, but that they are also remnants of older coastal sabkhahs of Pleistocene and Late Pliocene sea levels. This also is valid for the numerous sabkhahs on the east margin of Ar Rub' Al Khali. Special investigations on the specific characteristics of individual sabkhah deposits which may provide information about the development and maybe the age, have not yet been conducted.

The Quaternary marine sediments of the coastal regions of the Arabian Gulf cannot be examined without reference to the present sedimentary processes in the Gulf at the same time. In fact, not only the middle Holocene beach deposits and sabkhah formations show an interfingering with the sedimentation caused by Holocene transgression, but also the underlying sand dunes—having developed during the regression of the Würm glacial era—can still be identified far out in the Gulf. The submerging of the "Bahr Formation" in the Gulf near Ras Tannurah could also be proved by drillings. Summaries, especially about the Holocene sedimentary process in the Gulf area, were published by B. H. PURSER (1973). Unfortunately, there are only general references to Pleistocene sedimentation processes; till now a more detailed investigation of the stratigraphy and the sequence of strata has not been carried out.

## 3.1.5. Chronological and Stratigraphic Classification of Processes of Erosion and Sedimentation During the Quaternary

### *3.1.5.1. Development of the drainage system*

The drainage system of the central and eastern part of the Arabian Peninsula is directed towards the east, according to the general geologic conditions. From the distribution of the thick complex of sedimentary strata it becomes evident that this direction of drainage has been mainly responsible for the transport of material at least since the Late Paleozoic, if not since the Cambrian. The general situation of the denudation areas in the west and of the accumulation regions in east is confirmed impressively by the thickness of the strata complex. The lower tectonic units with their metamorphic rocks crop out in the west of the Arabian Peninsula, whereas the overlying strata of sediments increase in thickness towards the east. In the Gulf coastal area the strata complex of terrigenous and marine deposits reaches a thickness of over 4,000 meters (G. F. BROWN, 1972). The opening up of the Red Sea Graben with the uplift of the graben flanks and the young inclination of the Arabian Peninsula together with the Alpine orogenic events of the Tertiary have favoured this tendency and have raised the absolute gradient again.

The constancy of the general direction of the gradients can also be observed in the existing wide wadi systems which represent morphogenetically the result of an early-formed drainage system. Under the arid conditions of today they only effect a local drainage or serve as local lake basins without outlet for the episodic rainfal. Separated by large dune belts they have become hydrologically independent areas, the original connection of which often can be recognized only by the morphological uniformity. Their enormous channels which cross the Central Arabian cuesta landscape, as well as the ample gravel fans with their high percentage of crystalline rocks that extend towards the Gulf, show that they have, under humid climatic conditions, transported considerable quantities of water and sediments from the areas of the crystalline Shield to the accumulation areas in the east.

The most important of these wadi systems, whose recharge areas reach back to the Shield, are the systems of Wadi Ar Rimah with its continuation Wadi Al Batin in the north (sect. 2.3.), Wadi As Sah'ba (sect. 2.1.3.4.). Wadi Al Luhy and Wadi Hanifah (sect. 2.4.) in the central region and the system of Wadi Ad Dawasir (sect. 2.5.) in the south. About the age of their origin only incomplete details can be given. Situated in a relatively straight centrifugal line, the wadis which break across the Central Arabian cuesta landscape in consequent channels must be older than the cuesta landscape and must have undergone further development during its formation.

The distribution of the fluviatile or lacustrine sediments of the Miocene Kharj Formation (R. W. POWERS et al., 1966) gives a reference time especially in the area of the basin in front of the escarpment of Cretaceous strata. The sediments demonstrate a clear relation with Wadi As Sah'ba system. As these Miocene sediments can be found on the present wadi floor, the drainage system with the breaching through the Jurassic and Cretaceous cuestas must have existed even at that time. There may also be a similar relationship between the Kharj Formation south of Al Bijad area and Wadi Ad Dawasir as well as between a few gravel fans in the Miocene and Lower Pliocene Hofuf Formation and Wadi Atk 150 kilometers north of Ar Riyadh.

Large areas of the crystalline Shield were subject to denudation throughout the Mesozoic. Its surface drainage was directed towards the east into the adjacent marine area. According to the above-mentioned results it can be assumed that the drainage has, since the Mesozoic, been progressing eastward in step with the gradual regression of the sea.

Wadi As Sah'ba illustrates this fact. It was in action as early as the Miocene reaching to the east of Al Kharj plain, contributed to the sedimentation of the Hofuf Formation, and later cut into it during Late Pliocene, as it migrated eastwards. There it produced a delta fan and then, in Late Pliocene/Early Pleistocene, eroded a channel into this gravel as it followed the regressing sea towards the east.

Regression of the sea combined with the humid phase of the Late Pliocene/Early Pleistocene led to down-cutting erosion which reached back to the headwater regions. During this time the older accumulations were obviously washed out of the narrow breachings and the wadi floor partly became considerably deeper. In the area of the vast accumulation plain relatively narrow channels developed.

As there are accumulation processes practically all along these wadis today and as they consist for the most part of similar sediments built under arid conditions, it is exceedingly important to place the date of development of these erosion channels in a period of much wetter climate. Concerning the sediments of the wadi fills only the upper zone can be dated (cf., sect. 3.1.5.3.). The beginning of the next accumulation in the wadi channels after the last important erosion period is not yet known.

Investigations in Wadi Ranyah with the datings of two basalt flows (cf., sect. 2.5.2.) placed the last intensive precipitation period, with the morphodynamically still widely spread effective runoffs in the center of the Arabian Peninsula, in the period of 3 to 1 million years B.P.; a time which, lacking a closer definition, is called at the moment "Late Pliocene/Early Pleistocene". The parallelism of this wet period to the above-mentioned latest erosion and washing out of the large drainage channels can be validated on the basis of the conditions in Wadi Ranyah. During the indicated time a vertical erosion of at least 20 meters took place. Such erosional processes on the eastern margin of the crystalline Shield must have caused the sediments to be transported beyond the Central Arabian cuesta landscape. The recent channel deposits of these parts consist almost entirely of components of local strata, whereas unquestioned allochthonous components, such as found in the Wadi Birk detritus derived from the crystalline hinterland, occur only at the base of the wadi deposit.

As the development of the channels took place in the Late Pliocene/Early Pleistocene, the depositional processes which led to partial filling of the wadi channels must be assigned to the Quaternary. The sediments dissected by these channels must accordingly be older. This seems to be of importance with regard to the also partially eroded, vast gravel fans. The deeper erosion of these channels behind the gravel fans, as well as their decreasing gradient and their disappearance towards the east, coincide with the washing out of upstream wadi sections and the gravel accumulations. The lowering of the base level of erosion during this time, which also runs parallel to the Late Pliocene/Early Pleistocene marine regression, favoured the further extension of these gravel fans together with erosion into the newly accumulated section in the back. Evidence of these facts can be found in the eastern region of Wadi As Sah'ba.

### *3.1.5.2. Processes of marine transgressions*

The Quaternary sea level fluctuations and resulting transgressions and regressions in the coastal region of the Arabian Gulf are thoroughly described in section 2.1.2., and their correlation discussed. As the respective sea level is very important as a base level of erosion and the extension of the sea for sedimentation, the most relevant results may be listed once more in a summary (cf. also Table 50).

During the Late Tertiary, the eastern margin of the Arabian Shelf was for the most time above sea level. The terrestrial sedimentation was interrupted only by two short marine transgressions. During the middle Miocene the first transgression took place in the course of which the sediments of the marine Dam Formation were deposited. The second occurred during the Late Pliocene and caused the formation of the scarp on the eastern margin of the As Summan Plateau. It is remarkable that both transgressions proceeded as far as 70 to 100 kilometers west of the present shore line. The Late Pliocene transgression marks are found in the area of Al Hasa, about 150 meters above the present sea level.

The regression went parallel with the accumulation of deltaic fans. This permits the correlation of this second regression phase with the Late Pliocene/Early Pleistocene. In comparison with the Upper Pliocene maxima sea level was lowered by 50 meters. Further details on the extent of the regression cannot be given now, because the plain referred to ends at the edge of the steep coast near Salwah.

On the western coast of the Bay of Salwah, which is a down-faulted block of Early Pleistocene age, there are, at altitudes between 5 and 70 meters above sea level, 7 erosional terraces or erosion ridges (cf., Table 6). They are partly unambiguous shore lines and conform with the mostly eustatic fluctuations of the sea level during the Pleistocene. A detailed chronological classification cannot be provided at the moment, as there are no reliable age dates for the individual terraces available. An interpretation of their elevation above present sea level compared with the level of the Sizil, Milazzo and Thyrrhenian terraces in the near Mediterranean area would be too speculative with regard to the present stage of investigation.

The worldwide regression in connection with the Würm period is the first that can be dated reliably. In its course the sea withdrew completely from the shallow Arabian Gulf, as has been documented by eolian deposits spread far into the Gulf. The transgression between Würm I and Würm II on the Arabian Gulf coast can be proved by absolute age dates. The assignment of the 10 meter terrace in the area of Salwah to this interstadial transgression seems to be possible but cannot be given with confidence.

About the area of the Arabian Gulf and its shores, there is detailed information, supported by many carbon-14 datings, concerning the process of the Flandrian transgression—the rising of the sea level after its minimum during the Würm period. The rising of the level started about 15,000 years ago and reached its widest expansion between 6,000 and 4,000 years ago (according to the Atlanticum stage). The sea level then lay 2 meters above the present one and caused the formation of the 1.5 to 2.5 meters terraces with old beach sand barriers and small serpulid reefs.

### *3.1.5.3. Age determination of the sequence of sediments*

It is obvious that the time classification and correlation of mainly clastic sediments of the vast, arid terrestrial areas has proved to be extremely difficult. Generally, the process of sedimentation takes place in more or less isolated areas, which strongly depend on local factors concerning the sources of material as well as the process of sedimentation itself. Furthermore, the arid regions represent basically an extremely unfavourable living space and the arid conditions prevent the fossilization of zoogenic and even phytogenic material to a great extent. Thus, lithostratigraphic methods can be applied only in some cases and biostratigraphic methods practically cannot be used.

Guiding facts for an extensive correlation as well as for the stratigraphic division in arid regions are provided by climatic fluctuations. Changes of the arid conditions, even caused by only an irregular and seasonal precipitation, usually lead to a clear change of the exogene geological processes. In vast areas the geological processes resulting from climatic changes do not, however, yield the same responses. They may cause weathering here, erosion there, and sedimentation in other places. However, this does not alter the value of stratigraphic indicators of such climatic fluctuations which facilitate quite useful subdivisions fixed by single absolute age datings.

This investigation of the Quaternary sediments of the Shelf of the Arabian Peninsula, which does not claim to be a complete presentation, was defined to identify first of all the climatically formed key layers. It was necessary to conduct a chronostratigraphic division, by means of as many absolute age dates as possible, which should not only inform about the Quaternary sedimentation processes but also about the cycles of climatic change during the

Quaternary. This—and it must be anticipated—could only be carried out to a limited extent. The reasons for this are the following:

1. A lack of natural outcrops which involves vast complexes of strata.

As the sedimentation in the accumulation areas of the Quaternary is at present very active, most of the lower sequences are covered. Therefore, it will be necessary in the future to examine the results of drillings which are conducted in connection with different technical measures.

2. A lack of appropriate material for absolute age dating which facilitates the subdivision of the pre-Würm sections of the Pleistocene. It may be assumed that a more precise chronological division will be possible in the course of the investigations in the western part of the Arabian Peninsula where in many areas volcanism existed till historic times.

3. The Quaternary climatic changes themselves, which probably were less intense on the Arabian Shelf platform than have been assumed up till now with respect to the worldwide climatic changes.

Of all, in general maybe short humid periods of the Quaternary (cf. sect. 3.2.), only those of the Holocene and of the Würm interstadial could be supported geochronologically by carbon-14 dates. On the whole, 26 sediment samples were examined for carbon-14 dates. Apart from this a larger number of already published data were available especially for the Arabian Gulf coastal area and Ar Rub' Al Khali (H. A. McClure, 1976). For the earlier humid phases appropriate material for dating is lacking. Furthermore, these phases are morphologically and sedimentologically only partially provable by adequate deformation, as well as partly by destruction of the corresponding sediments and formations.

The intense humid phase of the Late Pliocene/Early Pleistocene, which caused a marked change in all the exogenous processes, produced an unequivocal "key horizon" observable all over the Arabian Peninsula. It could be delimited first to a period of about 1.1 to 3.5 million years B.P. by K-Ar age-datings of basalts.

In the following an age definition of the most important Quaternary sediments will be attempted. Table 50 contains a schematic outline, too.

*Deltaic fans*

The deltaic fans and the comparable gravel plains of the northern and western margin of Ar Rub' Al Khali—formerly having been correlated with various pluvials of the Pleistocene owing to their fluviatile origin (cf., D. A. Holm, 1960; R. W. Powers et al., 1966; W. Fuchs et al., 1968)—must be assigned to the "Late Pliocene/Early Pleistocene" as defined in section 3.2. The massive gravel deposits in the area of the Arabian Shelf platform, with a high percentage of gravels from the far distant crystalline Shield as well as the parallel erosional processes in the hinterland which washed out the basins and channels are geological indicators of the climatostratigraphic key horizon caused by the above mentioned intense humid phase. Its precise time classification and age dating was discussed in section 2.1.3.4.

Undoubtedly, the period of about 1.5 to 3.5 million years B.P., defined by the two K-Ar dates represents a very long span of time which wants further investigation and additional subdivision. In reference to the complete sedimentation process correlation could be done based on the cutting down of the large gravel fans and the later Late Pliocene regression.

During the humid phases of the Pleistocene the gravel plains underwent certain changes along their margins due to channel-forming processes as well as local accumulation processes. In those parts where mainly materials from the adjoining hinterland contributed to the accumulation of the large gravel plains, e.g., on the southern margin of Ar Rub' Al Khali, the distinction of the plain sections from the younger gravel fans is very difficult.

*Sediments of the large wadi channels*

The lowering of the base level of erosion during the regression of the Late Pliocene/Early Pleistocene humid period was connected with the process of deep vertical erosion. In all the river systems distinctive narrow channels were built. Their refill set in after a climatic change under primarily arid conditions. Even today a fine clastic, partly local sedimentation (cf. sect. 3.1.2.) dominates in the whole channel system. The thickness of the channel deposits amounts to 70 or more meters. Concerning the strata sequences within individual channels, only incomplete data are available from boreholes and wells. The sedimentation at the base of the channels consisted at the beginning of non-local gravels. Therefore, they must be assigned to the terminal phase of the Late Pliocene/Early Pleistocene humid period. On top of these follow, according to the present arid accumulation conditions, fine clastic sediments which show single layers of gravels consisting of local material. They are assigned to the humid phases of the Pleistocene. Neither for the beginning of the sedimentation nor for these intercalated gravel layers are there methods for a precise geochronological classification.

The uppermost strata of the channel deposits were investigated more thoroughly in Wadi Al Luhy, Wadi Hanifah and Wadi Birk where there are accumulation terraces up to 11 meters high (cf. sect. 2.4.2. and 2.4.3.). In Wadi Al Luhy the gastropods in lake deposits, 2.5 meters below the upper edge of the terrace, could be dated at 8,400 ± 140 years B.P.; in the lower Wadi Birk, a layer of charcoal in mainly eolian sands and silts about 6 meters below the terrace edge was dated at 2,170 ± 130 years.

Redeposition within the large erosion channels has today levelled up again the Late Pliocene/Early Pleistocene vertical erosion channels to a great extent. Consequently, the present floor of Wadi Al Atwa, the actual erosion channel of Wadi Ad Dawasir east of the Tuwayq breach, is situated only few meters below the surface of the terrace deposited during Late Pliocene/Early Pleistocene. Similar conditions can be observed in the uppermost part of Wadi Ar Rimah where an old terrace plain, which is covered with basalts near Hulayfah, is higher than the wadi floor in places by two meters. These small differences in altitude and partly sharp terrace edges characterize an essentially younger channel erosion. On the other side of the wadi in the relatively narrow transverse valleys the sedimentation occupies the total width of the wadi floor. A similar situation exists in the main channel of Wadi Ad Dawasir west of the Tuwayq Escarpment where the recent sediments obviously fill the deep vertical erosion channels. It is unknown to what extent older "terrace bodies" of the pre-Late Pliocene/Early Pleistocene already have been covered by more recent deposits in these sections.

*Eolian deposits of the vast dune areas*

Eolian sand dunes especially characterize the arid conditions of the concerned sedimentation area. The estimation of the duration of accumulations in young dune areas usually is very difficult as in those very places there are extremely adverse conditions for life which therefore precludes biostratigraphic division in any way. The dune material is continuously shifted furthermore and so the disposition of the dunes reflects the wind system of the respective period.

In the east and south of the Arabian Peninsula useful data on the origin of the vast dune areas can be drawn, however, from the age of the basal surface. Al Jafurah, the wide dune area of the Arabian Gulf occupies the very Late Pliocene/Early Pleistocene regression plain as well as the deltaic fan of Wadi As Sah'ba which is of the same origin.

The dunes in the western part of Ar Rub' Al Khali, the largest complex dune area of all also rest on the above mentioned Late Pliocene/Early Pleistocene gravel plains. The basal surface of the dunes in the east of Ar Rub' Al Khali, according to its altitude, must have been covered by the sea during the Upper Pliocene. Numerous extensive sabkhahs, obviously remains of this marine phase, are probably proof of this.

From these results it can be gathered that the first dunes were built during the change to an arid climate at the end of the Late Pliocene/Early Pleistocene humid period. The following more humid stages of the arid Quaternary hardly influenced the eolian processes as can be concluded from the development of the drainage system. So the dune formation proceeded throughout the Quaternary with perhaps changing intensity.

The material of the sand dunes mainly consists of fine clastic particles blown out of the wide, accumulated gravel plains. The eolian sands up to 30 meters thick on the bottom of the western Arabian Gulf manifest how rapidly the dune formation proceeds, provided the necessary sands are available. They were accumulated during the geologically relatively short Würm regression, when the Gulf was dry.

*Marine sediments*

Between the margin of the Upper Pliocene transgression and the present shore line can often be found young marine sediments on top of various old Tertiary rocks. Although this has been known for a long time, a detailed stratigraphic investigation is lacking except for the uppermost Holocene sections. Such a research would facilitate a better subdivision by means of micropaleontologic investigation of these deposits which often contain a high percentage of biogenic material.

The stratigraphic classification of sections for which carbon-14 age-dates are still lacking is effected today mainly by correlation with the recent transgression and regression processes (cf. sect. 3.1.5.2.). Some beach sediments can be correlated directly with marine deposits. This refers, for instance, to the marine sediments in the area of Al Hasa, which originated during the Upper Pliocene transgression. Other beach sediments, e.g., in the region west of Salwah, are related with Pleistocene marine terraces. As their precise position has not been determined yet, their sediments cannot be classified either.

Fig. 101. Pleistocene calcarenite in the Gulf coastal region north of Ras Tannurah. (Photo: V. Maurin, 1973)

The regionally spread calcarenites of the "Bahr Formation" (cf., sect. 3.1.4.) cannot be directly connected with the marine terraces at the moment. They presumably represent the period of the Pleistocene sea level fluctuations, which could probably be proved by an interbedding of terrestrial strata. Significant factors are the nature of subdeposits and above all, the overlaying eolian sediments. By means of the unambiguous correlation of the overlaying dunes and the Würm regression a pre-Würm Pleistocene age can be assumed for the higher situated calcarenites. It cannot be denied that the calcarenites located about 10 meters above the present sea level originated partly during the Würm interstadial transgression (Fig. 101). This fact is also proved by a few carbon-14 dates of marine organic remnants, limestones and dolomites of the Gulf coast (cf. sect. 2.1.2.). A final chronostratigraphic classification of these calcarenites must be postponed, however, till the detailed investigation of sequences of strata.

The beach sediments of the middle Holocene 2- to 3-meter terraces are for the most part distinctly separated from the similar Pleistocene deposits by eolian sand accumulations. Their precise time classification results not only from the strata overlaying the Würm dunes but is supported also by a large number of absolute age dates. These give a clear picture of the Flandrian transgression process (cf. sect. 2.1.2.3.).

*Calcareous crusts*

For the reconstruction of the exogenous processes, soil formations are of great importance. They especially admit conclusions on the climate at the time of origin and, referring to the morphology, they represent morphodynamic stability. Such residuals of weathering are the calcareous crusts, which occupy vast areas of the Arabian Shelf platform (cf. sect. 3.1.3.4.). Therefore special attention was paid to their age classification.

R. W. Powers et al. (1966) assumed at first a Quaternary age because of the crustification of Miocene and Pliocene (Hofuf Formation) sediments. R. W. Chapman (1974) supposed them to be formation of several phases during Pleistocene interpluvials. His classification was based on the existence of duricrust younger than the erosion terraces underneath the escarpment of As Summan Plateau near Al Hofuf. He ranged these erosional terraces at the beginning to fluviatile erosion phases of a different age and thus, in turn, he connected them with the generally accepted pluvials of the Pleistocene. This concept could not be confirmed by the writers' investigations. In section 2.1.3. it was already clearly demonstrated that these erosional terraces were built by the Upper Pliocene marine transgression. An Upper Pliocene to Late Pliocene/Early Pleistocene age for the calcareous crusts must be assumed. The especially thick crust formation in some parts of As Summan Plateau which is the result of a high ground water level and the incomplete development in the area of the lower regression plain of the Late Pliocene/Early Pleistocene, indicate that the crust formation was under way already during the Upper Pliocene transgression. Its further development during the Late Pliocene/Early Pleistocene is manifested by the crustification of the side walls of the erosion channels in the Wadi As Sah'ba deltaic body, whereas the recent sediments were not included in this process. Besides, Pleistocene deposits are covered only by thin crusts, if at all.

During the mainly arid climate of the Pleistocene the devastating process of mechanical weathering, which led to a nodular disintegration of the uppermost zones of the crust, prevailed. This process must be considered in different ways depending on the sometimes more humid phases as well as on the respective local ground water levels. So, the crust can be destroyed on a terrace surface under totally arid conditions and at the same time in a gully only few meters away be built by evaporation of the capillary waters of these rocks.

Salt precipitations and authentic caliche formations caused by high ground water level can be found in coastal areas under arid conditions, as reported by P. A. Scholle, D. J. Kinsman (1974), of the area of Abu Dhabi. The crusts in the regions of some Pleistocene marine terraces, as can be observed west of the Bay of Salwah, may have been formed in a similar way.

## 3.2. Climatic Changes During the Quaternary Period

(H. Hötzl, J. G. Zötl)

The main factors producing climatic changes are altering conditions of precipitation (precipitation depth, intensity, kind of precipitation, annual distribution) and of air temperature. Other factors such as humidity, wind conditions, etc. are partly dependent on the main factors or else hard to ascertain.

The climate influences the exogenous processes of geomorphogeny as well as the conditions vital to flora, fauna and man. Working methods employed to explore past climatic epochs thus largely correspond to above-mentioned spheres of influences (morphogenetic and sedimentologic studies, palynology, treatment of fossils and anthropogenic discoveries). Most of these methods were applied—to varying extents, to be sure—to the present sections of the study of the Quaternary geology of selected areas of the Shelf of the Arabian Peninsula.

### 3.2.1. Holocene

What is true for most geological problems also applies to the field of paleoclimatic research, i.e., with increasing age of events their recording generally becomes ever more difficult. Methodically, it is preferable to describe the younger climatic changes first. Apart from local, short-term and unusual climatic differences that occurred in historical times or even in recent years, the youngest supraregional climatic fluctuations took place during the era from approximately 9,500 until 4,500 years B.P. (Table 50). A shifting of climatic conditions in central and eastern Saudi Arabia during this period has been confirmed time and again.

For the early and middle Holocene there are indications of a greater amount of precipitation and stronger surface runoff of wadis. Accumulation terraces (partly consisting of fluvially transported eolian sediments) as well as lacustrine sedimentation in backwaters of smaller wadis (Wadi Hanifah and Wadi Al Luhy) and at the margins of dune areas (Al Hasa and the southwest part of Ar Rub' Al Khali) are witnesses of a stronger surface runoff and thus heavier (seasonal?) precipitation. Absolute time determinations have been made in many cases. Firstly, there are shells collected from limnic sediments lying on the base of the young terrace in Wadi Al Luhy (2.4.2., Fig. 65). Their age (8,400 ± 140 years B.P.) corresponds to the stratigraphic position. Also peat and charcoal remnants can be found in the early and middle Holocene rocks on the flat hills rising above the areas of Al Hasa which have turned boggy ever since their formation (2.1.3.6., Table 7, No. 8). Sinter sediments dating back to the Atlanticum (corresponding with the "Neolithic pluvial"

Table 50. *Abstract of geologic processes and climatic changes in central and eastern Saudi-Arabia (Arabian Shelf) during the Quaternary period based on field trips 1972–1975 (cf. sections 3.1.5. and 3.2.)*

| Epoch | Erosion and deposition | Sea level fluctuations | Climate |
|---|---|---|---|
| Holocene | During the dry phases:<br>Eolian and ephemeral fluviatile deposition; sedimentation of clay, silt and fine sand in larger wadis and basins; further development of dunes.<br>During more humid phases:<br>Considerable fluviatile deposition in numerous wadis, partly coarse sand and gravels. Limnic sediments in individual (isolated) basins. Development of calcareous crusts. Vertical erosion in the upper part of tributaries but no re-establishment of the large river systems. | Regression (insignificant),<br>transgression, coastal terrace 2–3 m above the present sea level, maximum 4,000–6,000 years B.P. | dry<br>more humid than at present (about 4,500–7,000 years B.P. Atlanticum)<br>dry<br>humid<br>(about 8,000–9,000 years B.P.)<br>dry<br>humid?<br>(about 11,000 years B.P.) |
| Latest Pleistocene | Mainly deflation and eolian deposition.<br>Stronger fluviatile accumulation and limnic sedimentation in basins in humid phases.<br>Eolian and ephemeral fluviatile deposition in large wadis and basins without outlet. | Regression,<br>coastal terrace 7–10 m above the present sea level (maximum of the interstadial transgression) | extremly dry<br>(about 14,000–25,000 years B.P.)<br>humid<br>(about 25,000–30,000 years B.P.)<br>changing, mainly dry |
| Pleistocene in general | Mainly deflation and eolian deposition. Relative fine clastic sediments by ephemeral runoff. Destruction of calcareous crusts. Humid phases with fluviatile transport and local gravel deposits in wadis, partly development of soil. Insignificant fluviatile erosion. No continuous river systems from the Shield to the Arabian Gulf. | Eustatic sea level fluctuations with general regression trend. Marine terraces above the present sea level:<br>15–18 m<br>25 m<br>38–40 m<br>45–50 m<br>about 70 m<br>about 110 m | mainly dry with humid phases |
| Early Pleistocene/ Late Pliocene | Intensive vertical erosion in the upper and middle part of the recharge area of large W–E running river systems; development of narrow erosion channels.<br>Accumulation of large deltaic fans. Temporarily intensive chemical weathering and development of terra rossa and laterite in the west, and calcareous crusts in the east. | Regression,<br>sea level of the Upper Pliocene (maximum) 150 m above the present one | semiarid to subtropic<br>(changing with arid phases?)<br>between 1 and 3 million years B.P. |
| Upper Pliocene | Development of calcareous crusts. Marine sedimentation in the eastern regions. | Transgression | changing |

of K. KAISER et al., 1973) were discovered in the cave of Nidw Abu Khurjayn (As Sulb Plateau, 5,060 ± 250 years B.P. (cf. 2.2.2., Table 35, No. 9); near Sha'ib Hathag northwest of Al Jirthamiyah (6,880 ± 290 years B.P., Table 40), and in Wadi Birk. No recent sinter formations were observed under the present arid conditions despite episodic rainfalls.

The described sinter discoveries as well as their range between lat 21° N and lat 27° N give evidence of humid warm phases prevailing over the entire peninsula; another indirect argument speaking in favour of this theory are the finds of Neolithic tools between the southwestern part of Ar Rub' Al Khali and Qatar, a rather inhospitable area today (H. A. McCLURE, 1976).

The existing age determinations have not been drawn from a uniform profile but from samples from widely separated and isolated localities of the Arabian Shelf plateau area. As many of these data have been taken from calcareous crusts there are certain doubts about their reliability (cf. G. E. WILLIAMS, H. A. POLACH, 1971). The picture to be formed from these data of the climatic development during the past 10,000 years is still incomplete.

The above-mentioned data could—though they are spread very widely—still be assigned in the most comprehensive sense to the "Neolithic Pluvial", according to its definition by K. KAISER et al. (1973) for the basin of Damascus, representing an equivalent to the European "climatic optimum" in the Atlanticum. Investigations of the marine sedimentation area immediately adjoining the Arabian Peninsula (Gulf of Aden: E. OLAUSSON, I. O. OLSSON, 1969; Arabian Gulf: L. DIESTER-HAAS, 1973) as well as detailed researches in the contiguous African area (e.g., K. W. BUTZER, C. L. HANSEN, 1968; M. SERVANT, S. SERVANT, 1969; G. DELIBRIAS, 1974; M. A. WILLIAMS, 1976; E. T. DEGENS, R. E. HECKY, 1974) however, show multiple changes of humid and arid phases during the Holocene period.

There are reports describing two to four humid phases, the regional correlations of which have not yet been examined thoroughly enough. In accordance with the conclusions drawn from the sediments of the Arabian Gulf by L. DIESTER-HAAS (1973), there can be deduced from the Holocene carbon-14 data of the Arabian Shelf plateau an older humid phase between 8,000 and 9,000 years B.P. and a younger one between 7,000 and 4,500 years B.P. H. A. McCLURE (2.6., Fig. 84) assumes a hyper-arid climate from 6,000 B.P. to present for the Rub' Al Khali.

Not to be classified in this pattern is the age of a calcareous crust (11,000 years B.P.) from the terrace surface near the Bir Juqjuq on the southeast margin of the Arabian Shield. Especially for this crust a mixing of particles of different ages must be taken into account as mentioned in section 2.5.2., in that the age given might be too old. It may also be possible that there was another more humid phase at the end of the Pleistocene, as reported for the African area by a few authors. For the Arabian Peninsula, however, this has still to be established.

### 3.2.2. Climatic Fluctuations During the Würm

The sensible climatic changes in North and Central Europe as well as in North America were primarily due to falling temperatures resulting in enormous formations of inland ice, glaciers and a different form of precipitation. In an arid region as far south as the Arabian Peninsula an eventual drop in temperature surely left its marks behind (e.g., physical weathering); what still remains open is the question of altered conditions of precipitation.

For a long time there was the tendency to conclude for the subtropical regions pluvials and interpluvials parallel to the European glacial and interglacial eras. The intense researche conducted all over the world during recent years concerning the Quaternary climatic changes, however, show clearly that a simple system like the co-ordination of the humid phases of the lower latitudes to the glaciations of the higher latitudes is not possible. Also the differentiation between north and south pluvials cannot yet explain satisfactorily the obviously very widely differing conditions. In addition to this, the expression "pluvial" is too inaccurate for the sometimes only slightly increased quantities of precipitation.

Sure proofs of greater amounts of precipitation are—as mentioned before—sinter formations in terraces, caves, and soils, and weathering horizons. These phenomena have not been observed under the present arid climatic conditions. On the Arabian Peninsula, crust formations dating back to 25,000–30,000 years B.P. have been found in the northern and southern areas. Carbon-14 measurements made from calcite crusts collected from crystalline gravels below the surface of terraces in Wadi Sha'bah (a northern tributary of Wadi Ar Rimah, lat. 27° N), revealed ages of 28,900 ± 1,300 and 30,200 ± 1,300 years B.P. (cf. chapt. 2.3.1.), and the calcareous weathering horizons 0.7 to 1.0 meter below land surface in the Muqabil loam pit in Wadi Ranyah (lat. 21° N) showed ages of 26,000 and 29,000 years B.P. (chapt. 2.5.3., Table 44). Additionally, the number of measurements carried out in this field and the numerous examinations made independently in various laboratories, relying on the same results, also speak in favour of the above established theory. Collected sinter pebbles of an age of 28,000 years B.P. can precisely be fitted into this scheme.

Even though the climatological results will not be compared with the investigations of the adjacent areas before volume 2 of this work, it must be stated that this humid phase in the Arabian regions between 25,000 and 30,000 B.P. is in accordance with the increased precipitation in the south of the Sahara (A. T. Grove, A. Warren, 1968; M. Servant, S. Servant, 1969; G. Delibrias, 1974), in Sudan and Egypt (K. W. Butzer, 1961; R. Huckriede, 1972), as well as in the areas of Jordan and Lebanon (R. Huckriede, 1966; A. Leroi-Gourhan, 1974).

From 25,000 to 12,000 years B.P. arid conditions prevailed again in Arabia[1]. There do not exist, it is true, datable facts for the continental areas, but the absence of soil formations or adequate sediments in itself suggests a more arid period. Distinctive sand dune formations in the area of the Arabian Gulf, desiccated at that time (P. Kassler, 1973), as well as investigations in the neighbouring areas confirm such a hypothesis.

---

[1] H. A. McClure (2.6., Fig. 84) assumes hyper arid conditions from 17,000 to 9,000 B.P. for the Rub' Al Khali. A more precise correlation should be possible after investigations in the western part of the country (Vol. 2).

It seems, however, that some results of carbon-14 data on ground-water do not yet confirm this arid period. Today's ground-water reserves from greater depths , especially waters from the aquifer of the Umm er Radhuma limestones that are vital to large parts of the Shelf, mainly had their origin in more humid phases, when great quantities of water infiltrated into the subsurface. This hypothesis is supported by age determinations made for the big springs of Al Hasa and those in the area of Al Qatif.

The investigated carbon-14 ages lie between 22,000 and > 34,500 years B.P. and thus at first give rise to the conclusion of increased precipitation activity parallel to the whole Würm period. On the other hand, as infiltration of precipitation during the younger Holocene humid phases cannot be excluded, it may be assumed that the age data corresponding with the Würm maxima are due to mixed waters of older and younger humid phases. These waters, furthermore, prove that considerably lower temperatures must have prevailed at the time of their infiltration into the subsoil. As described in detail in chapter 2.1.8., numerous measurements of the stable isotopes of deuterium and oxygen-18, carried out in quite a number of water samples taken from Al Hasa and Al Qatif, show a clear climatic effect (Fig. 50) in comparison to younger, unconfined groundwaters occurring in wadi fills. In the diagrams, the younger waters are co-ordinated above the straight line $\delta D = 8 \cdot \delta ^{18}O + 10$. The parameter $t$ in the equation $\delta D = m \delta ^{18}O + t$, has in this case a higher value than that found in Central Europe. Parameter $t$ has frequently been described as climatic factor for various parts of the Middle East. What is new are the results gained through measurements of lower $t$-values in waters of this area which originated in the Würm period. There are not enough criteria available yet to state clearly by how many degrees centigrade the yearly average temperature was lower; however, studies in this field have not yet been carried out.

### 3.2.3. Prewürm Pleistocene

It appears that the longest phases of the Pleistocene on the Arabian Shelf plateau were marked by mainly arid conditions. The Late Pliocene/Early Pleistocene landscape, with its deep erosion channels and its ample gravel fans influenced by the enormous surface runoffs, was overprinted almost only by eolian processes. Deflation and dune accumulation, which created the present sand dunes, dominated in the areas of the old fluviatile accumulation plains. The erosion channels were filled up with eolian sands, fanglomeratic deposits and wadi sabkhah sediments. There was no soil formation at the time and old soil was partly destroyed or blown out.

Certainly, it would be a mistake to conclude that up to the Würm period no more humid phases occurred, just because the big river systems that existed in the early Pleistocene have never been regenerated again. A number of different conditions such as enormous eolian accumulations as well as the relatively small increase of precipitation during humid phases made a considerably stronger surface runoff impossible. The local fluviatile accumulation of ancient tributaries was caused by the obstruction of the main river channels not by their regeneration. Proof of more humid phases can be found in the fluviatile sands and gravels intercalated in the wadi fills (cf. sects. 2.4.3. and 2.5.3.). Unfortunately, there are insufficient detailed profiles to permit conclusions about the extent and number of these humid phases during the mainly arid climate.

A caution must be exercised in overestimating of these humid phases in the literal meaning of the word "pluvial", which easily leads to misconceptions especially regarding the area of the Arabian Shelf plateau. In contrast to the intense humid phase of the Late Pliocene/Early Pleistocene, which—though older—can still be substantiated in different, regionally extended morphological and sedimentological forms, the Pleistocene humid phases are manifested in the relatively thin fluviatile channel fillings. Their material mainly originates from very local recharge areas. This means that the precipitation during the latter humid phases increased only to a limited extent and in consequence of this, semiarid conditions might have prevailed on the Arabian Shelf plateau.

### 3.2.4. Geochronological Delimitation and Division of the Quaternary

The time extension of the Quaternary was in the past a matter for discussion and still is today. Up till now it has been impossible to unify the various points of view and arguments concerning climatology, marine vertebrate development, anthropology, paleobotany and paleomagnetics in order to find a universal delimitation. In the very latest years, however, can be noticed promising improvements in the time classification of Quaternary geology and climatology (W. A. Berggren, J. A. van Couvering, 1974).

Until now, considerably differing proposals for the geochronological delimitation of the Quaternary-Tertiary boundary have been based primarily on the close relationship of the Quaternary and cooler climates. Therefore, it was supposed that the beginning of the generally cooler period coincided with the first appearance of larger glaciation. The discovery of still older glacial periods and improved methods for dating showed that the Quaternary lasted longer than was originally supposed. The end of the Tertiary was therefore corrected in Europe from 0.6 million years B.P.—according to the solar irradation curve of M. Milankovic (1941)—to about 2.5 million years B.P. with regard to the time fixation of the cold "Praeteglium"-period (e.g., W. H. Zagwijn, 1975). There can be proved even older maxima of coldness not only in polar regions (e.g., T. Einarson et al., 1967), but also in the temperate latitudes of North America. There, the first glaciation of the Sierra Nevada appeared 3 million years ago (R. P. Curry, 1966; B. W. Birkeland et al., 1971). In the Antarctic a vast glacial cover had developed as early as 4 to 6 million years ago (J. H. Mercer, 1973). Since the first appearance of glaciation differs regionally, this partly explains the varying concepts concerning the duration of the Quaternary and shows that such facts are of no help for a universal time classification.

There are also difficulties in the division of the older phases of the Quaternary in accordance with some worldwide events. This concerns particularly the periods of less characteristic climatic changes in areas of lower latitude. The simple equation of the glacials to the pluvials of the subtropics and tropics, customary 15 years ago (R. BRINKMANN, 1966), has been replaced today by a more detailed comparison using the idea of zonal shiftings and changes. Postglacial climatic fluctuations, which are proved by numerous absolute age datings all over the world, show that a general cooling-off or a temperature increase respectively, has a worldwide effect, but with differing accentuation and intensity depending on numerous factors.

We have improved our knowledge of the geochronology and biography of the Quaternary in the last 15 years by increased investigations of marine deposits.

The main factors have been:

1. Progress in the intercontinental correlation of marine microbiostratigraphic zones by means of evolutionary series of planktonic organisms (Foraminifera and Radiolaria). The worldwide correlation of these Cenozoic microzones has been promoted by deep-sea drilling projects.

2. The growing number of absolute age dates by the K-Ar method provided an increasingly more compact pattern of geochronological data to classify the climatic, sedimentological and volcanological processes.

3. The improvement and completion of the paleomagnetic time scale with the data of geomagnetic changes of polarity (A. COX, 1969) provides an additional gauging and classification of biostratigraphic sequences.

An excellent summary has been published by W. A. BERGGREN and J. A. VAN COUVERING (1974). They give a survey of the planktonic microfaunal and biostratigraphic division of the last 15 million years and compare it with the biochronology of the mammalians which is important for the division of continental series. The geochronological standardization was developed by means of numerous absolute age determinations as well as by the paleomagnetic time scale. In Table 51 the most important results concerning the geochronological classification of the Quaternary divisions are listed according to the various aspects.

For the beginning of the Quaternary, the base of the Calabrian was used (acc. to M. GIGNOUX, 1910). This is in accordance with the advice of the 19th International Geologic Congress held in Algiers (1952), where the Calabrian was nominated to be the oldest terrace of the marine Pleistocene. The biostratigraphic delimitation is defined by a faunal boundary according to which the beginning of the Quaternary is marked by the appearance of *Globorotalia truncatulinoides* d'ORBIGNY (W. A. BERGGREN et al., 1967). The geochronological comparison shows that this boundary is placed in the Olduvai-Gilsa normal magnetic event (from 1.61 to 1.82 million years) of the Matuyama inverse magnetization epoch. The Olduvai-Gilsa event (C. S. GROMMÉ, R. L. HAY, 1971)—with its duration of 0.2 million years a geologically very short period—is therefore suggested for the Pliocene/Pleistocene delimitation in accordance with many other authors.

In contrast with this are delimitations which principally refer to terrigenous geological process. W. H. ZAGWIJN (1975) assumed the beginning of the Quaternary, on the basis of paleobotanic and climatic standards, at

2.5 million years B.P. Comparable data are also given by W. SIMON, H. J. LIPPOLT (1967) based on the evoluationary trend of the vertebrates. With regard to the significance of marine stratigraphy for the worldwide correlation, this study is based on the concept of the Quaternary as given by W. A. BERGGREN, J. A. VAN COUVERING (1974).

Difficulties with the delimitation of the Pliocene and Pleistocene undoubtedly arise for terrigenous deposits from the lack of fossils, as long as there are no other methods for dating. A similar situation exists with the wide ranging classification of the Villafranchian. It was considered for a long time to be the oldest Pleistocene and in Italy its interfingering with the marine Calabrian and with the Astian is known. Today this expression is used as the biochronological name for an evolutionary stage of the continental mammals in Europe. Its duration is about 1 to 3 million years. The lower Villafranchian, therefore, can be taken most by Pliocene.

Today there are still difficulties with the geochronological division of the continental Pleistocene. An insufficient number of exactly dated processes makes it difficult to correlate precisely. Thus, the data in Table 51 for the classification of the glacial eras and their parallelization are still provisional and require further detailed investigations. A comparison of the given data of individual glacial periods, e.g., Günz 0.9 to 0.7 million years B.P. with those of C. EMILIANI (1966), J. FRECHEN, H. J. LIPPOLT (1965) or P. EVANS (1972), who fix the age of the Günz glacial stage at 0.5 million years B.P. or even less according to K-Ar dates, makes evident that there are basic differences concerning the integration and classification of the individual cold periods and ice ages [1].

A detailed geochronological division of the climatic processes is only available for the uppermost Quaternary, so the climatic fluctuations of the Würm period can be used for a comparison all over the world. This is even more valid for the oscillating Holocene increase in temperature. Carbon-14 dates of carbonate rocks, remnants of shells, bone and organic substances facilitate a precise time classification of the processes of the last 30,000 to 40,000 years. The geochronological subdivision of the Quaternary can at present be applied only with great caution to the geological and climatic processes in the areas of the Arabian Shelf plateau. As there is frequently a lack of fossils in the terrigenous sediments, which were mainly deposited in arid climates, any biostratigraphic division is impossible for the moment. This is also true for the marine shoal-water deposits, the facies fossils of which practically correspond with present faunal associations.

For the Pleistocene Epoch on the Arabian Shelf a stratigraphic and geochronological subdivision cannot yet be worked out. Up till now there are only two competent K-Ar dates of basalts (3.5 and 1.1 million years B.P., sect. 2.5.2.2.). These two data limit the phase of more humid climate and consequently also of the different erosion and sedimentation processes to the

---

[1] For Central Europe at least 17 interglacials after the Olduvai Event have been identified in a study of J. FINK and G. J. KUKLA (1977) published after the correction of the second proof of this volume. This, and further research of J. FINK shall be discussed in volume II.

Table 51. *Geochronological correlation of the last 3 million years B.P.* (after W. A. Berggren, J. A. van Couvering, 1974)

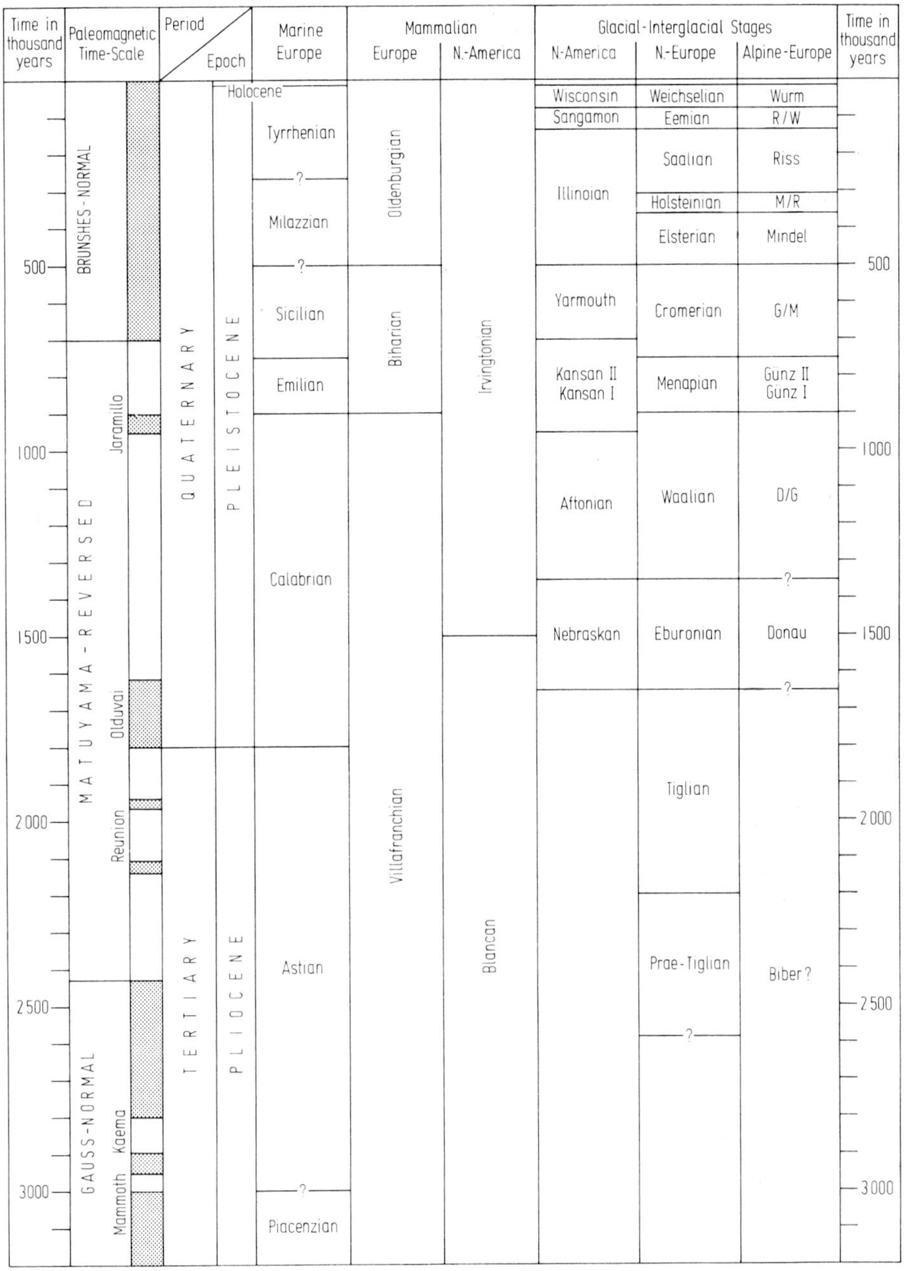

period of the Late Pliocene/Early Pleistocene in contrast to the following more or less arid periods.

Processes which can be correlated with this humid phase are classified in consequence with the "Late Pliocene/Early Pleistocene", although the duration of the phase is not known. This double definition is meant rather to express more the undetermined classification for the Late Pliocene or the Early Pleistocene. In contrast to this, all processes which took place between this period and Würm are classified simply as "Pleistocene". Processes older than the humid Late Pliocene/Early Pleistocene but younger than the sedimentation of the complex strata of the Hofuf Formation are placed in the "Upper Pliocene".

### 3.2.5. The Late Pliocene/Early Pleistocene Phase

An essential conclusion of our research is—as pointed out in the previous chapters—that the last intense long humid period took place in Late Pliocene/Early Pleistocene between 1.2 and 3.5 million years B.P. In contrast to the relatively short interruptions of the arid Quaternary climate, the durations of which generally were too short to influence morphologic and geologic events, the above mentioned long-lived climatic fluctuations caused a lasting alteration of the whole morphological processes. So, the Late Pliocene/Early Pleistocene differs by its intense chemical weathering and enormous fluviatile erosion and sedimentation processes from the mainly eolian processes of the Quaternary.

Our knowledge today is far from sufficient to give a more detailed picture of the Late Pliocene/Early Pleistocene conditions and totally inadequate to provide out a more detailed climatic-stratigraphic subdivision of the 2.5 million years in question. But today there are enough facts to justify in this case differentiating between a period of its own processes, marked by the humid climate, and a dominantly arid period of the remaining Quaternary.

Regarding the Arabian Peninsula the basalts of Wadi Ranyah must be mentioned once more (cf. sect. 2.5.2.). The intense lateritic weathering and deeply furrowed dissection of a basalt flow 3.5 million years old, and the absence of a corresponding weathering as well as the lack of erosion appearances of the younger basalt flow (1.2 million years old) provide, by means of K-Ar ages, a times classification of this more humid climatic period.

The surface runoff promoted by the copious precipitation led, in connection with the lowering of the general base level of erosion, to a distinct linear erosion (cf. e.g., sect. 2.4.3. and 2.5.3.). This is evident not only in the dissection of the basalts but also in the now mostly filled-up erosion channels of the ancient drainage system. Directly connected with them are the enormous gravel fans which have accumulated in the forefield of these river systems, partly interfingering with deposits of the regressing Late Pliocene/Early Pleistocene sea. These deltaic gravel fans were declared long ago to be remnants of an intense humid period. However, according to the former customary interpretation of the effect of the Pleistocene temperature decrease in subtropical areas, had been taken for Pleistocene pluvials.

In the eastern part of the Arabian Peninsula thick calcareous crusts—proof of semiarid conditions—must be assigned to this humid period. In contrast to the lateritic, i.e., humid tropical, weathering in the Shield area there are available good indications of regionally different climates on the Arabian Peninsula. This idea is also supported by the development of the ancient drainage system. Its tributaries mainly descended from the higher situated Shield area whereas there was no adequate runoff pattern developed in the east.

# References


ALIMEN, H., 1963: Considérations sur la chronologie du Quaternaire saharien. Soc. Géol. France, Bull. sér. 7, **5**, 627—634, Paris.

ALIMEN, H., 1965: The Quaternary era in the Northwest Sahara. Geol. Soc. Amer., Spec. Pap. **84**, 273—291, New York.

AL SINAWI, S., AL ANSARI, N., 1970: Glossary of earth science terms. 2nd. ed., 78 p. Geol. Soc. Iraq, Baghdad.

ANDERBERG, M. R., 1973: Cluster Analysis for Applications. 359 p. Harcourt, Brace, Jovanovich, Publishers. New York: Academic Press.

ANDERSON, E. C., LIBBY, W. F., WEINHOUSE, S., REICH, A. F., KIRSHENBAUM, A. D., GROSSE, A. V., 1974: Natural radiocarbon from cosmic radiation. Phys. Rev. **72**, 931—936, New York.

Arabian American Oil Company, 1935–1974: Oceanographic Meteorological Services, Survey Unit. Dhahran.

Arabian American Oil Company, 1968: ARAMCO Handbook, Oil and the Middle East. 279 p. Dhahran.

Arabian American Oil Company, 1973: Booklet on oil statistics. 45 p. Dhahran.

Arabian American Oil Company, 1974: Facts and figures, 1974. 22 p. Dhahran.

Arabian American Oil Company Staff, 1959: Ghawar oil field Saudi Arabia. Amer. Assoc. Petrol. Geologists, Bull. **43**, 434—454, Tulsa, Oklahoma.

ARISTARAIN, L. F., 1971: On the definition of caliche deposits. Z. Geomorph. N.F. **15**, 274—289, Berlin-Stuttgart.

ASSEMIEN, P., FILLERON, J. C., MARTIN, L., TASTET, J. P., 1971: Le Quaternaire de la zone littorale de Cote d'Ivoire. Quaternaria **15**, 305—316, Roma.

AWAD, H., 1963: Some aspects of the geomorphology of Morocco related to the Quaternary climate. Geogr. J. **129**, 129—139, London.

BAGNOLD, R. A., 1941: The physics of blown sand and desert dunes, 265 p. London: Methuen.

BAGNOLD, R. A., 1951: Sand formations in Southern Arabia. Geogr. J. **117**, 78—86, London.

BAKKER, J. P., 1958: Some observations in connection with recent Dutch investigations about granite weathering and slope development in different climates and climatic changes. Z. Geomorph. N.F. Suppl. **1**, 69—92, Berlin-Stuttgart.

BALOUT, L., 1952: Pluviaux interglaciaires et préhistoire saharienne. Inst. Rech. Sahar., Trav. **8**, 9—22, Alger.

BARBEAU, J., 1956: Notice explicative sur la feuille Fort Lamy, Carte géol. de reconnaissance 1 : 1,000,000. Gouvernement générale de l'Afrique équatorial français, Paris.

BARSCH, D., ROYSE, C. F., 1972: A model for development of Quaternary terraces and pediment terraces in the southwestern United States of America. Z. Geomorph. N.F. **16**, 54—75, Berlin-Stuttgart.

BARTH, H. K., 1976: Pedimentgenerationen und Reliefentwicklung im Schichtstufenland Saudi Arabiens. Z. Geomorph. N.F. Suppl. **24**, 11—119, Berlin-Stuttgart.

BAR-YOSEF, O., TCHERNOV, E., 1972: On the palaeo-ecological history of site of "Ubeidiya". 35 p. Jerusalem: Israel Academy of Sciences and Humanities.

BAUMGARTNER, A., REICHEL, E., 1975: Die Weltwasserbilanz. 179 p. München-Wien: Oldenbourg.

BENDER, F., 1968: Geologie von Jordanien. Beitr. Region. Geol. Erde **7**, 230 p. Berlin-Stuttgart: Borntraeger.

BENDER, F., 1975: Geology of the Arabian Peninsula: Jordan. U.S. Geol. Survey, Prof. Paper 560-I, 136 p. Washington.

BERGGREN, W. A., 1971: The Pliocene/Pleistocene boundary in deep-sea cores; its significance for the concept of the Quaternary. Quaternaria **15**, 77—78, Roma.

BERGGREN, W. A., COUVERING, J. A. VAN, 1974: The late Neogene: biostratigraphy, geochronology and paleoclimatology of the last 15 million years in marine and continental sequences. Palaeogeogr. Palaeoclimat. Palaeoecology **16**, 1/2, 1-216, Amsterdam.

BERGGREN, W. A., PHILLIPS, J. D., BERTELS, A., WALL, D., 1967: Late Pliocene-Pleistocene stratigraphy in deep-sea cores from the south central North Atlantic. Nature **216**, 253—254, London.

BERRY, L., WHITEMAN, A. J., BELL, S. V., 1966: Some radiocarbon dates and their geomorphological significance: emerged reef complex of Sudan. Z. Geomorph. **10**, 119—143, Berlin-Stuttgart.

BEYDOUN, Z. R., 1966: Geology of the Arabian Peninsula: Eastern Aden Protectorate and part of Dhufar. U.S. Geol. Survey, Prof. Paper 560-H, 49 p. Washington.

BEYDOUN, Z. R., GREENWOOD, J. E., 1968: Protectorat d'Aden et Dhufar. Lexique Stratigr. Intern. **3**, 10 b 2, 128 p. Paris.

BIRKELAND, P. W., CRANDELL, D. R., RICHMOND, G. M., 1971: Status of correlation of Quaternary stratigraphical units in the western conterminous United States. Quaternary Res. **1**, 208—227, New York.

BISHOP, W. W., 1967: Annotated lexicon of Quaternary stratigraphical nomenclature in East Africa. In: BISHOP, W. W., CLARK, J. D. (Eds.), Background to evolution in Africa, 375—395. Chicago-London: Univ. Chicago Press.

BISHOP, W. W., 1969: Pleistocene stratigraphy in Uganda. Geol. Surv. Uganda, Mem. **10**, 128, Entebbe.

BISHOP, W. W., CLARK, J. D. (Eds.), 1967: Background to evolution in Africa, Proceedings of the Symposium "Systematic Investigation of the African Later Tertiary and Quarternary", Burg Warthenstein, Austria 1965, 935 p. Chicago-London: Univ. Chicago Press.

BLANCHARD, R., 1926: Le Relief de l'Arabic Central. Rev. Geogr. Alpine **14**, 765—786, Grenoble.

BLANKENHORN, M., 1914: Syrien, Arabien und Mesopotamien. Handbuch der regionalen Geologie, 5, 4, 159 p. Heidelberg: Winter.

BLISSENBACH, E., 1954: Geology of alluvial fans in semiarid regions. Geol. Soc. Amer. Bull. **65**, 175—190, New York.

BLÜMEL, W. D., 1976: Kalkkrustenvorkommen in Südwestafrika. Basler Afrika Bibliogr., Mitt. **15**, 17—46, Basel.

BMDP Computer Programs: Written for Center for Health Science of the University of California at Los Angeles, 1973. Modified for use at Los Alamos Scientific Laboratory.

BOBEK, H., 1955: Klima und Landschaft Irans in vor- und frühgeschichtlicher Zeit. Geogr. Jbericht Österreich **25**, 1—42, Wien.

BOBEK, H., 1961: Die Salzwüsten Irans als Klimazeugen. Anz. Österr. Akad. Wiss., phil.-hist. Kl. 1961, **3**, 7—19, Wien.

BOBEK, H., 1963: Nature and implications of Quaternary climatic changes in Iran. UNESCO, Arid Zone Res. **20**, 403—422, Paris.

BÖTTGER, U., ERGENZINGER, P. J., JAECKEL, S. H., KAISER, K., 1972: Quartäre Seebildungen und ihre Mollusken. Inhalte im Tibestigebirge und seinen Rahmenbereichen der zentralen Ostsahara. Z. Geomorph. N.F. **16**, 2, 182—234, Berlin-Stuttgart.

BOGOCH, R., COOK, P., 1974: Calcite cementation of Quaternary conglomerate in southern Sinai. J. Sed. Petrology **44**, 3, 917—920, Tulsa, Oklahoma.

BONIFAC, E., MARS, P., 1959: Le Tyrrhénien dans le cadre de la chronologie méditeranéenne. Soc. Géol. France, Bull., sér. **7**, 1, 62—78, Paris.

BOWLER, J. M., 1976: Aridity in Australia: age, origins and expression in aeolian landforms and sediments. Earth-Science Reviews **12**, 279—310, Amsterdam.

BOWMAN, D., 1970: Geomorphology of the shore terrace of the Late Pleistocene Lisam lake (Israel). Palaeogeogr., Palaeoclimat., Palaeocology **9**, 183—209, Amsterdam.

BRAMKAMP, R. A., BROWN, G. F., 1948: Ground water in the Nejd, Saudi Arabia. New York Acad. Sci., Trans., Ser. **2**, 10, 236—237, New York.

BRAMKAMP, R. A., GIERHART, R. D., BROWN, G. F., JACKSON, R. O., 1956: Geologic map of the Southern Tuwayq quadrangle, Kingdom of Saudi Arabia. U.S. Geol. Survey, Misc. Geol. Investig. Map I-212 A, 1 : 500,000, Washington.

BRAMKAMP, R. A., GIERHART, R. D., OWENS, L. D., RAMIREZ, L. F., 1963: Geologic map of the western Rub' al Khali quadrangle, Kingdom of Saudi Arabia. U.S. Geol. Survey, Misc. Geol. Investig. Map. I-218 A, 1 : 500,000, Washington.

BRAMKAMP, R. A., RAMIREZ, L. F., 1958: Geologic map of the Northern Tuwayq quadrangle, Kingdom of Saudi Arabia. U.S. Geol. Survey, Misc. Geol. Investig. Map I-207 A, 1 : 500,000, Washington.

BRAMKAMP, R. A., RAMIREZ, L. F., 1959 a: Geologic map of the Wadi al Batin quadrangle, Kingdom of Saudi Arabia. U.S. Geol. Survey, Misc. Geol. Investig. Map I-203 A, 1 : 500,000, Washington.

BRAMKAMP, R. A., RAMIREZ, L. F., 1959 b: Geologic map of the Northwestern Rub' al Khali quadrangle, Kingdom of Saudi Arabia. U.S. Geol. Survey, Misc. Geol. Investig. Map I-213 A, 1 : 500,000, Washington.

BRAMKAMP, R. A., RAMIREZ, L. F., 1961 a: Geologic map of the Central Persian Gulf quadrangle, Kingdom of Saudi Arabia. U.S. Geol. Survey, Misc. Geol. Investig. Map I-209 A, 1 : 500,000, Washington.

BRAMKAMP, R. A., RAMIREZ, L. F., 1961 b: Geologic map of the Northeastern Rub' al Khali quadrangle, Kingdom of Saudi Arabia. U.S. Geol. Survey, Misc. Geol. Investig. Map I-214 A, 1 : 500,000, Washington.

BRAMKAMP, R. A., RAMIREZ, L. F., 1963: Geologic map of the Darb Zubaydah quadrangle, Kingdom of Saudi Arabia. U.S. Geol. Survey, Misc. Geol. Investig. Map I-202 A, Washington.

BRAMKAMP, R. A., RAMIREZ, L. F., BROWN, G. F., POCOCK, A. E., 1963 a: Geologic map of the Wadi Ar Rimah quadrangle, Kingdom of Saudi Arabia. U.S. Geol. Survey, Misc. Geol. Investig. Map I-206 A, 1 : 500,000, Washington.

BRAMKAMP, R. A., RAMIREZ, L. F., STEINECKE, M., REISS, W. H., 1963 b: Geologic map of the Jawf-Sakakah quadrangle, Kingdom of Saudi Arabia. U.S. Geol. Survey, Misc. Geol. Investig. Map I-210 a, 1 : 500,000, Washington.

BRINKMANN, R., 1966: Abriß der Geologie, 9. Aufl., Bd. 2, 345 p. Stuttgart: Enke.

BROSCHE, K. U., MOLLE, H. G., 1976: Geomorphologische und klimageschichtliche Studien in Süd- und Zentraltunesien. Z. Geomorph. N.F. Suppl. **24**, 149—159, Berlin-Stuttgart.

BROWN, F. H., HEINZELIN, J. DE, HOWELL, F. C., 1970: Pliocene/Pleistocene formations in the lower Omo Basin southern Ethiopia. Quaternaria **13**, 247—268, Roma.

BROWN, G. F., 1949: The geology and ground water of Al Kharj district, Nejd, Saudi Arabia. New York Acad. Sci., Trans. Ser. **2**, 10, 370—375, New York.

BROWN, G. F., 1960: Geomorphology of western and central Saudi Arabia. Intern. Geol. Congr., 21st., Rept. **21**, 150—159, Copenhagen.

BROWN, G. F., 1970: Eastern margin of the Red Sea and the coastal structures in Saudi Arabia. Royal Soc. London Philos. Trans. **A 267**, 75—89, London.

BROWN, G. F., 1972: Tectonic map of the Arabian Peninsula. Kingdom of Saudi Arabia, Dir. Gen. Mineral Resources, Map AP-2, 1 : 4,000,000, Jiddah.

BROWN, G. F., JACKSON, R. O., 1960: The Arabian Shield. Intern. Geol. Congr., 21st, Rept. **9**, 69—77, Copenhagen.

BROWN, G. F., JACKSON, R. O., BOGUE, R. G., MACLEAN, W. H., 1963: Geologic map of the Southern Hijaz quadrangle. U.S. Geol. Survey. Misc. Geol. Investig. Map I-210 A, 1 : 500,000, Washington.

BROWN, G. F., LAYNE, N., GOUDARZI, G. H., MACLEAN, W. H., 1963: Geologic map of the Northeastern Hijaz quadrangle, Kingdom of Saudi Arabia. U.S. Geol. Survey, Misc. Geol. Investig. Map I-205 A, 1 : 500,000, Washington.

BRÜCKNER, W., 1957: Laterite and bauxite profiles of West Africa as an index of rhythmical climatic variations in the tropical belt. With an appendix on the chief factors involved in the formation and preservation of laterite and bauxite. Eclogae Geol. Helv. **50**, 2, 239—250, Basel.

BRÜCKNER, W., 1960: Afrikanische Quartärprobleme. Regio Basiliensis **2**, 41—46, Basel.

BRUMLEY, B., 1977: Not only old bones. Teheran Journal, Feb. 12, 1977.

BRUNNER, H., 1970: Pleistozäne Klimaschwankungen im Bereich des östlichen Mysore-Plateaus (Südindien). Geologie **19**, 1, 72—82, Berlin.

BUCHBINDER, B., BEGIN, Z. B., FRIEDMAN, G. M., 1974: Pleistocene algal tufa of Lake Lisan, Dead Sea area. Israel J. Earth-Sci. **23**, 131—138, Jerusalem.

BÜDEL, J., 1952: Bericht über klimamorphologische und Eiszeitforschungen in Nieder-Afrika auf Grund einer Forschungsreise 1950/51. Erdkde. **6**, 104—132, Bonn.

BÜDEL, J., 1955: Reliefgenerationen und plio-pleistozäner Klimawandel im Hoggar-Gebirge. Erdkde. **9**, 100—115, Bonn.

BÜDEL, J., 1957: Die „Doppelten Einebnungsflächen" in den feuchten Tropen. Z. Geomorph. N.F. **1**, 201—228, Berlin-Stuttgart.

BÜDEL, J., 1963: Die pliozänen und quartären Pluvialzeiten der Sahara. Eiszeitalter und Gegenwart **14**, 161—187, Öhringen.

BULL, W. B., 1964: Geomorphology of segmented alluvial fans in western Fresno County, California. U.S. Geol. Survey, Prof. Paper 352 E, 89—129, Washington.

BUNKER, D. G., 1953: The southwest borderland of the Rub' al Khali. Geogr. J. **119**, 420—430, London.

BURDON, D. J., ÖTKUN, G., 1967: The groundwater potential of karst aquifers in Saudi Arabia. Intern. Assoc. Hydrogeologists, Istanbul Meeting, Istanbul.

BURDON, D. J., ÖTKUN, G., 1968: Hydrogeological control of development in Saudi Arabia. Intern. Geol. Congr., 23rd, **12**, 145—153, Prag.

BUREK, P. J., 1974: Plattentektonische Probleme in der weiteren Umgebung Arabiens sowie der Danakil-Afar-Senke. Geotekton. Forsch. **47**, 1/2, 1—100, Stuttgart.

BUSCHE, D., 1973: Die Entstehung von Pedimenten und ihre Überformung untersucht an Beispielen aus dem Tibesti-Gebirge, République du Tchad. Berliner Geogr. Abh. **18**, 110 p. Berlin.

BUSH, P. R., 1970: Chloride rich brines from sabkha sediments and their possible role in ore formation. Inst. Mining Metallurgy, Trans., sec. B. **79**, 137—144, New York.

BUSH, P., 1973: Some aspects of the diagenetic history of the sabkha in Abu Dhabi, Persian Gulf. In: PURSER, B. H. (Ed.), The Persian Gulf, 395—406, Berlin-Heidelberg-New York: Springer.

BUTLER, G. P., 1969: Modern evaporite deposition and geochemistry of co-existing brines, the sabkha, Trucial Coast, Arabian Gulf. J. Sed. Petrology **39**, 70—89, Tulsa, Oklahoma.

BUTLER, G. P., 1971: Origin and controls on distribution of arid supratidal (sabkha) dolomite, Abu Dhabi, Trucial Coast (abs.). Amer. Assoc. Petrol. Geologists, Bull. **55**, 332, Tulsa, Oklahoma.

BUTLER, G. P., KENDALL, C. G., KINSMAN, D. J., SHEARMAN, D. J., SKIPWITH, S. P., 1965: Recent anhydrite from the Trucial coast of the Arabian Gulf. Geol. Soc. London, Circ. **120**, 3, London.

BUTZER, K. W., 1957 a: Late glacial and postglacial climatic variation in the Near East. Erdkde. **11**, 21—35, Bonn.

BUTZER, K. W., 1957 b: The recent climatic fluctuations in lower latitudes and the general circulation of the Pleistocene. Geogr. Annaler **39**, 2/3, 105—113, Stockholm.

BUTZER, K. W., 1958: Quaternary stratigraphy and climate in the Near East. Bonner Geogr. Abh. **24**, 1—157, Bonn.

BUTZER, K. W., 1961: Climatic change in arid regions since the Pliocene. UNESCO, Arid Zone Res. **17**, 13—56, Paris.

BUTZER, K. W., 1963: The last "pluvial" phase of the Eurafrican subtropics. UNESCO, Arid Zone Res. **20**, 211—221, Paris.

BUTZER, K. W., 1966: Climatic changes in the arid zones of Africa during early to mid-Holocene times. In: SAWYER, J. S. (Ed.), World climate from 8,000 to 0 B.C., 72—83, London: Royal Meteor. Soc.

BUTZER, K. W., HANSEN, C. L., 1967: Upper Pleistocene stratigraphy in Southern Egypt. In: BISHOP, W. W., CLARK, J. D. (Eds.), Background to evolution in Africa, 329—356, Chicago-London: Univ. Chicago Press.

BUTZER, K. W., HANSEN, C. L., 1968: Desert and river in Nubia. 562 p. Madison, London: University of Wisconsin Press.

CATON-THOMPSON, G., 1938: Geology and archaeology of the Hadhramaut, South West Arabia. Nature **142**, 3586, 139—142, London.

CATON-THOMPSON, G., 1964: Flint tools from Southern Arabia. Nature **201**, 872, London.

CHALLINOR, J., 1973: Dictionary of geology, 4th ed., 350 p. Cardiff: Univ. Wales Press.

CHAPMAN, R. W., 1971: Climatic changes and the evolution of landforms in the eastern province of Saudi Arabia. Geol. Soc. Amer., Bull. **82**, 2713—2728, Boulder, Colorado.

CHAPMAN, R. W., 1974: Calcareous duricrust in Al-Hasa, Saudi Arabia. Geol. Soc. Amer., Bull. **85**, 119—130, Boulder, Colorado.

CLARK, J. D., WILLIAMS, M. A. J., SMITH, A. B., 1973: The geomorphology and archaeology of Adrar Bous, Central Sahara: a preliminary report. Quaternaria **17**, 245—297, Roma.

COLEMAN, J. M., 1969: Brahmaputra River: Channel processes and sedimentation. Sediment. Geol. **3**, 129—239, Amsterdam.

COOMER, P. G., BARRETT, W. L., DODSON, M. H., 1970: Natural variations in sulfur isotope abundances: a bibliography. J. Earth Sci. Leeds **8**, 75—99, Leeds.

COQUE, R., 1955: Notes morphologiques sur les grands chottes Tunisiens. Ass. Géogr. Franc., Bull. **253/4**, 174—185, Paris.

COQUE, R., JAUZEIN, R., 1967: The geomorphology and Quaternary geology of Tunisia (transl. by DE COSTER, G. L.). In: MARTIN, L. (Ed.), Guidebook to the geology and history of Tunisia. Petrol. Explor. Soc. Libya, Ann. Field. Conf., 9th, 227—257, Tripoli.

CORNELIUS, R. F. S., FALCON, N. L., South, D., VITA-FINZI, C., 1973: The Musandam expedition 1971—1972. Scientific results, part 1. Geogr. J. **139**, 400—425, London.

CORNWALL, P. B., 1946: Ancient Arabia: Explorations in Hasa 1940—1941. Geogr. J. **107**, 28—50, London.

COX, A., 1969: Geomagnetic reversals. Science **163**, 237—245, Washington.

CRAIG, H., 1961 a: Isotopic variations in meteoric waters. Science **133**, 1702—1703, Washington.

CRAIG, H., 1961 b: Standard for reporting concentrations of deuterium and oxygen-18 in natural waters. Science **133**, 1833—1834, Washington.

CURRAY, J. R., 1961: Late Quaternary sea level. A discussion. Geol. Soc. Amer., Bull. **72**, 1707—1712, New York.

CURRY, R. P., 1966: Glaciation about 3,000,000 years ago in the Sierra Nevada. Science **154**, 770—771, Washington.

CURTIS, R., EVANS, G., KINSMAN, D. J., SHEARMAN, D. J., 1963: Association of dolomite and anhydrite in the recent sediments of the Persian Gulf. Nature **197**, 4868, 679—680, London.

Daily and monthly climate reports. Kingdom of Saudi Arabia, Ministry of Defence and Aviation, Met. Dept., Climate Section. Ar Riyadh.

DANSGAARD, W., 1964: Stable isotopes in precipitation. Tellus **16**, 436—468, Stockholm.

DAVIES, O., 1971: Sea-level during the past 11,000 years (Africa). Quaternaria **14**, 195—204, Roma.

DAVIS, J. C., 1973: Statistics and Data Analysis in Geology. 550 p. New York: J. Wiley & Sons, Inc.

DEGENS, E. T., HECKY, R. E., 1974: Paleoclimatic reconstruction of late Pleistocene and Holocene based on biogenic sediments from the Black Sea and a tropical African lake. Colloques Internationaux du C.N.R.S., 219, Les méthodes quantitatives d'étude des variations du climat au cours du Pleistocène, 13—24, Paris.

DELFOUR, J., 1966: Report on the mineral resources and geology of the Hulayfah-Musayna'ah region (sheet 78, zone 1 North). Bur. Recherch, Géol. Minières, Jeddah, SG 66 A 8, Saudi Arabian Dir. Gen. Mineral Resources, open file report 64 p. Jeddah.

DELIBRIAS, C., 1974: Variation du niveau de la mer sur la côte ouest africaine, depuis 26.000 ans. Colloques Internationaux du C.N.R.S., 219, Les méthodes quantitatives d'étude des variations du climat au cours du Pleistocène, 127—133, Paris.

DELIBRIAS, C., LABOREL, J., 1971: Recent variations of the sea level along the Brazilian coast. Quaternaria **14**, 45—49, Roma.

DIESTER, L., 1972: Zur spätpleistozänen und holozänen Sedimentation im zentralen und östlichen Persischen Golf. „Meteor" Forsch.-Ergebnisse, Reihe C, **8**, 37—83, Berlin-Stuttgart.

DIESTER-HAASS, L., 1973: Holocene climate in the Persian Gulf as deduced from grain-size and pteropod distribution. Marine Geol. **14**, 207—223, Amsterdam.

DINÇER, T., AL-MUGRAIN, A., ZIMMERMANN, V., 1974: Study of the infiltration and recharge through the sand dunes in arid zones with special reference to the stable isotopes and thermonuclear tritium. J. Hydrology **23**, 79—109, Amsterdam.

DINÇER, T., NOORY, M., JAVED, A. R. K., NUTI, S., TONGIORGI, E., 1973: Study of groundwater recharge and movement in shallow and deep aquifers in Saudi Arabia with stable isotopes and salinity data. Manuscript SM-182/17, Ar Riyadh.

DOEGLAS, D. J., 1962: The structure of sedimentary deposits of braided rivers. Sedimentology **1**, 167—190, Amsterdam.

DORAN, J. E., HODSON, F. R., 1975: Mathematics and Computers in Archeology. 381 p. Cambridge, Massachusetts: Harvard University Press.

DOUGHTY, C. M., 1964: Travels in Arabia Deserta. 2nd ed., 349 p. London: J. Cape.

DRESCH, J., 1965: Le « désert » de Thar. Ass. Géogr. Franç., Bull. **332/333**, 36—47, Paris.

DROST, W., MOSER, H., NEUMAIER, F., RAUERT, W., 1974: Isotope methods in groundwater hydrology. Eurisotop Office Information Booklet **61**, 176 p. Brussels.

DUBIEF, J., 1959: Le climat du Sahara. Inst. Rech. Sahar., Mém. **1**, Alger.

DUBIEF, J., 1963: Le climat du Sahara. Inst. Rech. Sahar., Mém. **2**, Alger.

DURAN, B. S., ODELL, P. L., 1974: Cluster Analysis, A Survey. Lecture Notes in Economic and Mathematical Systems, 137 p. Berlin-Heidelberg-New York: Springer.

DURY, G. H., 1971: Relict deep weathering and duricrusts in relation to the palaeoenvrionments of middle latitudes. Geogr. J. **137**, 511—522, London.

EINARSON, T., HOPKINS, D. M., DOELL, R. R., 1967: The stratigraphy of Tjornes, Northern Iceland and the history of the Bering Land Bridge. In: HOPKINS, D. M. (Ed.), The Bering Land Bridge, 312—324, Stanford, Calif.

EINSELE, G., HERM, D., SCHWARZ, H. V., 1974: Holocene eustatic sea level fluctuation at the coast of Mauritania. "Meteor" Forsch.-Ergebnisse, Reihe C **18**, 43—62, Berlin-Stuttgart.

EL RAMLY, I. M., 1971: Shorelines changes during the Quaternary in the western desert Mediterranean Coastal Region (Alexandria-Salhum), U.A.R. Quaternaria **15**, 285—292, Roma.

EMERY, K. O., 1956: Sediments and water of Persian Gulf. Amer. Assoc. Petrol. Geologists, Bull. **40**, 2354—2383, Tulsa, Oklahoma.

EMILIANI, C., 1966: Paleotemperature analysis of the Caribbean cores P 6304–8 and P 6304–9 and a generalized temperature curve for the past 425,000 years. J. Geol. **74**, 109—126, Chicago.

EPSTEIN, S., MAYEDA, T., 1953: Variations of the $^{18}O$ content of waters from natural sources. Geochim. Cosmochim. Acta **4**, 213—224, London.

EVANS, G., 1967: Persian Gulf. In: FAIRBRIDGE, R. W. (Ed.), Encyclopedia of Oceanography, 684—694. New York: Reinhold.

EVANS, G., KENDALL, C. G., SKIPWITH, P. A., 1964: Origin of the coastal flats, the sabkha of the Trucial Coast, Persian Gulf. Nature **202**, 4934, 759—761, London.

EVANS, G., KINSMAN, D. J., SHEARMAN, D. J., 1964: Reconnaissance survey of the environment of recent carbonate sedimentation along the Trucial Coast, Persian Gulf. In: VAN STRAATEN, L. M. (Ed.), Deltaic and shallow marine deposits, 12—135. Amsterdam-London-New York: Elsevier.

EVANS, G., SCHMIDT, V., BUSH, P., NELSON, H., 1969: Stratigraphy and geologic history of the sabkha, Abu Dhabi, Persian Gulf. Sedimentology **12**, 145—159, Amsterdam.

EVANS, G., SHEARMAN, D. J., 1964: Recent celestite from the sediments of the Trucial Coast of the Persian Gulf. Nature **202**, 385—386, London.

EVANS, P., 1972: The present status of age determination in the Quaternary (with special reference to the period between 70,000 and 1,000,000 years ago). Intern. Geol. Congr., 24th, Section 12, 16—21, Montreal.

Everitt, B., 1974: Cluster Analysis. 122 p. New York: Halsted Press, division of John Wiley & Sons, Inc.

Fairbridge, R. W., 1961: Eustatic changes in sea level. Physics and Chemistry of the Earth **4**, 99—185, London.

Fairbridge, R. W., 1962: New radiocarbon dates of Nile sediments. Nature **196**, 108—110, London.

Fairbridge, R. W., 1963: Nile sedimentation above Wadi Halfa during the last 20,000 years. Kush **11**, 96—107, Khartoum.

Fairbridge, R. W., 1964: African ice-age aridity. In: Nairn, A. E. (Ed.), Problems in palaeoclimatology. Proc. NATO-Conf. on Palaeoclimate, 1963 Newcastle upon Tyne, 356—360, London.

Fairbridge, R. W., 1965: Eiszeitklima in Nordafrika. Geol. Rdsch. **54**, 399—414, Stuttgart.

Fairbridge, R. W., 1970: World paleoclimatology of the Quaternary. Rev. Geogr. phys. Geol. dyn. (2), **12**, 97—104, Paris.

Fairbridge, R. W., 1972: Climatology of a glacial cycle, Quaternary Res. **2**, 283—302, New York.

Fairbridge, R. W., 1972: Quaternary sedimentation in the Mediterranean region controlled by tectonics, paleoclimates and sea level. In Stanley, D. J. (Ed.), The Mediterranean Sea: a natural sedimentation laboratory, 99—113. Stroudsburg, Pa.: Dowden, Hutchinson & Ross.

Fairbridge, R. W., Richards, H. G., 1967: The INQUA Shorelines commission. Z. Geomorph. N.F. **11**, 2, 205—215, Berlin-Stuttgart.

Falcon, N. L., 1947: Raised beaches and terraces of the Iranian Makran coast. Geogr. J. **109**, 149—151, London.

Falcon, N. L., 1967: The geology of the north-east margin of the Arabian basement shield. Advancement Sci. **24**, 31—42, London.

Falcon, N. L., 1973: The Musandam (Northern Oman) expedition 1971–1972. Geogr. J. **139**, 1, 1–19, London.

Faltings, V., Harteck, P., 1950: Der Tritiumgehalt der Atmosphäre. Z. Naturforsch. **5 a**, 438—439, Wiesbaden.

Faure, H., 1969: Lacs quaternaires du Sahara. Mitt. Intern. Limnolog.

Faure, H., Elouard, P., 1967: Schéma des variations du niveau du l'océan Atlantique sur la côte de l'Ouest de l'Afrique depuis 40.000 ans. Acad. Sci. Paris, Comptes Rendus, Série D **265**, 784—787, Paris.

Faure, H., Hoang, C., Lalou, C., 1974: Le récifs soulevés à l'ouest du Golfe d'Aden (T.F.A.I.) et les hauts niveaux de coraux de la dépression de l'Afar (Ethiopie), géochronologie et paléoclimats interglaciaires. Colloques Internationaux C.N.R.S., Les méthodes quantitatives d'étude des variations du climat au cours du Pleistocène, 103—114, Paris.

Felber, H., 1976: Vienna, Radium Institute Radiocarbon Dates VII. Radiocarbon **18**, in press.

Felix, J., 1904: Studien über tertiäre und quartäre Korallen und Riffkalke aus Ägypten und Sinai-Halbinsel. Z. dt. geol. Ges. **56**, 168—206, Berlin.

Field, H., 1955: New stone age sites in the Arabian Peninsula. Man **121**, 93—94, London.

Field, H., 1958: Stone implements from Rub' al-Khali, Southern Arabia. Man **145**, 136—138, London.

Fink, J., Kukla, G. J., 1977: Pleistocene Climates in Central Europe: At Least 17 Interglacials after the Olduvai Event. Quaternary Research **7**, 363—371, New York.

Fleck, R. J., Coleman, R. G., Cornwall, H. R., Greenwood, W. R., Hadley, D. G., Prinz, W. C., Ratté, J. C., Schmidt, D. L., 1973: Potassium-argon geochronology of the Arabian Shield. U.S. Geol. Survey, Saudi Arabian Project Report 165, 40 p. Jiddah.

Fleck, R. J., Coleman, R. G., Cornwall, H. R., Greenwood, W. R., Hadley, D. G., Schmidt, D. L., Prinz, W. C., Ratté, J. C., 1976: Geochronology of the Arabian Shield, western Saudi Arabia: K-Ar results. Geol. Soc. Amer., Bull. **87**, 9—21, Boulder, Colorado.

FLINT, R. F., 1959: Pleistocene climates in eastern and southern Africa. Geol. Soc. Amer., Bull. **70**, 343—360, New York.

FLINT, R. F., 1969: Pleistocene sand ridges and pans in western Rhodesia. Geol. Soc. Amer., Bull. **79**, 299—314, Boulder, Colorado.

FLINT, R. F., 1971: Glacial and Quaternary geology, 892 p. New York: Wiley & Sons.

FLOHN, H., 1952: Allgemeine atmosphärische Zirkulation und Paläoklimatologie. Geol. Rdsch. **40**, 153—178, Stuttgart.

FLOHN, H., 1958: Bemerkungen zum Problem der globalen Klimaschwankungen. Archiv Meteor. Geophys. Bioklimatol. **9**, 1—13, Wien.

FLOHN, H., 1969: Ein geophysikalisches Eiszeit-Modell. Eiszeitalter und Gegenwart **20**, 204—231, Öhringen.

FRANZ, H., 1958: Beitrag zur Kenntnis der Stratigraphie und Klimatologie des Quartärs im Tschadbecken. Geol. Ges. Wien, Mitt. **51**, 19—65, Wien.

FRANZ, H., 1967: On the stratigraphy and evolution of climate in the Chad Basin during the Quaternary. In: BISHOP, W. W., CLARK, J. D. (Eds.), Background to evolution in Africa, 273—283. Chicago-London: Univ. Chicago Press.

FRECHEN, J., LIPPOLT, H. J., 1965: Kalium-Argon-Daten zum Alter des Laacher Vulkanismus, der Rheinterrassen und der Eiszeiten. Eiszeitalter und Gegenwart **16**, 5—30, Öhringen.

FRENZEL, B., 1967: Die Klimaschwankungen des Eiszeitalters, 296 p. Braunschweig: Vieweg.

FRIEDMAN, G. M., 1972: Significance of Red Sea in problem of evaporites and basinal limestones. Amer. Assoc. Petroleum Geologists, Bull. **56**, 6, 1072—1086, Tulsa, Oklahoma.

FRIEDMAN, I., REDFIELD, A. C., SCHOEN, B., HARRIS, J., 1964: The variation of the deuterium content of natural waters in the hydrologic cycle. Rev. Geophys. **2**, 117—224, Washington.

FUCHS, W., GATTINGER, T. E., HOLZER, H. F., 1968: Explanatory text to the synoptic geologic map of Kuwait, 87 p. Vienna: Geol. Survey Austria.

FUJII, S., LIN, C. C., TJIA, H. D., 1971: Sea level changes in Asia during the past 11,000 years. Quaternaria **14**, 211—216, Roma.

GARDNER, L. R., 1972: Origin of the Mormon Mesa caliche, Clark Country, Nevada. Geol. Soc. Amer., Bull. **83**, 143—156, Boulder, Colorado.

GARSON, M. S., KRS, M., 1976: Geophysical and geological evidence of the relationship of Red Sea transverse tectonics to ancient features. Geol. Soc. Amer. Bull. **87**, 169—181, Boulder, Colorado.

GARY, M., MCAFEE, R., JR., WOLF, C. L., 1972: Glossary of geology, 805 p. Washington: Amer. Geol. Inst.

GASSE, F., 1975: L'évolution des lacs de l'Afar central (Ethiopie et T.F.A.I.) du Plio-Pleistocène à l'actuel. Doctorate d'Etat, thesis University of Paris.

GAT, J. R., 1971 Comments on the stable isotope method in regional groundwater investigations. Water Resources Res. **7**, 980—993, Washington.

GEUKENS, F., 1966: Geology of the Arabian Peninsula: Yemen. U.S. Geol. Survey, Prof. Paper 560-B, 23 p. Washington.

GEYH, M. A., 1971: Die Anwendung der $^{14}$C-Methode. Clausthaler Tektonische Hefte **11**, 118 p. Clausthal-Zellerfeld.

GEYH, M. A., JÄKEL, D., 1974 a: Late glacial and holocene climatic history of the Sahara desert derived from a statistical assay of $^{14}$C dates. Palaeogeogr., Palaeoclimat., Palaeoecology **15**, 205—208, Amsterdam.

GEYH, M. A., JÄKEL, D., 1974 b: Spätpleistozäne und holozäne Klimageschichte der Sahara auf Grund zugänglicher $^{14}$C-Daten. Z. Geomorph. N.F. **18**, 82—98, Berlin-Stuttgart.

GEYH, M. A., OBENAUF, K. P., 1974: Zur Frage der Neubildung von Grundwasser unter ariden Bedingungen. Ein Beitrag zur Hydrologie des Tibesti Gebirges. Freie Universität Berlin, Pressedienst Wissenschaft **5**, 70—91, Berlin.

GHOSE, B., PANDEY, S., SINGH, S., 1966: Geomorphology of the Central Luni Basin, Western Rajasthan. Ann. Arid Zone **5**, 1, 10—25, Jodpu, Rajastan, India.

GIGNOUX, M., 1910: Sur la classification du Pliocène et du Quaternaire de l'Italie du Sud. Acad. Sci. Paris, Comptes Rendus **150**, 841—844, Paris.

GILL, E. D., 1967: Criteria for the description of Quaternaria shorelines. Quaternaria **9**, 237—243, Roma.

GILL, E. D., 1968: Multiple factors in the description of Quaternary shorelines. Proceedings VII Congress INQUA **8**, 375—379, Boulder, Colorado.

GILL, E. D., 1971: The Paris symposium on World sea levels of the past 11,000 years. Quaternaria **14**, 1—6, Roma.

GILL, E. D., HOPLEY, D., 1972: Holocene sea levels in eastern Australia. Marine Geology **12**, 223—242, Amsterdam.

GLENNIE, K. W., 1970: Desert sedimentary environments. Developments in Sedimentology **14**, 222. Amsterdam: Elsevier.

GLENNIE, K. W., EVAMY, B. D., 1968: Dikaka: plants and plantroot structures associated with aeolian sands. Palaeogeogr., Palaeoclimat., Palaeoecology **4**, 77—87, Amsterdam.

GODWIN, H., 1962: Half-life of Radiocarbon. Nature **195**, 984, London.

GOELLNER, W. A., 1972: The Bill Goellner Centennial Hofuf Tour. 18 p., Dhahran.

GOLDBERG, M., YARON, F., 1977: Absolute dating of some raised marine terraces on the Island of Tiran and the South Eastern coast of Sinai. Z. Geomorph. N.F. **21**, in press.

GOSH, A., 1952: The Rajputana desert: its archaeological aspects. Proc. Symp. Rajputana Desert, 37—42, Jaipur, India.

GOUDIE, A., 1973: Duricrusts in tropical and subtropical landscapes. Oxford research studies in Geography, 174 p. Oxford: Clarendon.

GREENWOOD, J. E., BLEACKLEY, D., 1967: Geology of the Arabian Peninsula: Aden Protectorate. U.S. Geol. Survey, Prof. Paper, 560-C, 96 p. Washington.

GREENWOOD, W. R., 1973: The Hail arch—a key to deformation of the Arabian shield during evolution of the Red Sea rift. Dir. Gen. Mineral Resources, Min. Resources Bull. 7, 4 p. Jiddah.

GREENWOOD, W. R., BROWN, G. F., 1972: Petrology and chemical analysis of selected plutonic rocks from the Arabian Shield. U.S. Geol. Survey, Saudi Arabian Proj. Rep. **147**, 21, Jiddah.

GROMMÉ, C. S., HAY, R. L., 1971: Geomagnetic polarity epochs: new data from Olduvai Gorge, Tanganyika. Earth Planet. Sci. Lett. **2**, 111—115, Amsterdam.

GROSS, H., 1966: Der Streit um die Geochronologie des Spätpleistozäns und sein Ausgang. Forschungen und Fortschritte **40**, 165—168, Berlin.

GROSSE, A. V., JOHNSTON, W. M., WOLFGANG, R. L., LIBBY, W. F., 1951: Tritium in nature. Science **113**, 1—2, Washington.

GROVE, A. T., 1959: The former extent of Lake Chad. Geogr. J. **125**, 465—467, London.

GROVE, A. T., PULLAN, R. A., 1964: Some aspects of the Pleistocene palaeogeography of the Chad Basin. In: HOWELL, F. C., BOURLIERE, F. (Eds.), African ecology and human evolution. Viking Fund Publ. Anthrop. **36**, 230—245, London.

GROVE, A. T., WARREN, A., 1968: Quaternary landforms and climate on the south side of the Sahara. Geogr. J. **134**, 194—208, London.

GROVE, A. T., GOUDIE, A. S., 1971: Late Quaternary lake levels in the Rift Valley of Southern Ethiopia and elsewhere in tropical Africa. Nature **234**, 403—405, London.

HAGEDORN, H., 1967: Studien über den Formenschatz der Wüste an Beispielen aus der Südost-Sahara. Dt. Geogr. Tag Bad Godesberg, Tag.-Ber. u. wiss. Abh., 401—411, Wiesbaden.

HAGEDORN, H., 1971: Untersuchungen über Relieftypen arider Räume an Beispielen aus dem Tibestigebirge und seiner Umgebung. Z. Geomorph., Suppl. **11**, 251 p. Stuttgart.

HARTIGAN, J. A., 1975: Clustering Algorithms. 351 p. New York: J. Wiley & Sons, Inc.

HARTMANN, M., LANGE, H., SEIBOLD, E., WALGER, E., 1971: Oberflächensedimente im Persischen Golf und Golf von Oman, I, Geolog. hydrogeologischer Rahmen und erste sedimentologische Ergebnisse. „Meteor"-Forschungsergebnisse, Reihe C **4**, 1—76, Berlin-Stuttgart.

HEBRARD, L., ELOUARD, P., FAURE, H., 1971: Quaternaire du littoral mauritanien entre Nouak-chott et Nouadhibon (Port-Etienne) (18°–21° Latitude Nord). Quaternaria **15**, 297—304, Roma.

HEIM, A., 1928: Die artesischen Quellen der Bahrein-Inseln im Persischen Golf. Eclogae Geol. Helv. **21**, 1—6, Basel.

HEIM, W., 1914: Südarabische Itinerare. Mitt. k.k. Geogr. Ges. Wien, Heft 1—3, 32—58, Wien.

HEINE, K., 1974: Bemerkungen zu neueren chronostratigraphischen Daten zum Verhältnis glazialer und pluvialer Klimabedingungen. Erdkde. **28**, 303—312, Bonn.

HEINE, K., 1975: Studien zur jungquartären Glazialmorphologie mexikanischer Vulkane mit einem Ausblick auf die Klimaentwicklung. Das Mexiko-Projekt der DFG **7**, 1—178, Wiesbaden.

HEINE, K., 1976: Auf den Spuren der Eiszeit in Mexiko. Natur und Museum **106**, 10, 289—298, Frankfurt.

HEINZELEIN, J. DE, 1967: Pleistocene sediments and events in Sudanese Nubia. In: BISHOP, W. W., CLARK, J. D. (Eds.), Background to evolution in Africa, 313—328. Chicago-London: Univ. Chicago Press.

HEY, R. W., 1971: Quaternary shorelines of the Mediterranean and Black Seas. Quaternaria **15**, 273—284, Roma.

HÖVERMANN, J., 1967: Hangformen and Hangentwicklung zwischen Syrte und Tschad. Congr. et Coll. l'Univ. Liége **40**, 139—156, Liége.

HOLM, D. A., 1953: Dome-shaped dunes of Central Nejd, Saudi Arabia. Intern. Geol. Congr., 19th, Alger 1952, Compt. Rend., Section **7**, 107—112, Alger.

HOLM, D. A., 1960: Desert geomorphology in the Arabian Peninsula. Science **132**, 1369—1379, Washington.

HOLSER, W. T., KAPLAN, I. R., 1966: Isotope geochemistry of sedimentary sulfates. Chem. Geol. **1**, 93—135, Amsterdam.

HOOKE, R. L., 1967: Processes on arid region alluvial fans. J. Geol. **75**, 438—460, Chicago.

HORN, P., LIPPOLT, H. J., TODT, W., 1972: Kalium-Argon-Altersbestimmungen an tertiären Vulkaniten des Oberrheingrabens; I. Gesamtgesteinsalter. Eclogae Geol. Helv. **65**, 131—156, Basel.

HOUBOLT, J. J., 1957: Surface sediments of the Persian Gulf near the Qatar Peninsula. Doctoral thesis, University of Utrecht, Den Haag.

HSÜ, K. J., SCHNEIDER, J., 1973: Progress report on dolomitization-hydrology of Abu Dhabi sabkhas, Arabian Gulf. In: PURSER, B. H. (Ed.), The Persian Gulf, 409—422. Berlin-Heidelberg-New York: Springer.

HUCKRIEDE, R., 1966: Das Quartär des arabischen Jordan-Tales und Beobachtungen über „Pebble Culture" und „Prä-Aurignac". Eiszeitalter und Gegenwart **17**, 211—212, Öhringen.

HUCKRIEDE, R., 1972: Ein $^{14}$C-Wert für den Pluvial-Kalk von Sodiri Kordofan, Republik Sudan. Geologica Paleontologica **6**, 173—175, Marburg.

HUCKRIEDE, R., VENZLAFF, H., 1962: Über eine pluvialzeitliche Molluskenfauna aus Kordofan (Sudan). Paläontolog. Z., H.-SCHMIDT-Festband, 93—109, Stuttgart.

HUCKRIEDE, R., WIESEMANN, G., 1968: Der jungpleistozäne Pluvial-See von El Jafr und weitere Daten zum Quartär Jordaniens. Geologica Palaeontologica **2**, 73—95, Marburg.

HUDSON, R. G. S., EAMES, F. E., WILKINS, G. L., 1957: The fauna of some recent marine deposits near Basrah, Iraq. Geol. Magazine **94**, 5, 393—401, Hertford.

ILLING, L. V., WELLS, A. J., TAYLOR, J. C., 1965: Penecontemporary dolomite in the Persian Gulf. In: PRAY, L. G., MURRAY, R. O. (Eds.), Dolomitization and limestone diagenesis. Soc. Econ. Palaeont. Mineral., Spec. Pub. **13**, 89—111, Tulsa, Oklahoma.

ISSAR, A., PICARD, L., 1971: On Pleistocene shorelines in the coastal plain of Israel. Quaternaria **15**, 267—272, Roma.

Italconsult, 1969: Water and agricultural development studies for area IV. Final report. Drilling investigations, discharge tests, water point inventory. Unpublished report, Rome.

JACKSON, R. O., BOGUE, R. G., BROWN, G. F., GIERHART, R. D., 1963: Geologic map of the Southern Najd quadrangle, Kingdom of Saudi Arabia. U.S. Geol. Survey, Misc. Geol. Investig. Map I-211 A, 1 : 500,000, Washington.

JÄKEL, D., 1971: Erosion und Akkumulation in Enneri Bardague-Araye des Tibestigebirges (Zentrale Sahara) während des Pleistozäns und Holozäns. Berliner Geogr. Abh. **10**, 1—55, Berlin.

JAECKEL, S. H., 1972: Quartäre Mollusken aus Seeabsätzen des Tibestigebirges. In: BÖTTCHER, V., ERGENZINGER, P. J., JAECKEL, S. H., KAISER, K., Quartäre Seebildungen und ihre Mollusken-Inhalte im Tibestigebirge und seinen Rahmenbereichen der zentralen Ostsahara. Z. Geomorph. N.F. **16**, 2, 182—234, Berlin-Stuttgart.

JAMES, N. P., 1972: Holocene and Pleistocene calcareous crust (caliche) profiles: criteria for subaerial exposure. J. Sed. Petrology **42**, 4, 817—836, Tulsa, Oklahoma.

JELGERSMA, S., 1966: Sea level changes during the last 10,000 years. In: SAWYER, J. S. (Ed.), World climate from 8,000 to 0 B.C., 54—71. London: Royal Meteor. Soc.

KABELAC, F., 1962: Krustenkalkbildungen in Nordostsyrien. Ber. naturf. Ges. Freiburg i. Br. **52**, 103—109, Freiburg.

KAISER, E., 1923: Was ist eine Wüste? Geogr. Ges. München, Mitt. **16**, 3, 1—20, München.

KAISER, K., 1963: Die Ausdehnung der Vergletscherungen und „periglazialen" Erscheinungen während der Kaltzeiten des quartären Eiszeitalters innerhalb der syrisch-libanesischen Gebirge und die Lage der klimatischen Schneegrenze zur Würmeiszeit im östlichen Mittelmeergebiet. INQUA, 6th, Warschau 1961, Rep. 3, 127—148, Warschau.

KAISER, K., KEMP, E. K., LEROI-GOURHAN, A., SCHÜTT, H., 1973: Quartärstratigraphische Untersuchungen aus dem Damaskus-Becken und seiner Umgebung. Z. Geomorph. N.F. **17**, 3, 263—353, Berlin, Stuttgart.

KAPEL, H., 1967: Atlas of the Stone-age cultures of Qatar. Reports of the Danish Archeological Expedition to the Arabian Gulf, Vol. 1. Jutland Archeolog. Soc., Publ. **4**, 43, Aarhus.

KASSLER, P., 1973: The structural and geomorphic evolution of the Persian Gulf. In: PURSER, B. H. (Ed.), The Persian Gulf, 11—32. Berlin-Heidelberg-New York: Springer.

KENDALL, C. G., 1969: Holocene shallow-water carbonate and evaporite sediments of Khor Bazam, Abu Dhabi, SW Persian Gulf. Amer. Assoc. Petrol. Geologists, Bull. **53**, 841—869, Tulsa, Oklahoma.

KENDALL, C. G., SKIPWITH, S. P., 1968: Recent algal mats of a Persian Gulf lagoon. J. Sed. Petrology **38**, 1040—1058, Tulsa, Oklahoma.

KENT, P. E., 1970: The salt plugs of the Persian Gulf region. Leicester lit. phil. Soc. **64**, 56—88, Leicester.

KETTANI, M. A., 1973: Climatological effects on heliohydroelectric (HHE) power generation. Compless **2**, 26—32.

KETTANI, M. A., LAM, E. Y., 1974: Attempts at mapping the solar intensity distribution for the Arabian Peninsula. Compless **3**, 3—7.

KIILSGAARD, T. H., 1974: Summary of activities. U.S. Geological Survey Saudi Arabian Project 1950 to 1974. U.S. Geol. Survey, Saudi Arabian Proj. Rep. **168**, 53, Jiddah.

KINSMAN, D. J., 1964: Recent carbonate sedimentation near Abu Dhabi, Trucial Coast, Persian Gulf, Ph.D. thesis, Univ. London.

KINSMAN, D. J., 1967: Huntite from a carbonate-evaporite environment. Amer. Mineralogist **52**, 1332—1340, Lancaster, Pennsylvania.

KINSMAN, D. J., 1969: Modes of formation, sedimentary associations, and diagnostic features of shallow-water and supratidal evaporites. Amer. Assoc. Petrol. Geologists, Bull. **53**, 830—840, Tulsa, Oklahoma.

KLITZSCH, E., 1970: Die Strukturgeschichte der Zentralsahara. Geol. Rdsch. **59**, 459—527, Stuttgart.

KNETSCH, G., SHATA, A., DEGENS, E., MÜNNICH, K. O., VOGEL, J. C., SHAZLY, M. M., 1962: Untersuchungen an Grundwässern der Ost-Sahara. Geol. Rdsch. **52**, 587—610, Stuttgart.

KÖPPEN, W., 1900: Versuch einer Klassifikation der Klimate, vorzugsweise nach ihren Beziehungen zur Pflanzenwelt. Geogr. Z. **6**, 593—611, 657—679, Leipzig.

KRUMBEIN, W. E., 1969: Über den Einfluß der Mikroflora auf die exogene Dynamik (Verwitterung und Krustenbildung). Geol. Rdsch. **58**, 333—363, Stuttgart.

KRUMBEIN, W. E., 1971: Sedimentmikrobiologie und ihre geologischen Aspekte. Geol. Rdsch. **60**, 438—471, Stuttgart.

KU, T. L., THURBER, D. L., MATHIEU, G. G., 1969: Radiocarbon chronology of Red Sea sediments. In: DEGENS, E. T., ROSS, D. A. (Eds.), Hot brines and recent heavy metal deposits in the Red Sea, 348—359. Berlin-Heidelberg-New York: Springer.

KUBIENA, W., 1957: Neue Beiträge zur Kenntnis des planetarischen und hypsometrischen Formenwandels der Böden Afrikas. Stuttgarter Geogr. Stud. **69**, 50—64, Stuttgart.

LAMARE, P., 1924: L'Arabie Heureuse: le Yémen. La Géographie **47**, Paris.

LAMB, H. H., 1970: Climatic changes and variations in the atmospheric and ocean circulations. Geol. Rdsch. **54**, 1, 486—504, Stuttgart.

LEES, G. M., 1928: The geology and tectonics of Oman and parts of South Eastern Arabia. Quat. J. Geol. Soc. London **84**, 4, 585—670, London.

LEES, G. M., 1928: The physical geography of South Eastern Arabia. Geogr. J. **71**, 1, 441—452, London.

LEES, G. M., FALCON, N. L., 1952: The geographical history of the Mesopotamian plains. Geogr. J. **118**, 24—39, London.

LEIDLMAIR, A., 1962: Klimamorphologische Probleme in Hadramaut. Herrmann v. Wissmann-Festschr., 162—179, Tübingen.

LEROI-GOURHAN, A., 1971: Pollens et terrasses marines Liban. Quaternaria **15**, 249—259, Roma.

LEROI-GOURHAN, A., 1974: Analyses polliniques préhistoire et variations climatiques quaternaires. Colloques Internationaux du C.N.R.S., 219, Le méthodes quantitatives d'étude des variations du climat au cours du Pleistocène, 61—66, Paris.

LESCH, W., 1931: Arabien. Geogr. Ges. München, Mitt. **24**, 4—153, München.

LIBBY, W. F., 1955: Radiocarbon dating. 2nd ed., 175 p. Chicago-London: Univ. Chicago Press.

LOBECK, A. K., 1945: Physiographic diagram of Asia. Geographical Press. Maplewood, N.J.: Hammond and Comp.

LOUIS, H., 1938: Eiszeitliche Seen in Anatolien. Z. Ges. Erdkde. **1938**, 268—285, Berlin.

LOZEK, V., 1964: Quartärmollusken der Tschechoslowakei. Rozpravy, Ustredniho ustavu geologického **31**, 374, Prag.

MACFADYEN, W. A., 1930: The undercutting of coral reef limestone on the coasts of some islands in the Red Sea. Geogr. J. **75**, 27—34, London.

MADGIN, W. M., SWALES, D. A., 1956: Solubilities in the system $CaSO_4$—NaCl—$H_2O$ at 25° and 35°. J. Appl. Chemistry **6**, 482—487, London.

MANDAHL-BARTH, G., 1957: Intermediate hosts of Schistosoma. African Biomphalaria and Bulinus. Wld. Hlth. Org., Bull. **16**, 1103—1163, and **17**, 1—65, Genf.

MCCLURE, H. A., 1974: The Arabian Peninsula and Prehistoric Populations. Field Research Projects, Study No. 58, Miami.

MCCLURE, H. A., 1976: Radiocarbon chronology of late Quaternary lakes in the Arabian Desert. Nature **263**, 755—756, London.

MCCLURE, H. A., SWAIN, F. M., 1974: Fresh water and brackish water fossil Quaternary Ostracoda from the Rub' al Khali, Saudi Arabia. Paper read at the 6th African Micropalaeontological Colloquium, Tunis. In print.

MCGINNIES, W. G., GOLDMAN, B. J., PAYLORE, P., 1968: Desert of the world, 788 p. Tuscon: Univ. Arizona Press.

MCKEE, E. D., CROSBY, E. J., BERRYHILL, H.-L., 1967: Flood deposits Bijon Creek, Colorado, June 1965. J. Sed. Petrology **37**, 3, 829—851, Tulsa, Oklahoma.

MELGUEN, M., 1973: Correspondence analysis for recognition of facies in homogeneous sediments of an Iranian river mouth. In: PURSER, B. H. (Ed.), The Persian Gulf, 99—113. Berlin-Heidelberg-New York: Springer.

MENSCHING, H., 1958: Glacis, Fußfläche, Pediment. Z. Geomorph. N.F. **2**, 165—186, Berlin-Stuttgart.

MENSCHING, H., 1968: Bergfußflächen und das System der Flächenbildung in den ariden Subtropen und Tropen. Geol. Rdsch. **58**, 62—82, Stuttgart.

MENSCHING, H., 1970: Flächenbildung in der Sudan- und Sahel-Zone (Ober-Volta und Niger). Z. Geomorph. N.F. Suppl. **10**, 1—29, Berlin-Stuttgart.

MENSCHING, H. G., 1974: Landforms as a dynamic expression of climatic factors in the Sahara and Sahel—a critical discussion. Z. Geomorph. N.F. Suppl. **20**, 168—177, Berlin-Stuttgart.

MERCER, J. H., 1973: Cainozoic temperature trends in the Southern Hemisphere: Antarctic and Andean glacial evidence. In: ZINDEREN-BAKKER, E. M. VAN (Ed.), Palaeoecology of Africa **8**, 87—114. Rotterdam-Capetown.

MEULEN, D. VAN D., WISSMANN, H. V., 1932: Hadramaut. 248 p., Leiden.

MICHEL, P., 1971: Chronologie du Quaternaire des bassins des fleuves Sénégal et Gambie. Ass. Franç. Etude Quaternaire, Bull., Suppl. **4**, 49—61, Paris.

MIENIS, H. K., 1970: Some note on freshwater mollusca from the marshes of Kurdani, Galilee. Argamon, J. Israel Malakol. Soc. **1**, 51—54, Jerusalem.

MILANKOVIČ, M., 1941: Kanon der Erdbestrahlung. 389 p. Kgl. serb. Akad. Belgrad.

MILLER, R. P., 1937: Drainage lines in bas-relief. J. Geol. **45**, 432—438, Chicago.

MILLIMAN, J. D., EMERY, K. O., 1968: Sea levels during the past 35,000 years. Science **162**, 3858, 1121—1123, Washington.

MILTON, D. I., 1967: Geology of the Arabian Peninsula: Kuwait. U.S. Geol. Survey, Prof. Paper 560-F, 7 p. Washington.

MITCHELL, R. C., 1957: Notes on the Geology of Western Iraq and Northern Saudi Arabia. Geol. Rdsch. **46**, 476—493, Stuttgart.

MONOD, TH., 1964: The late Tertiary and Pleistocene in the Sahara and adjacent southerly regions. In: HOWELL, C., BOURLIERE, F. (Eds.), African ecology and human evolution, 117—229. London: Methuen.

Monthly climatic data for the world, January 1966 to December 1974. U.S. National Oceanic and Atmospheric Administration, sponsored by the World Meteorological Organization (WMO).

MORGAN, J. P., 1971: Depositional processes and products in the deltaic environment. Soc. Econ. Paleont. Mineral., Spec. Publ. **15**, 31—47, Tulsa, Oklahoma.

MORITZ, B., 1923: Arabien. 133 p. Hannover: Orient-Buchh. H. Lafaire.

MORTENSON, H., 1933: Die „Salzsprengung" und ihre Bedeutung für die regionalklimatische Gliederung der Wüsten. Peterm. Geogr. Mitt. **79**, 130—135, Berlin.

MOSER, A. V., STICHLER, W., 1971: Verwendung des Deuterium- und Sauerstoff-18-Gehaltes bei hydrologischen Untersuchungen. Geologica Bavaria **64**, 7—35, München.

MOVIUS, H. L., 1944: Early man and Pleistocene stratigraphy in southern and eastern Asia. Peabody Museum. Harvard Univ., Pap. **55**, 3, 1—125, Boston.

MÜLLER, G., 1964: Methoden der Sedimentuntersuchung. In: ENGELHARDT, W., FÜCHTBAUER, H., MÜLLER, G., Sediment-Petrologie, Teil 1, 303 p. Stuttgart: Schweizerbart.

MÜNNICH, K. O., 1957: Messungen des $^{14}$C-Gehaltes von hartem Grundwasser. Naturwiss. **44**, 32—34, Berlin.

MÜNNICH, K. O., 1968: Isotopendatierung von Grundwasser. Naturwiss. **55**, 158—163, Berlin.

MÜNNICH, K. O., ROETHER, W., 1963: A comparison of Carbon-14 and Tritium ages of groundwater. Radioisotopes in Hydrology, IAEA, 397—406, Wien.

MÜNNICH, K. O., VOGEL, J. C., 1959: $^{14}$C-Altersbestimmungen von Süßwasserkalkablagerungen. Naturwiss. **46**, 168—169, Berlin.

MÜNNICH, K. O., VOGEL, J. C., 1962: Untersuchungen an pluvialen Wässern der Ost-Sahara. Geol. Rdsch. **52**, 611—624, Stuttgart.

MUSIL, A., 1926: The Northern Hegaz. Amer. Geogr. Soc., Publ., Oriental Explor. Stud. **1**, New York.

MUSIL, A., 1927: Arabia Deserta. Amer. Geogr. Soc., Publ., Oriental. Explor. Stud. **2**, New York.

NAIMI, A. I., 1965: The groundwater of Northeastern Saudi Arabia. Fifth Arab Petroleum Congress March 16–23, Cairo.

NAQIB, K. M. AL, 1967: Geology of the Arabian Peninsula: Southwestern Iraq. U.S. Geol. Survey, Prof. Paper 560-G, 54 p. Washington.

NIELSEN, H., RICKE, W., 1964: Schwefel-Isotopenverhältnisse von Evaporiten aus Deutschland; ein Beitrag zur Kenntnis von $\delta^{34}S$ im Meerwassersulfat. Geochim. Cosmochim. Acta **28**, 577—591, London.

NILSSON, E., 1941: Die Eiszeit in Indien. Stockholm Hogskolan. Geol. Inst., Medd. **54**, 1—23, Stockholm.

NILSSON, E., 1963: Pluvial lakes and glaciers in East-Africa. Stockholm Univ. Acta, Stockh. Contrib. Geol. **11**, 21—57, Stockholm.

NIR, D., 1970 a: Marine terraces of Southern Sinai. Geogr. Review **61**, 32—50, New York.

NIR, D., 1970 b: Les lacs quaternaires dans la région de Feiran (Sinai central). Rev. géogr. phys. et géol. dyn. **12**, 335—346, Paris.

OBENAUF, P., 1967: Beobachtungen zur spätpleistozänen und holozänen Talformung im Nordwest-Tibesti. Berliner Geogr. Abh. **5**, 27—38, Berlin.

OLAUSSON, E., OLSSON, I. V., 1969: Varve stratigraphy in a core from the Gulf of Aden. Palaeogeogr., Palaeoclimat., Palaeoecology **6**, 87—103, Amsterdam.

ÖTKUN, G., 1969: Outlines of groundwater resources of Saudi Arabia. Intern. Conf. on arid lands in an changing world. Unpublished report, Tucson, Arizona.

ÖTKUN, G., 1972: Observations on Mesozoic sandstone aquifers in Saudi Arabia. Intern. Geol. Congr., 24th, Sect. 11, 28—35, Montreal.

OVEJERO, G., ZAZO, C., 1971: Niveles marinos pleistocenos en Almeria (S. E. de Espana). Quaternaria **15**, 145—159, Roma.

PAK, E., FELBER, H., 1974: Massenspektrometrische Präzisionsmessung der Schwefelisotopenverhältnisse. Österr. Akad. Wiss., math.-nath. Kl., Sitzungsberichte **183**, 295—308, Wien.

PENCK, A., BRÜCKNER, E., 1901—1909: Die Alpen im Eiszeitalter. 3 vls., 1189 p., Leipzig.

PFANNENSTIEL, M., 1952: Das Quartär der Levante. Teil 1: Die Küste Palästina-Syriens. Akad. Wiss. Lit. Mainz, Abh. math.-nat. Kl. 1952, **7**, 375—475, Mainz.

PFANNENSTIEL, M., 1953: Das Quartär der Levante. Teil 2: Die Entstehung der ägyptischen Oasendepressionen. Akad. Wiss. Lit. Mainz, Abh. math.-nat. Kl. 1953, **7**, 335—411, Mainz.

PHILBY, H. S., 1925: Das geheimnisvolle Arabien, 2 Bde., 365 p., 320 p. Leipzig: Brockhaus.

PHILBY, H. S., 1933: The empty quarter. 389 p. London.

PHILLIPS, J. H., BUTZER, K. W., 1974: A "Silsilian" occupation site (GS-2 B-11) of the om Ombo Plain, Upper Egypt: geology, archeology, and paleoecology. Quaternaria **17**, 343—386, Roma.

PIAS, J., 1960: Sédimentation au Quaternaire dans l'Est de la cuvette tchadienne. Acad. Sci. Paris, Comptes Rendus **250**, 8, 1414—1516, Paris.

PICARD, L., 1937: On the structure of the Arabien Peninsula. Hebrew Univ. Bull. Geol. Dept. **3**, 1—11, Jerusalem.

PICARD, L., 1965: The geological evolution of the Quaternary in the central-northern Jordan Graben, Israel. Geol. Soc. Amer., Spec. Paper **84**, 337—366, New York.

PICARD, L., BAIDA, V., 1966: Geological report on the lower Pleistocene deposits of the 'Ubeidiya excavations, 39 p. Jerusalem: Israel Acad. Sci. Human.

PILGRIM, G. E., 1908: The geology of the Persian Gulf and the adjoining portions of Persia and Arabia. India Geol. Survey, Mem. **34**, 1—177, New Delhi.

PILSBRY, H. A., BEQUAERT, J., 1927: The aquatic mollusces of the Belgian Congo. With a geographical and ecological account of Congo malacology. Amer. Mus. Nat. Hist., Bull. **53**, 69—602, New York.

POWERS, R. W., 1968: Arabie Saoudite. Lexique Stratigr. Internat. **3**, 10 b, 1, 177 p. Paris.

POWERS, R. W., RAMIREZ, L. F., REDMOND, C. D., ELBERG, E. L., 1966: Geology of the Arabian Peninsula. Sedimentary Geology of Saudi Arabia. U.S. Geol. Survey, Prof. Paper 560-D, 1—147, New York.

PRAMANIK, S. K., 1952: Hydrology of the Rajasthan Desert. Rainfall, humidity and evaporation. Proc. Symp. Rajputana Desert, 183—197, Jaipur, India.

PURDY, E. G., 1965: Diagenesis of recent marine carbonate sediments (abs). In: PRAY, L. C., and MURRAY, R. O. (Eds.), Dolomitization and limestone diagenesis. Soc. Econ. Palaeont. Mineral., Spec. Pub. **13**, 169, Tulsa, Oklahoma.

PURSER, B. H., 1973 (Ed.): The Persian Gulf, 471 p. Berlin-Heidelberg-New York: Springer.

Quezel, P., Martinez, G., 1959: Le dernier interpluvial au Sahara central. Libyca Anthrop. Préhist. Ethnogr. **5**, 211—225, Alger.

Raikes, R. L., Dyson, R. H., 1961: The prehistoric climate of Baluchistan and the Indus Valley. Amer. Anthropol. **63**, 265—281, Lancaster.

Rathjens, C., 1957: Physisch-geographische Beobachtungen im Nordwestindischen Trockengebiet. Erdkde. **11**, 49—58, Bonn.

Rathjens, C., 1959: Menschliche Einflüsse auf die Gestalt und Entwicklung der Tharr. Geogr. Inst. Univ. Saarland, Arb. **4**, 1—36, Saarbrücken.

Rathjens, C., 1968: Schichtflächen und Schnittflächen im Trockenklima. Regio Basiliensis **9**, 1, 162—169, Basel.

Rathjens, C., Wissmann, H. v., 1931—1934: Rathjens-v.-Wißmannsche-Südarabien-Reise. Geographische Ergebnisse einer Reise in Yemen. Hamburg. Univ. Abh. Gebiet Auslandskunde **36**, 270 p., **38**, 212 p., **40**, 229 p. Hamburg.

Rathjens, C., Wissmann, H. v., 1947: Landschaftskundliche Beobachtungen im südlichen Hedjaz. Erdkde. **1**, 61—89 und 200—205, Bonn.

Reeves, C. C., 1970: Origin, classification and geologic history of caliche on the southern High Plains, Texas and eastern New Mexico. J. Geol. **78**, 352—362, Chicago.

Reineck, H. E., Singh, I. B., 1973: Depositional sedimentary environments, 439 p. Berlin-Heidelberg-New York: Springer.

Reinhardt, B. M., 1969: On the genesis and emplacement of ophiolites in the Oman Mountains geosyncline. Schweiz. Mineral. u. Petrog. Mitt. **49**, 1, 1—30, Bern.

Richards, H. G., 1971: Sea level during the past 11,000 years as indicated by data from North and South America. Quaternaria **14**, 7—15, Roma.

Ritter, C., 1846: Die Erdkunde. 12. und 13. Teil: Vergleichende Erdkunde von Arabien, 184 p. Berlin: Reimer.

Roether, W., 1967 a: Estimating the tritium input to groundwater from wine samples: Groundwater and direct run-off contribution to Central European surface waters. Isotopes in Hydrolog., IAEA, 73—90, Wien.

Roether, W., 1967 b: Tritium im Wasserkreislauf. Dissertation Univ. Heidelberg.

Rohdenburg, H., 1970: Morphodynamische Aktivitäts- und Stabilitätszeiten statt Pluvial- und Interpluvialzeiten. Eiszeitalter und Gegenwart **21**, 81—96, Öhringen.

Rohdenburg, H., Sabelberg, U., 1972: Quartäre Klimazyklen im westlichen Mediterrangebiet und ihre Auswirkungen auf die Relief- und Bodenentwicklung. Z. Geomorph. N.F. Suppl. **15**, 87—92, Berlin, Stuttgart.

Rossignol-Strick, M., 1974: Palynologie de sapropèles Méditerranéens du Villafranchien à l'holocène. Colloques Internat. du C.N.R.S., 219, Les méthodes quantitatives d'étude des variations du climat au cours du Pleistocène, 91—102, Paris.

Rubin, M., 1963: Simultaneity of glacial and pluvial episodes from $^{14}$C-chronology of the Wisconsin glaciation. UNESCO, Arid Zone Research **20**, 223—228, Paris.

Rumney, G. R., 1968: Climatology and the world's climates. 656 p. New York: McMillian Comp.

Sanlaville, P., 1971: Sur la Tyrrhénien libanais. Quaternaria **15**, 239—248, Roma.

Sarnthein, M., 1971: Oberflächensedimente im Persischen Golf und Golf von Oman. II. Quantitative Komponentenanalyse der Grobfraktion. „Meteor" Forschungsergeb., Reihe C **5**, 1—113, Berlin-Stuttgart.

Sarnthein, M., 1972: Sediments and history of the postglacial transgression in the Persian Gulf and northwestern Golf of Oman. Marine Geol. **12**, 245—266, Amsterdam.

Saxen, A., 1967: Situation der bewässerten Landwirtschaft in der Ostprovinz Saudi Arabiens. Z. Kulturtechnik Flurbereinigung **8**, 6, 321—344, Berlin-Hamburg.

Schäfer, A., 1975: Kalkkrusten auf der Insel Christiani, Santorin-Gruppe, Ägäis (Griechenland). N. Jb. Geol. Paläont., Abh. **150**, 1, 1—18, Stuttgart.

Scharlau, K., 1958: Zum Problem der Pluvialzeiten in Nordost-Iran. Z. Geomorph. N.F. **2**, 258—277, Berlin-Stuttgart.

Scheffers, F., Meyer, B., Kalk, E., 1963: Biologische Ursachen der Wüstenlackbildung. Z. Geomorph. N.F. **7**, 112—119, Berlin-Stuttgart.

SCHICK, A. P., 1974: Formation and obliteration of desert stream terraces—a conceptual analysis. Z. Geomorph. N.F. Suppl. **21**, 88—105, Berlin-Stuttgart.

SCHMIDT, D. L., HADLEY, D. G., GREENWOOD, W. R., GONZALEZ, L., COLEMAN, R. G., BROWN, G. F., 1973: Stratigraphy and tectonism of the southern part of the Precambrian shield of Saudi Arabia. Dir. Gen. Mineral Resources, Min. Res. Bull. **8**, 13 p. Jiddah.

SCHOLLE, P. A., KINSMAN, D. J., 1974: Aragonitic and high-Mg calcite caliche from the Persian Gulf. J. Sed. Petrology **44**, 904—916, Tulsa, Oklahoma.

SCHÜTT, H., 1973: Die Mollusken eines jungpleistozänen Seeprofils im Becken von Damaskus. In: KAISER, K., KEMPF, E. K., LEROI-GOURHAN, A., SCHÜTT, A., Quartärstratigraphische Untersuchungen aus dem Damaskus-Becken und seiner Umgebung. Z. Geomorph. N.F. **17**, 3, 263—353, Berlin-Stuttgart.

SEIBOLD, E., DIESTER, L., FÜTTERER, D., LANGE, H., MÜLLER, P., WERNER, F., 1973: Holocene sediments and sedimentary processes in the Iranian part of the Persian Gulf. In: PURSER, B. H. (Ed.), The Persian Gulf, 57—80. Berlin-Heidelberg-New York: Springer.

SERVANT, M., 1973: Séquences continentales et variations climatiques: évolution du bassin du Tchad au Cenozoique supérieur. Paris: Orstrom.

SERVANT, M., SERVANT, S., 1969: Chronologie du Quaternaire récent des bases du Tchad. Acad. Sci. Paris, Compt. Rendus, Série D **269**, 1603—1606, Paris.

SHEARMAN, D. J., 1963: Recent anhydrite, gypsum, dolomite and halite from the coastal flats of the Arabian shore of the Persian Gulf. Geol. Soc. London, Proc. **1607**, 63, London.

SHEARMAN, D. J., 1966: Origin of marine evaporites by diagenesis. Inst. Mining Metallurgy, Trans., sec. B **75**, 208—215, New York.

SHEPARD, F. P., CURRAY, J. R., 1967: Carbon-14 determination of sea level changes in stable areas. Progr. Oceanogr. **4**, 283—291. New York: Pergamon.

SHINN, E. A., 1969: Submarine lithification of Holocene carbonate sediments in the Persian Gulf. Sedimentology **12**, 109—144, Amsterdam.

SIMON, W., LIPPOLT, H. J., 1967: Geochronologie als Zeitgerüst der Phylogenie; In: HEBERGER, G. (Ed.): Die Evolution der Organismen, 3. Aufl. **1**, 161—237, Stuttgart.

SKIPWITH, P., 1973: The Red Sea and coastal plain of the Kingdom of Saudi Arabia. Dir. Gen. Min. Resources, Techn. Rec. TR-1973-1, 149 p. Jiddah.

SKOCEK, V., SAADALLAH, A. A., 1972: Heavy minerals in eolian sands, southern desert, Iraq. Sedimentary Geology **8**, 29—45, Amsterdam.

SMALLEY, I. J., VITA-FINZI, C., 1968: The formation of fine particels in sandy desert and the nature of "desert" loess. J. Sed. Petrology **38**, 766—774, Tulsa, Oklahoma.

SMITH, P. E., 1962: Two Neolithic collections from Saudi Arabia. Man **17**, 21—23, London.

SOLLE, G., 1966: Rezente und fossile Wüste. Notizbl. hess. L.-Amt Bodenforsch. **94**, 54—121, Wiesbaden.

SPARKS, B. W., GROVE, A. T., 1961: Some Quaternary fossil nonmarine Mollusca from the Central Sahara. J. Linnean Soc. London, Zoology **44**, 355—364, London.

STEINEKE, M., HARRISS, T. F., PARSONS, K. R., BERG, E. L., 1958: Geologic map of the Western Persian Gulf quadrangle Kingdom of Saudi Arabia. U.S. Geol. Survey, Misc. Geol. Investig. Map I-208 A, 1 : 500,000, Washington.

STONE, R. O., 1967: A desert glossary. Earth Sci. Rev. **3**, 211—268, Amsterdam.

STRATIL-SAUER, G., 1957: Die pleistozänen Ablagerungen im Innern der Wüste Lut. Festschr. Hundertjahrfeier Geogr. Ges. Wien **2**, 460—488, Wien.

STRATIL-SAUER, G., WEISE, O. R., 1974: Zur Geomorphologie der südlichen Lut und zur Klimageschichte Irans. Würzb. Geogr. Arb. **41**, 29, Würzburg.

STRINGFIELD, V. T., LEGRAND, H. E., 1966: Hydrology of limestone terraces in the coastal plain of the southeastern United States. Geol Soc. Amer., Spec. Papers, 93—46, Washington.

SUGDEN, W., 1962: Structural analysis and geometrical prediction for change of form with depth of some Arabian plains-type folds. Amer. Assoc. Petrol. Geologists, Bull. **46**, 2213—2218, Tulsa, Oklahoma.

SUGDEN, W., 1963: The hydrology of the Persian Gulf and its significance in respect to evaporite deposition. Amer. J. Sci. **261**, 741—755, New Haven.

SUGDEN, W., STANDRING, A. J., 1975: Qatar Peninsula. Lexiq. Stratigr. Intern. **3**, 10 b, 3, 120, Paris.

SUESS, H. E., 1970: Bristlecone pine calibration of the radiocarbon time scale 5,200 B.C. to the present. In: OLSSON, I. V. (Ed.), Radiocarbon Variations and Absolute Chronology. Stockholm: Almquist and Wiksell.

TAYLOR, J. C. M., ILLING, L. V., 1969: Holocene intertidal calcium carbonate cementation Quatar, Persian-Gulf. Sedimentology **12**, 69—107, Amsterdam.

TERRA, H. DE, PATERSON, T., 1939: Studies on the ice-age in India and associated human cultures. Carnegie Inst. Wash., Publ. **493**, 354, Washington.

THATCHER, L., RUBIN, M., BROWN, G. F., 1961: Dating desert groundwater. Science **134**, 105—106, Washington.

THESIGER, W., 1948: Across the Empty Quarter. Geogr. J. **111**, 1—21, London.

THESIGER, W., 1949: A further journey across the Empty Quarter, Geogr. J. **113**, 21—46, London.

THODE, H. G., 1970: Sulfur isotope geochemistry. Mineral. Soc. Amer., Spec. Paper **3**, 133—144, Washington.

THODE, H. G., MONSTER, J., 1965: Sulfur isotope geochemistry of petroleum, evaporites and ancient seas. Amer. Assoc. Petrol. Geologists, Mem. **4**, 367—377, Tulsa, Oklahoma.

THORNTHWAITE, C. W., 1943: Problems in the classification of climates. Geogr. Rev. **33**, 3, 233—255, London.

THORNTHWAITE, C. W., 1948: An approach toward a rational classification of climate. Geogr. Rev. **38**, 55—94, London.

THRUSH, P. W., 1968: Dictionary of mining, mineral and related terms, 1296 p. Washington: U.S. Bur. Mines.

TLEEL, J. W., 1973: Surface geology of Dammam dome, Eastern Province, Saudi Arabia. Amer. Assoc. Petrol. Geologists, Bull. **57**, 558—576, Tulsa, Oklahoma.

TOTHILL, J. D., 1946: The origin of the Sudan Gezira Clay. Sudan Notes Rec. **27**, 153—183, Khartoum.

TRICART, J., 1956: Tentative de corrélation des périodes pluviales africaines et des périodes glaciaires. Soc. Geol. France, Compt. Rendus **10**, 164—167, Paris.

TRICART, J., MICHEL, P., VOGT, J., 1957: Oscillations climatiques quaternaires en Afrique occidentale. INQUA, 5th, Madrid 1957, Madrid.

TROLL, C., PAFFEN, K. H., 1964: Karte der Jahreszeitenklimate der Erde. Erdkde. **1**, 5—28, Bonn.

TSCHERNOV, E., 1973: On the Pleistocene molluscs of the Jordan Valley. The Pleistocene of the central Jordan Valley, the excavations at Ubeidiya, 50 p. Jerusalem: Israel Acad. Sci. Human.

TUREKIAN, K. K. (Ed.), 1971: Late Cenozoic glacial ages, 606 p. New Haven, London: Yale Univ. Press.

TURNER, W. M., 1971 a: Quaternary sea levels of Western Cyprus. Quaternaria **15**, 197—202, Roma.

TURNER, W. M., 1971 b: The Calabrian Touloupos formation of the Khrysokhou basin of Western Cyprus. Quaternaria **15**, 203—209, Roma.

UHLIG, D., 1965: Das Be- und Entwässerungsprojekt der Al-Hassa-Oase in Saudi Arabien. Die Wasserwirtschaft **55**, 53—57, Stuttgart.

UHLIG, D., 1971: Anwendung von Stahlbetonfertigteilen für das Bewässerungssystem des Al-Hassa-Projektes. Die Bautechnik **48**, 109—113, Berlin-München-Düsseldorf.

United Nations Educational Scientific and Cultural Organization, 1968. International tectonic map of Africa. Sheet 3, 1 : 5,000,000. Paris: Inst. Geogr. Nat.

U.S. Geological Survey Arabian American Oil Company, 1963: Geologic map of the Arabian Peninsula, 1 : 2,000,000. U.S. Geol. Survey, Misc. Geol. Investig. Map I-270 A, Washington.

U.S. Geological Survey Saudi Arabian Project, 1972: Topographic map of the Arabian Peninsula. Arabian Peninsula Series Map AP-1, 1 : 4,000,000, Washington.

VAN DAALHOFF, H., 1974: Mineral locality map of the Arabian Shield. Metalliferous minerals. Saudi Arabian Dir. Gen. Min. Res., Geol. Map GM-15, Plate 1, 1 : 2,000,000, Jiddah.

VAN DAALHOFF, H., 1974: Mineral locality map of the Arabian Shield. Industrial minerals and dimension stone, marble and others. Saudi Arabian Dir. Gen. Min. Res., Geol., Map GM-15, Plate 2, 1 : 2,000,000, Jiddah.

VAN STRAATEN, L. M., 1960: Some recent advances in the study of deltaic sedimentation. Liverpool-Manchester Geol. J. **2**, 411—442, Liverpool.

VERHAGEN, B. T., SELLSHOP, J. P., JENNINGS, C. M., 1970: Contribution of environmental tritium measurements to some geohydrological problems in Southern Africa. IAEA, Isotope Hydrology 1970, 289—313, Wien.

VERSTAPPEN, H. TH., 1970: Aeolian geomorphology of the Thar Desert and palaeo-climates. Z. Geomorph., Suppl. **10**, 104—120, Berlin-Stuttgart.

VINCENT, P., 1963: Les volcans tertiaires et quaternaires du Tibesti occidental et central. Bur. Rech. Géol. Min., Mém. **23**, 307, Paris.

VINCK, F., 1972: Die Auswertung von Niederschlagsmessungen in ariden Gebieten. Wasser und Boden **1972**, 15—16, Hamburg.

VITA-FINZI, C., 1969: Late Quaternary alluvial chronology of Iran. Geol. Rdsch. **58**, 951—973, Stuttgart.

VOGEL, J. C., 1970: $^{14}C$ dating of groundwater. IAEA, Isotope Hydrology **1970**, 225—237, Wien.

VOGEL, J. C., EHHALT, D., 1963: The use of carbon isotopes in groundwater studies. IAEA, Radioisotopes in Hydrology, 383—396, Wien.

VOGEL, J. C., EHHALT, D., ROETHER, W., 1963: A survey of the natural isotopes of water in South Africa. IAEA, Radioisotopes in Hydrology, 407—415, Wien.

VOUTE, C., 1959: Geology of the Assab area (Southern Eritrea, Ethiopia). Intern. Geol. Congr., 20th, Mexico 1956, Asoc. Serv. Geol. Africanos, 455—468, Mexico.

VOUTE, C., 1960: Climate and landscape in the Zagros mountains (Iraq). Intern. Geol. Cong., 21st, Copenhagen 1960, Rep. 4, 81—87, Copenhagen.

VOUTE, C., WEDMAN, E. J., 1963: The Quaternary climate as a morphological agent in Iraq. UNESCO, Arid Zone Research **20**, 395—402, Paris.

Wakuti-Ingenieurbüro, 1960—1964: Studies for the project of improving irrigation and drainage in the region of Al-Hassa, Saudi Arabia. 2 vols., **1** (1960), 156 p. and **2** (1964), 322 p., unpublished report, Siegen, Westf.

WENDT, I., 1970: Carbon isotopes in hydrological processes. In: OGATA, K., HAYAKAWA, T. (Eds.), Recent Developments in Mass Spectroscopy. Tokyo: Univ. Tokyo Press.

WENDT, I., STAHL, W., GEYH, M. A., FAUTH, F., 1967: Model experiments for $^{14}C$ water-age determinations. IAEA, Isotopes in Hydrology, 331—336, Wien.

WENZ, W., 1943: Land- und Süßwassermollusken aus fluvioäolischen Ablagerungen von Hadramaut (Südarabien), Archiv für Molluskenkunde **75**, 148—151 und 240—241, Frankfurt.

WHITEMAN, A. J., 1971 a. The geology of the Sudan Republic. 290 p. Oxford: Oxford Univ. Press, Clarendon.

WHITEMAN, A. J., 1971 b: Pleistocene and Recent stratigraphy of Sudan Republic. Assoc. Senegolaise Quatern. Ouest Africa, Bull. **29**, 29, Dakar.

WHITTEN, D. G., BROOKS, J. R., 1972: Penguin dictionary of geology, 516 p. Penguin Books Ltd.

WILHELMY, H., 1958: Klimamorphologie der Massengesteine. 238 p. Braunschweig: Westermann.

WILHELMY, H., 1968: Indusdelta und Rann of Kutch. Erdkde. **22**, 177—191, Bonn.

WILLIAMS, G. E., 1970: Piedmont sedimentation and late Quaternary chronology in the Biskra region of the northern Sahara. Z. Geomorph. N.F. Suppl. **10**, 40—63, Berlin-Stuttgart.

WILLIAMS, G. E., POLACH, H. A., 1971: Radiocarbon dating of arid-zone calcareous paleosols. Geol. Soc. Amer. Bull. **82**, 3069—3086, Boulder, Colorado.

WILLIAMS, M. A., 1976: Radiocarbon dating and late Quaternary Sahara climates: a discussion. Z. Geomorph. N.F. **20**, 3, 361—362, Berlin-Stuttgart.

WISSMANN, H. v., 1932: Übersicht über Aufbau und Oberflächengestaltung Arabiens. Z. Ges. Erdkde. Berlin **9/10**, 335—357, Berlin.

WISSMANN, H. v., 1957: Karsterscheinungen in Hadramaut. Peterm. Geogr. Mitt., Erg.-H. **262**, 259—268. Berlin: VEB. H. Haack.

WISSMANN, H. v., RATHJENS, C., KOSSMAT, F., 1943: Beiträge zur Tektonik Arabiens. Geol. Rdsch. **33** (1942), 221—353, Stuttgart.

WOLDSTEDT, P., 1958—1965: Das Eiszeitalter, Bd. 1, 3. Aufl. 1961, 374 p., Bd. 2, 2. Aufl. 1958, 438 p., Bd. 3, 2. Aufl. 1965, 328 p. Stuttgart: Enke.

WOLDSTEDT, P., 1966: Ablauf des Eiszeitalters. Eiszeitalter und Gegenwart **17**, 153—158. Öhringen.

WOLDSTEDT, P., 1969: Quartär. Handb. Stratigr. Geol. **2**, 263 p. Stuttgart: Enke.

WOLFART, R., 1961 a: Geologic-hydrogeologic research for the utilization of groundwater in the Wadi Nisah area. A supplement to the water supply of Ar Riyadh, Saudi Arabia. Unpub. report, Bundesanstalt für Bodenforschung, Hannover.

WOLFART, R., 1961 b: Hydrogeology of the Central Tuwaiq mountains and adjoining regions (Saudi Arabia). Intern. Ass. Sci. Hydrology, Publ. **57**, 98—112, Gentbrugge.

WOLFART, R., 1965: Geologie und Hydrogeologie des mittleren Tuwaiq-Gebirges und der angrenzenden Gebiete (Saudi Arabien). Geol. Jb. **83**, 149—180, Hannover.

WOLFART, R., 1966: Zur Geologie und Hydrogeologie von Syrien. Geolog. Jb. Beihefte **68**, 129, Hannover.

WOLFART, R., 1967: Geologie von Syrien und dem Libanon. Beitr. Reg. Geol. Erde **6**, 326. Berlin: Borntraeger.

WRIGHT, H. E., 1961: Pleistocene glaciation in Kurdistan. Eiszeitalter und Gegenwart **13**, 131—164, Öhringen.

YAALON, D. H., SINGER, S., 1974: Vertical variation in strength and porosity of calcrete (nari) on chalck, Shefela, Israel and interpretation of its origin. J. Sed. Petrology **44**, 4, 1016—1023, Tulsa, Oklahoma.

ZAGWIJN, W. H., 1975: Variations in climate as shown by pollen analysis especially in the Lower Pleistocene of Europe. In: WRIGHT, A. E., MOSELEN, F. (Eds.), Ice Ages: ancient and modern. Geol. J., Spec. Issues **6**, 137—152, Liverpool.

ZEIST, W. VAN, 1967: Late Quaternary vegetation history of western Iran. Rev. Palaeobot. Palynol. **2**, 303—311, Amsterdam.

ZEIST, W. VAN, WRIGHT, H. E., JR., 1963: Preliminary pollen studies at Lake Zeribar, Zagros Mountains, southwestern Iran. Science **140**, 65, Washington.

ZEUNER, F. E., 1952: Pleistocene shore-lines. Geol. Rdsch. **40**, 39—50, Stuttgart.

ZEUNER, F. E., 1953: Das Problem der Pluvialzeiten. Geol. Rdsch. **41**, 242—253, Stuttgart.

ZEUNER, F. E., 1954: Neolithic Sites from Rub' al Khali. Man **209**, 133—136, London.

ZEUNER, F. E., 1958: Dating the past. 4th ed., 516 p. London: Methuen.

ZEUNER, F. E., 1959: The Pleistocene period. 3rd. ed., 447 p. London: Hutchinson & Co.

ZIEGERT, H., 1967: Zur Pleistozän-Gliederung in Nordafrika. Afrika-Spectrum **3**, Pfaffenhofen/Ilm: Afrika Verlag (in Komm.).

ZILCH, A., 1959: Gastropoda, Teil 2. Euthyneura, Lieferung 1—4. In: SCHINDEWOLF, O., Handbuch der Palaeozoologie **6**, 834.

ZINDEREN BAKKER, E. M. VAN, 1962: A late glacial and postglacial climatic correlation between East Africa and Europe. Nature **194**, 201—203, London.

ZINDEREN BAKKER, E. M. VAN, 1967: Upper Pleistocene and Holocene stratigraphy and ecology on the basis of vegetation changes in sub—Saharan Africa. In: BISHOP, W. W., CLARK, J. D. (Eds.), Background to evolution in Africa, 125—148. Chicago-London: Univ. Chicago Press.

# Index of Geographical Names

As there is no universally agreed on scheme for the transliteration of Arabic words into English, we have found some Arabic place names spelled in as many as five different ways in a series of various maps and reports. It was therefore decided to use a different approach when assigning place names.

The spelling of Arabic names in this volume aims at producing correct or nearly correct pronunciation by a person unfamiliar with Arabic but used to common Latin spelling. As it is not possible to transliterate directly from Arabic into English, the names were spelled phonetically according to their Latin pronunciation. The results may differ in some cases from other versions but it is believed that the spelling adopted here will result in pronunciation closer to the original Arabic than is usual with other schemes.

# Transcription of Names Used in Some Figures

| | |
|---|---|
| p. 31, Fig. 3: | Musqat, correct: Masqat |
| p. 61, Fig. 12: | Ayn, correct: 'Ayn |
| p. 94, Fig. 25: | ayn, correct: 'ayn |
| p. 117, Fig. 37: | W. Miyah, correct: Wadi Al Miyah |
| p. 122, Fig. 39: | As Sha'bah, correct: Ash Sha'bah<br>Ayn, correct: 'Ayn<br>'Ayn Sha'bah, correct: 'Ayn Shabah<br>Jabal Buraiqa, correct: Jabal Burayqa |
| p. 147, Fig. 48: | Qatif, correct: Al Qatif |
| p. 167, Fig. 53: | Dahl Abu Marwa, correct: Dahl Abu Marwah |
| p. 175, Fig. 56: | Ar Ras, correct: Ar Rass<br>Buraidah, correct: Buraydah<br>Ha'il, correct: Hail<br>Nafud Ath Thuwayrat, correct: Nefud Ath<br>R. Al Khabra, correct: Riyadh Al Khabra<br>Thuwayrat<br>Wadi Ash Sha'ba, correct: Wadi Sha'ba |
| p. 183, Fig. 60: | Ar Ras, correct: Ar Rass<br>Unayzah, correct: 'Unayzah |
| p. 217, Fig. 72: | Wadi Nisha, correct: Wadi Nisah<br>Wadi Sulaij, correct: Wadi As Seley |
| p. 227, Fig. 77: | As Sulayyil, correct: As Sulayil<br>Jabal Tuwayq, correct: Jebal Tuwayq<br>Nafud Ad Dahi, correct: Nefud Adh Dhai<br>Rawdah, correct: Rawdhah<br>Wadi Tathlit, correct: Wadi Tathlith |
| p. 247, Fig. 80: | As Sulayyil, correct: As Sulayil |
| p. 255, Fig. 83: | Al Arid, correct: Al 'Aaridh<br>Hadramaut, correct: Hadhramawt<br>Jebal Tuwaiq, correct: Jebal Tuwayq |

Druck: Adolf Holzhausens Nfg., A-1070 Wien